AF816525

The SURFACE TEXTURE answer book

by

Carl Musolff

and

Mark C. Malburg

Edited by Mike Zecchino

Cover Design by Todd Adkins, Born to Design

ISBN: 978-1-7368468-2-7

TABLE OF CONTENTS

TABLE OF FIGURES

TABLE OF TABLES

PREFACE

During our careers we've been asked hundreds of questions about surface texture. "What does this parameter mean?" "Which cutoff should I use?" "What is a 'cutoff'?'" "Why can't I get rid of this raised metal?" Often, the follow-up question was, "Where can I read about this?"—a question for which we had no good answer. This book was written to change that.

The questions come from product designers, product engineers, manufacturing engineers, quality personnel, laboratory technicians—people that need a quick answer to a relatively straight-forward question, not a full course on surface metrology. With that in mind, this book is written in a question-and-answer format. There is no need to read a whole book, or even a whole chapter; the answer is given in one or two pages. Over 280 figures illustrate these answers.

Most laboratories have charts on their walls showing visual representations of surface texture concepts. There are also numerous books and journal articles on surface metrology that go into extensive detail about the underlying mathematics and research behind surface analysis. This book occupies a middle ground, providing more information than the pictures on the wall charts, but avoiding the lengthy, sometimes highly technical descriptions found in the metrology literature.

This book documents the exchange between people that have questions and the two of us attempting to provide answers. It could not have happened without people being inquisitive enough to ask those questions. If that is you . . . thank you.

INTRODUCTION

Many of today's most challenging technical design problems involve the interaction of surfaces. Wear reduction, friction control, energy efficiency, noise and vibration reduction, sealing, contract stress capacity, and other phenomenon are of interest in many industries. Optical properties, fluid flow characteristics, and electrical behavior are critical to others. The manufacture and assembly of the physical components is important to all. Design, development, and manufacturing all require measurement—you won't fully understand things you can't measure.

The discipline of measuring surfaces is called surface metrology, and it provides the foundation for this book. But this isn't a book for surface metrologists. It's a book for the rest of the technical community—those of you who deal with specifying and creating functional surfaces. The primary goal is to "help you with your job" rather than "teach you to be a metrologist."

Surface interactions are complex because every surface has a texture—a micro-topography with peaks and valleys of varying heights, depths, slopes, and curvatures lying at various distances from each other. This topography, or surface texture, results from the mechanical history of the surface, with possible added effects from its thermal and chemical processing. The manufacturing methods create surface features through the action of the cutting tools, vibration in the equipment or environment, errors in the tool guides, and deformations from clamping or fixturing. Once the surface is placed into service it may undergo further changes through contact with mating components or from exposure to its environment.

Of all the features that exist at any point in time, some will impact the surface's future usefulness, some won't. Surface metrology provides the tools to detect, define, and quantify this surface texture. This information provides the basis for determining which characteristics do and don't matter, and to what extent. This can then be fed back into the design specifications and the manufacturing process to optimize product performance and cost. An understanding of the specification, measurement, or influence of surface texture is therefore an integral part of the whole life cycle of many products (see **Figure 1**).

This book is intended for practicing technicians or engineers who are collecting, analyzing, and applying surface texture data in the course of their professional activities. This can include designing a component or system, designing and monitoring a manufacturing process, working with a component supplier to verify or improve product quality, or analyzing a part or system that does not function as intended. The material applies to a wide range of industries—automotive, aerospace, optics, heavy equipment, bio-medical, electronics, machine tools, manufacturing, etc.—and to disciplines such as tribology, contact mechanics, hydrodynamics, machining, quality control, bioengineering, failure analysis, and many, many more.

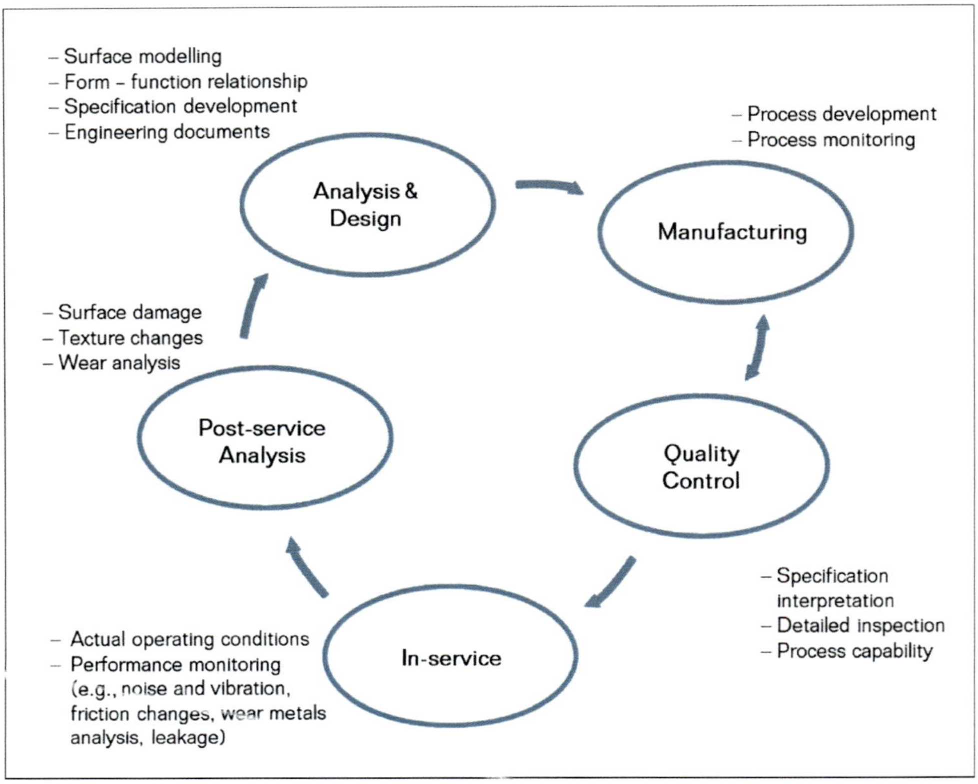

Figure 1: Surface texture in a product's life cycle

The book is designed to provide answers to the specific questions you may have. It is organized into the following sections:

Measurement

How do I measure? What do I measure with? What can go wrong?

Data analysis

I've measured it—now what? Can I put numbers on the things I need to control? What do I do with all these filters and cutoffs? Why are there so many options?

Data interpretation

Is my part OK? Can I trust the results?

Documentation

How do I put this stuff on a technical drawing? What parameters should I be using?

Applications

All this information is great, but what is it good for? How can I use it to solve real problems?

Definitions and terminology

What do all these words and symbols mean? Does everyone use the same ones?

References and resources

Where did this information come from? Where do I go to find out more?

The goal of this structure and the question/answer format is for the reader to quickly find the needed information, without having to wade through material that isn't relevant.

In order to keep the scope of this book within manageable limits and speak to the broadest audience we've made the following decisions about the subject matter:

- We primarily deal with the surface texture terminology and concepts contained in the ASME and ISO standards. It is recognized that this leaves out numerous other national and industry standards that address this topic. However, most of the concepts presented here are universally applicable, even if the terminology isn't.

- Some parameters are discussed that are not yet included in the national or international standards. These have been useful in solving problems and in specifying critical surfaces so they are presented for your consideration.

- We focus on 2D contact (stylus-based) measurements because this is the more common technology in a majority of product development and manufacturing environments. This is not to minimize the value of 3D and non-contact techniques, which are very powerful and will no doubt become more common as familiarity increases and cost decreases. 3D examples are occasionally included to illustrate their advantages in specific situations.

The national and international standards referenced in this book are continually being revised. Changes may be made to the parameter definitions, analysis algorithms, default settings, specification syntax, or surface texture terminology. These revisions may result in changes in how a design specification is interpreted, how your measurement data is analyzed, or the numerical value of your surface texture parameters. Most measurement systems and analysis software lag these changes. It may not be clear from the instrument's purchase date or the software revision level which specification version is being followed. Consult your instrument's documentation to determine the specification revision to which your instrument conforms. Your company should also document the source or derivation of any critical functional parameter limits in order to understand the impact of any future specification changes on those parameters.

Note: At the time it was written (2019–2020), this book was consistent with the currently released ASME and ISO documents.

Off-line analysis software

This book is about acquainting the practicing engineer or technician with the principles behind surface texture measurements. While it attempts to provide answers to the more common questions, the only way to really develop an understanding of the subject is to analyze data and examine the results. If you work in a laboratory and have measurement equipment readily available, you have that opportunity. However, if you work outside of a lab and are a recipient rather than a creator of this data, experimentation can be difficult. This is where off-line data analysis software is useful.

With your own analysis software you can process raw data files and experience how changes in settings affect the data, in real time. This provides a better grasp of concepts such as profile type, reference geometry or form removal, cutoff selection, filter type, and parameter relationships. You can creatively explore aspects of the surface not explicitly addressed by a technical drawing or other product documentation. Access to aspects of the data such as the *primary* profile instead of just the *roughness* profile will help in the resolution of product and process issues. If you are responsible for products design you will be able to more knowledgeably specify filters, parameters, and limits that truly define the functional requirements of the component.

Offline analysis affords several benefits to productivity, including:

- Measurement technicians can spend their time measuring parts rather than analyzing data
- Technicians can focus on accuracy rather than interpreting results, thus better utilizing these valuable lab resources.
- Instead of providing processed results (e.g., parameters), the lab focuses on providing accurate data (i.e., raw profiles) that can be interpreted using various analysis settings and methodologies.
- Many of the common, repeat measurements can be avoided; the part doesn't have to be re-fixtured and re-measured just because the wrong filter was used, or because a parameter wasn't calculated. A "measure once, analyze often" approach can be used to get the most information from each measurement.

One such example of an off-line analysis package is OmniSurf from Digital Metrology Solutions, Inc. (www.digitalmetrology.com/). OmniSurf, and its 3D counterpart OmniSurf3D, were used to make many of the images and analyses in this book.

One important service provided by software developers such as Digital Metrology is the creation of new parameters to address specific needs. While parameter proliferation is to be avoided whenever possible, there are some situations where the existing tools do not adequately describe some functional, or dysfunctional, characteristic of a surface. This is, of course, how all the parameters were originally developed: to address a need. National and international standards organizations and instrument manufacturers usually cannot respond to these situations as quickly as desired, but third party software developers may be able to do so. This book includes several examples of such parameters contained within OmniSurf.

TERMINOLOGY AND DISPLAY CONVENTIONS

Throughout this book the following conventions will apply:

- ISO surface texture coordinate system:

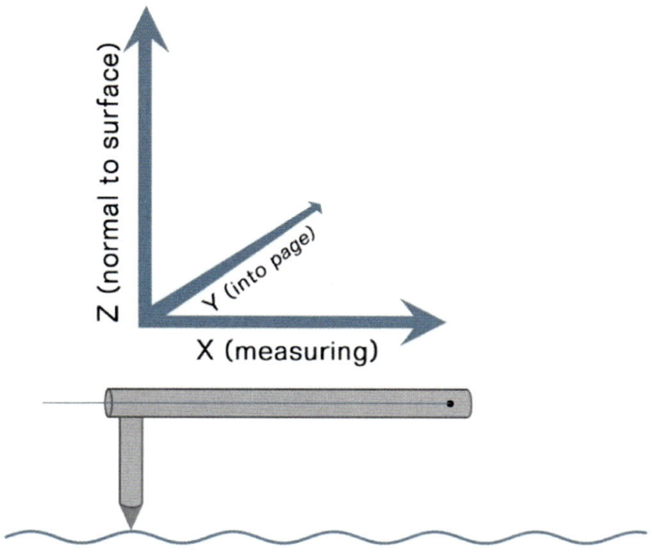

Figure 2: ISO surface texture coordinate system

  - The traversing direction for a measurement will be called the x-axis.
  - The height of surface features will be considered to be along the z-axis.
  - The y-axis is based on a right-handed coordinate system.

- The standard parameter naming convention will be used wherever possible. The first (capitalized) letter will indicate the profile and the following (lowercase) letters will indicate the characteristic. For example, the "Ra" parameter indicates the "R" (roughness) profile with the "a" (arithmetic average) characteristic.

- Parameter names will generally follow the historical convention of calculation based on evaluation length rather than sampling length. This is to avoid confusion in an environment where both naming conventions exist. For example, Rp (roughness peak height) and Rpm (roughness average maximum peak height) are both covered here, with Rp being the highest peak in the evaluation length and Rpm being the average of the highest peak in each sampling length. This is consistent with ASME standards.

- "RzDIN" and "RzJIS" will occasionally be used to distinguish between the two definitions of Rz (10-point roughness height). RzDIN and RzJIS are not official names, but they may be found in some software packages.

- Measurement graphics will typically be displayed as *primary* profiles with a least squares line suppressed. It will be noted when different profiles or reference geometry options are used.

- *Primary* profiles will be displayed in blue; *waviness* profiles will be displayed in red; *roughness* profiles will be displayed in black; other profiles such as those created by morphological filters will be displayed in black.

- No attempt is made to use consistent x- and z-axis scales in the figures and graphs. Scales are chosen to emphasize the features of interest or the concept being discussed.

MEASUREMENT

MEASUREMENT

It all starts with a measurement. While most users of surface texture data will not be directly involved in making a measurement, it is important to understand the process. Data analysis, interpretation, and application are only as good as the information obtained by an instrument interacting with the workpiece surface. Numerous factors can affect the integrity of this information, including:

- **Sample condition**
  - Cleanliness
  - Surface uniformity
  - Presence of defects
- **Equipment**
  - Stylus geometry and condition
  - Vertical resolution
  - Horizontal resolution (data density)
  - Frequency response
  - Datum straightness and cleanliness (for skidless instruments)
  - Calibration
  - Internal mechanical vibration
- **Environment**
  - External mechanical noise or vibration
  - Electromagnetic noise
- **Measuring process**
  - Sample support and alignment
  - Measuring speed
  - Traverse length
  - Positioning of stylus

This section addresses topics such as:

- Measurement equipment and fixture characteristics
- How the instrument interacts with the workpiece surface
- How to identify factors which have affected a measurement
- How to tell if an instrument is working correctly

As a starting point, it is important to remember the size scale for surface texture measurements. This is a world of micron and sub-micron features, at least 100 times smaller than the limit of human vision. **Figure 3** is a schematic of a micron relative to a typical human hair (larger circle) and a 0.5 mm pencil lead (gray circular segment). **Table 1** lists the dimensions of common objects. This perspective can give an appreciation not only of the challenge in measuring features this small, but also how things such as airborne debris and mechanical vibration can negatively affect the data being collected.

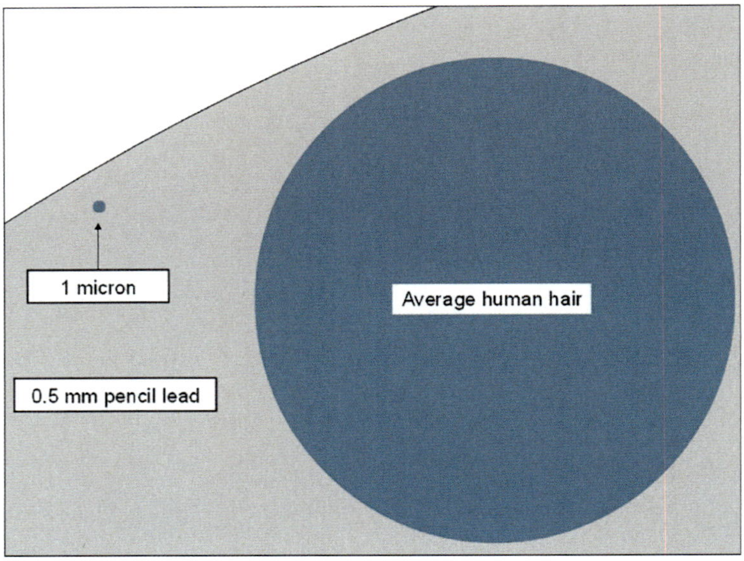

Figure 3: Illustration of a micron compared to a typical human hair and a 0.5 mm pencil lead

Table 1: Approximate dimensions of common objects

Object	Size (μm)
Width of hand	75,000 – 100,000
Thickness of window glass	2000 – 4000
Thickness of copier / printer paper	~ 100
Diameter of human hair	20 – 200
Diameter of silkworm silk thread	~ 30
Diameter of a strand of a spider's web	2 – 5
Smoke particle	0.5 - 5

How does a roughness measurement system work?

This discussion will focus on the operation of a contact instrument. Consult other references for a description of non-contact instruments.

Figure 4 is a schematic of the mechanical components of a common, skidless surface texture instrument. In the simplest terms, a motor moves a carriage along a datum surface. Attached to the carriage is a stylus on a pivot. One end of the stylus contacts the workpiece and the other end is attached to a sensor. The sensor translates z-axis motion of the stylus into displacements relative to the datum. The sensor output is typically fed through an amplifier, analog-to-digital convertor, and signal conditioning electronics to a computer which controls the instrument, saves and analyzes the digital data, and displays the results. The x-axis component of the digital data is supplied by a timing circuit or by an x-axis scale or encoder.

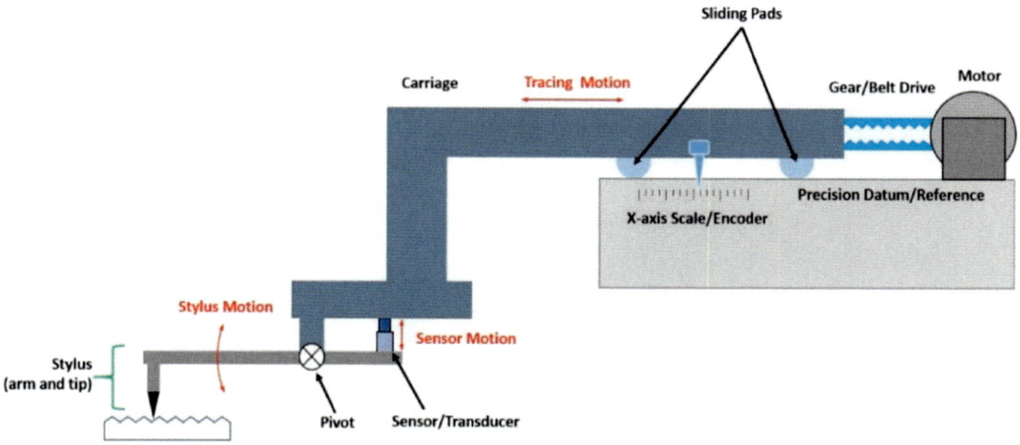

Figure 4: Schematic of a skidless contact surface texture instrument

Several key design and operational considerations are evident from this schematic:

- Form errors or waviness on the datum surface will be incorporated into the measured data. These errors can be inherent to the datum or due to the buildup of contamination.
- Vibration in the motor, gears, drive, and other moving components will be seen as noise in the measured data. This may include other factors such as stick-slip between the pads and the datum due to dirt or contamination.
- Backlash in the motor and drive mechanism will appear as low level noise or a discontinuity at the beginning of a trace as the sensor gathers data before the stylus moves.
- Pivot friction, stylus mass, and sensor mechanical resistance all contribute to the stylus tip force which, if large enough, can damage the workpiece surface.

- The length of the arm between the stylus and pivot directly influences the z-axis resolution in the measured data.

A skidded instrument operates similarly, with the exception of adding a skid adjacent to the stylus tip. The skid acts as the reference surface so a precision internal datum is not needed.

The stylus provides the interface between the instrument and the surface being measured. Since the stylus tip geometry interacts with the surface features, the stylus must have the correct shape and be in good condition. Various shapes are available, including spherical, truncated pyramid, chisel point, chisel knife edge, and conical. The spherical tipped cone is the most common (see **Figure 5**), with common cone angle options of 90, 60, and 40°. The spherical tip radius is typically 2, 5, or 10 µm, and must be sized to meet the characteristics of the workpiece surface and the short filter settings (see **Table 6** on page **38**).

There are various types of sensors for detecting the stylus motion, including piezo-electric crystals, LVDTs (linear variable differential transformer), and laser interferometers. In the past, the sensor type could have a significant influence on the instrument's capabilities. For example, an LVDT sensor had good resolution but a limited range, while a laser interferometer had poorer resolution but a larger range. However, advances in sensor technologies have reduced these differences so that considerations other than sensor design will most likely determine which instrument to use.

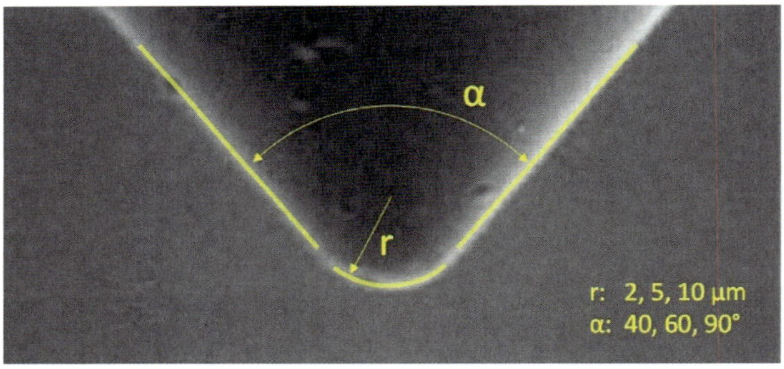

Figure 5: Conical diamond stylus tip geometry

Is the world moving to optical 3D measurement systems?

Are stylus-based, 2D measurement systems obsolete?

3D/areal surface texture data can provide a great deal of useful information about a surface, as well as providing a visually appealing representation of that surface—a representation that is often more intuitive than a 2D graph (see **Figure 6**). 3D data facilitates the application of new tools to the characterization and study of surfaces, potentially providing an improved understanding of a surface's function and a better language for communicating certain design requirements to the manufacturing process.

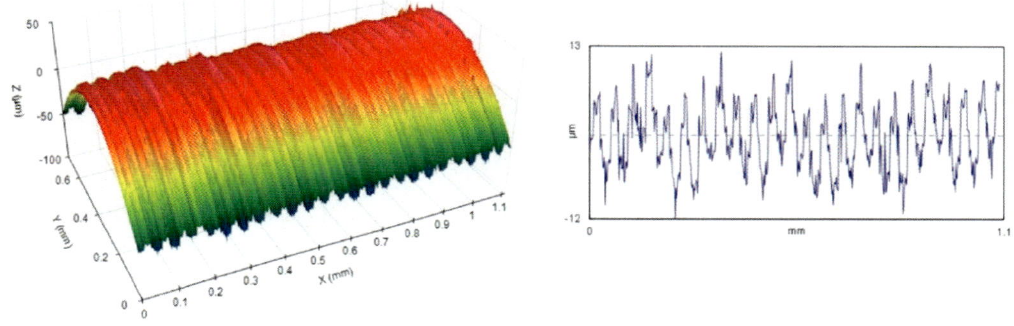

Figure 6: 3D (left) and 2D representations of a turned surface

Much of the current surface metrology research is directed at non-contact 3D (areal) systems. However, 2D stylus instruments will continue to be used for many industrial applications for the foreseeable future. There are several reasons for this:

- Many component specifications and manufacturing process controls have been derived from stylus instrument data. While some roughness parameters correlate well between 2D and 3D data, others don't. Switching to a non-contact 3D type of instrument could require revisiting those specifications—potentially a very costly endeavor.

- Most manufacturing facilities use stylus instruments, with minimal incentive to spend the money to change.

- Table-mounted contact instruments can be purchased for as little as $11,000, while hand-held units can cost as little as $1000. Non-contact instruments can cost significantly more for the actual equipment. There may also be additional costs for training, setup, and support systems.

- At a comparable lateral resolution to a stylus system, optical instruments typically evaluate an area of the surface on the order of a square millimeter or less. Many systems now incorporate "stitching" which allows multiple images to be combined into one large image. However, stitching can be quite time consuming, depending on the image size. In contrast, contact instruments can be purchased with traverse lengths of 200 mm. Units with traverse lengths on the order of 50 mm are very common. In many situations, a long profile measurement actually provides a better assessment of surface function than a small area measurement. For example, for a plateau honing operation, a profile measurement with a stylus can incorporate 20 or more valleys, while an optical 3D (areal) measurement may only include one.

- The small measured area with an optical instrument provides roughness data but does not provide information about the waviness or form characteristics. For many applications, the waviness and form play a key role in the component function. Without extensive stitching, optical data may not adequately characterize a surface.
- Many of the 3D (areal) parameters are analogous to 2D (profile) parameters—they are measured relative to a mean plane rather than to a mean line. There may be more incentive to consider 3D/areal analyses of surfaces as more descriptive parameters are developed, parameters that utilize more of the unique information available from 3D data.

Table 2: Advantages and disadvantages of stylus-based 2D instruments

2D (Stylus) Advantages	2D (Stylus) Disadvantages
Accepted in industry.	Two dimensional data cannot distinguish between a "ridge" and a "single peak" or a "valley" and a "pit."
Relatively easy to use.	Not suitable for relatively soft surfaces.
Not sensitive to surface color and/or reflectivity.	The stylus can damage the surface.
Horizontal resolution: A typical stylus can have a 2 μm tip radius, with only a small portion contacting the surface. This results in a very small contact area.	Cannot assess connectivity of features or "leak paths" along a surface.
Vertical resolution: Many instruments can resolve single-digit nanometers; some can resolve fractions of a nanometer.	

Table 3: Advantages and disadvantages of optical 3D instruments

3D/Areal (Optical) Advantages	3D/Areal (Optical) Disadvantages
Can distinguish between a "ridge" and a "single peak" or a "valley" and a "pit." Can assess connectivity between surface features.	Some systems have problems with high slopes.
Good vertical resolution on the nanometer level. Some instruments are good down in the Angstrom range.	Horizontal resolution is typically similar to that with a stylus instrument (despite many instrument claims).
Some highly engineered surfaces incorporate features that can only be evaluated in 3D (see **Figure 7**).	Some systems are sensitive to material optical properties and can react poorly to material boundaries (e.g., inclusions, partial coatings, joined materials).
Some systems provide very fast data collection for small areas.	Computationally expensive due to the large data sets (less of an issue as computing speeds increase).
Ability to measure actual volumes of features	
Can measure soft or fragile materials (e.g., the paper towel in the figure below).	

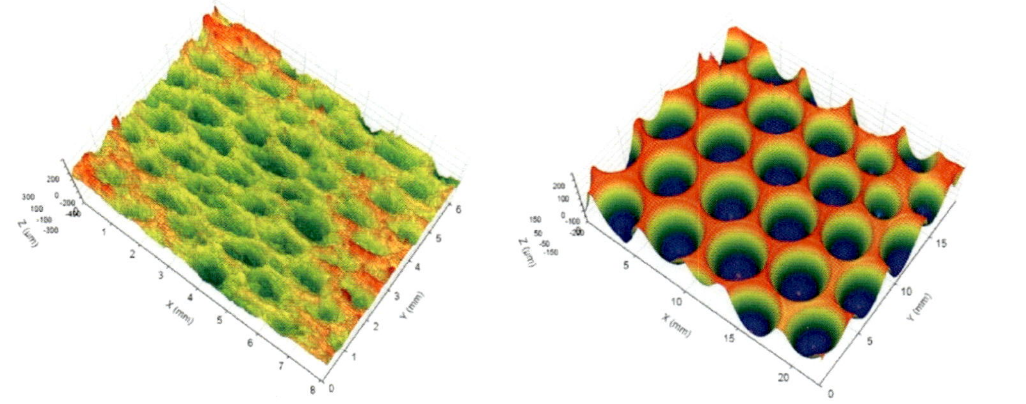

Paper towel

Golf ball with spherical form removed

Figure 7: Examples of surfaces best suited for 3D (areal) rather than 2D (profile) measurements

When should I use 3D (areal) instead of 2D (profile) measurements?

2D and 3D surface texture measurement systems both offer advantages and disadvantages relative to the other. Several of these differences are listed in the previous section. These will contribute to your decision about which type of system to use. However, two key factors for selecting the measurement system should be the manufacturing process used to create the surface and the functional sensitivity of the surface to those manufactured features. Many common machining processes generate surfaces that are relatively consistent in one direction, including turning, boring, broaching, cylindrical grinding, and tape lapping. **Figure 8** (a) shows that any randomly-placed sectioning plane that is transverse to the surface lay will provide a reasonable estimate of the topography of this type of surface. A 2D assessment should be adequate.

Other manufacturing processes generate a much more random surface, either simply as a consequence of that process or to meet specific design goals. Shot peening, tumbling, texturing, some coating processes, as well as as-cast and as-forged surfaces, all fall into this category. Randomly-placed sectioning planes from a 2D measurement can provide widely varying profiles, as in **Figure 8** (b). If the surface's function is sensitive to the pits, peaks, nodules, or other characteristics that might not be captured by a sectioning plane, a 3D areal assessment should be considered.

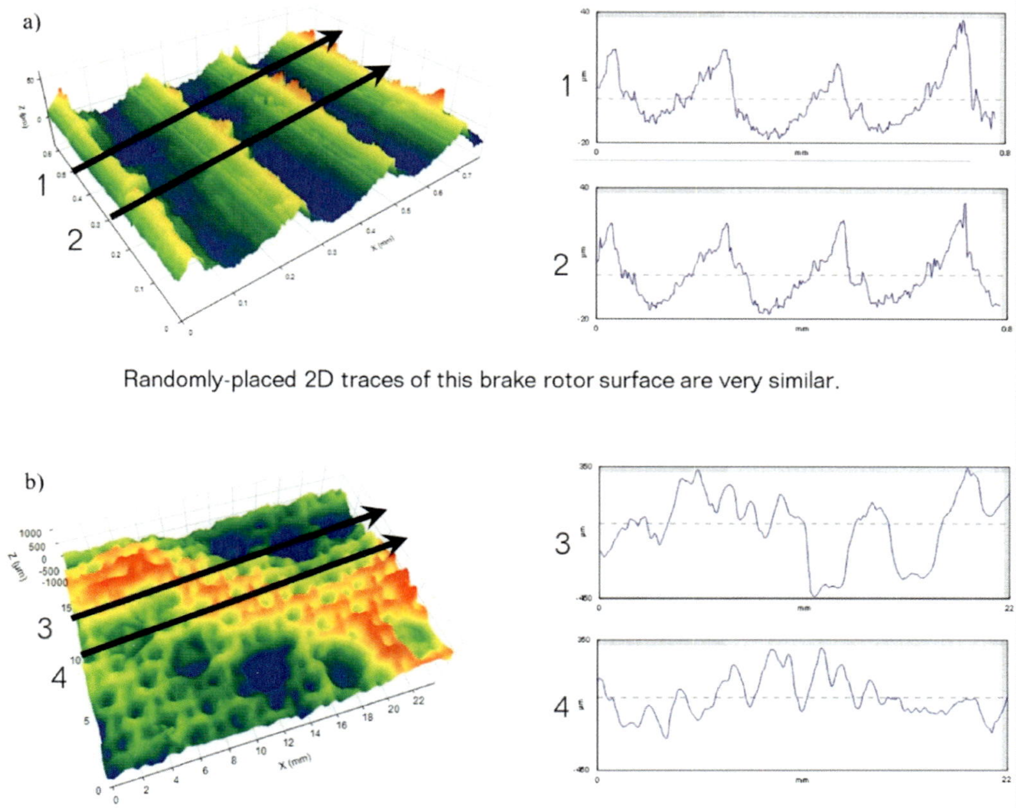

Randomly-placed 2D traces of this brake rotor surface are very similar.

Randomly-placed 2D traces of this textured surface are quite different.

Figure 8: 2D and 3D measurements of two types of surfaces

Another consideration when choosing 2D versus 3D measurements is the amount of length/area sampled by that measurement, and the sizes (wavelengths) of interest. For example, the measurement of roughness typically requires a maximum x-axis data spacing of 0.5 µm. This is very typical for a stylus-based instrument. However, to achieve that resolution optically, a 1024 x 1024 pixel camera can only measure a 0.5 mm square area. The same workpiece may have a waviness specification based on controlling wavelengths of several millimeters. This waviness assessment would obviously require a much larger image. Stitching of adjacent images is possible, but it can take a great deal of time. A lower resolution image can be taken for waviness, but that can also require additional high resolution measurements for roughness. A single, long 2D measurement provides an assessment of both roughness and waviness. **Figure 9** shows one example comparing the size of the measured regions.

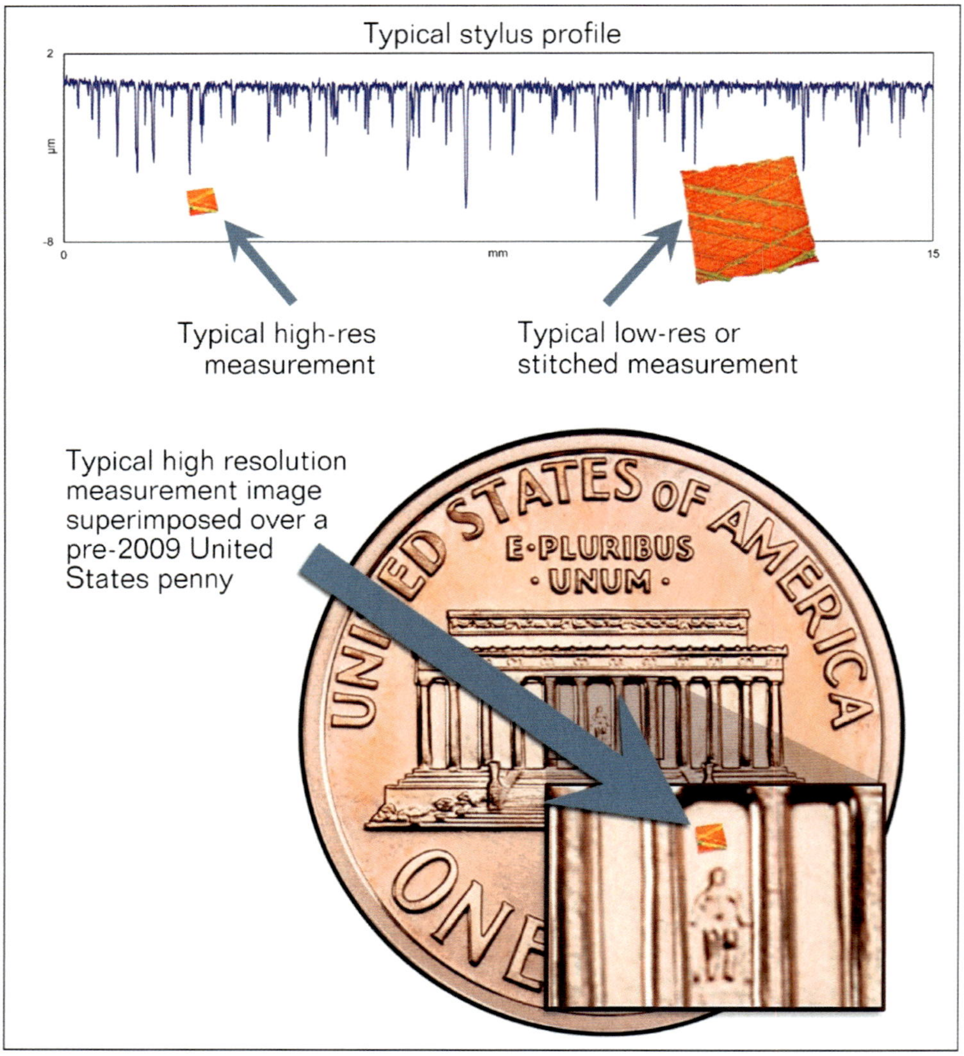

Figure 9: Comparison between a 2D profile and a typical 3D (areal) measurement

Will my 2D stylus system correlate with my 3D optical system?

Many studies have shown very good correlation between 2D (stylus) and 3D/areal (optical) measurement systems. Agreement is made possible when both systems are configured to measure similar wavelength domains. This includes not only filtering but instrument-specific items such as pre-processing of the data, optical objectives, mechanical tip radii, data point spacing, etc.

The degree of correlation depends on the parameters being considered. Amplitude parameters such as average roughness (Ra, Sa) and RMS roughness (Rq, Sq) have shown good correlation (correlation coefficient ≥ 0.98)[1]. Parameters with a large inherent uncertainty, such as skewness (Rsk, Ssk), may not show good correlation in an absolute sense, but may still correlate within their respective levels of uncertainty. Parameters such as RMS Slope (Rdq, Sdq) tend to show poor correlation.

Correlation will also be a function of the type of 3D system being used and the absolute surface roughness. For example, a 2005 NIST study[2] found that phase shifting interferometers correlated with stylus instruments better than did white light interferometers, especially at an average roughness in the range of 0.05 to 0.3 µm. Much has changed in this regard since 2005, however, and today good correlation can occur with any of these technologies. More recent improvements in hardware and software[3] have addressed some of the issues of switching between different optical measurement modes. As a general rule, phase shifting instruments correlate better for roughness below ~1 µm, while white light interferometers are better suited when it comes to roughness above ~1 µm.

Comparisons of 2D to 3D areal data sets (rather than parameters) are more problematic, especially for extreme value parameters, and especially on surfaces containing peaks and pits rather than ridges and grooves. For example, 3D data will be able to identify the actual peak or valley, while a 2D stylus will almost certainly not traverse the highest or lowest points.

---

[1] D. K. Cohen, R. P. Garibary, J. McNeill, and J. Huang, "Correlation of Optical Surface Profilometry with Contact Profilometry on Precision Peened Surfaces," *SME Conference on 3D Gaging and Technology*, Detroit, MI, March 4, 1997

[2] H.G. Rhee, T.V. Vorburger, J.W. Lee and J. Fu, "Discrepancies between roughness measurements obtained with phase shifting interferometry and white-light interferometry," *Applied Optics* (2005);44:5919.

[3] Vivek G. Badami et al. "Evaluation of the measurement performance of a coherence scanning microscope using roughness specimens," *Proceedings of ASPE Annual Meeting* (2011).

What should I consider when buying a surface measurement system?

There are numerous considerations when buying a stylus-based surface texture measurement system, most of them driven by the following questions:

- What are the size and geometry of the component(s) that will be measured?
- Will you be evaluating surface form and/or waviness or only surface roughness?
- Will the instrument be used on the shop floor or in a laboratory?
- What are the parameters and parameter ranges of interest?
- What is your budget?
- Where (geographically) will the instrument be located?
- Who will be using your data?

These questions lead to the following list of specific instrument characteristics and supplier considerations:

Traverse length

This must be as long or longer than the longest workpiece feature you will be measuring. Systems are available with traverse lengths of 200 mm or more. However, these longer drive units cost more, so the traverse length shouldn't be unnecessarily large.

Traverse speed

The traverse speed must be compatible with the inspection lab productivity expectations. This is especially important if you will be measuring long features or multiple features. Note that higher traverse speeds may lead to a sacrifice in frequency response or data density, so uncertainty and data integrity studies must be performed at the anticipated traverse speeds.

Column height

The column must be able to raise the drive unit above the tallest feature that will be measured.

Skidded versus skidless

If the device will be used for quick surface roughness checks on large components on the shop floor, a skidded design should be considered. This has several limitations (see pages 25 and 29) but the ease of use, relative insensitivity to vibration, and reduced risk of stylus damage may make it the right choice. However, form and waviness data and many of the more advanced analyses are only available with skidless instruments.

Probe configuration

The availability of break-away probes and a range of probe shapes and designs may be worth considering.

Probe limiter

Will you be measuring interrupted features (for example, a surface with a groove) or do you need to start and/or stop your measurement off the edge of a feature? If so, you will need some type of probe limiter, which mechanically stops the probe from falling out of range.

Datum straightness

On skidless instruments the straightness of the datum is key to obtaining accurate form and waviness analyses. Straightness specifications on the order of 0.2 μm/100 mm are usually appropriate. More stringent requirements will impact cost.

Horizontal resolution/data density

Is the instrument's data spacing consistent with your anticipated data analysis parameters? (see **Table 4**; reference ISO 3274:1996, Table 1).

Table 4: Maximum horizontal data spacing as a function of lambda-s (λs)

λs (μm)	Maximum data spacing (μm)
2.5	0.5
8	1.5
25	5

Vertical resolution

The instrument's vertical resolution must be consistent with the specification limits you'll be working to. This is especially important when the surfaces are smooth and the acceptable ranges are narrow. One rule of thumb is that the vertical resolution should be no larger than 10% of the Ra values being measured. As noted below under "Vertical range," the range and resolution are typically related, so the profile height (z-axis height) of the workpiece must be considered when assessing the resolution. See page **359** for more information about range and resolution.

Vertical range

The device must be capable of measuring the total vertical extent of the feature (the range of the feature in the z-axis). Note that the range can be increased by using a longer stylus, but this sacrifices vertical resolution, so it is usually not recommended. The interaction of range and resolution is very important. Many people wrongly choose an instrument based on its largest range and its finest resolution. They neglect the fact that the finest resolution cannot be obtained at the largest range. The resolution of most measuring systems is based on dividing the total range into a fixed number of increments; thus, the smaller ranges will typically have the finest resolution.

Analysis algorithms

Does the system software meet the requirements of the latest versions of the relevant national and international standards? Does the system software incorporate the parameters of interest for your surface?

Special analysis requirements

Features such as bandpass waviness or robust filters are not available on some commercial instruments. In some cases these limitations can be overcome by incorporating third-party software. If so, check for compatibility between the internal software and the third-party program. Make sure the data files are compatible and the right data is being written to the files (e.g., unfiltered vs filtered). Include in your budget any additional software licensing and training costs. Also, be sure to understand the conditions concerning software upgrades: how much do they cost, how frequently are they available, how easy are they to do?

Special parameters

Several parameters documented in this book are not yet included in the standards (e.g., waviness convexity (Wcvx), waviness void area (Wvoid). If your application requires these parameters, you may need to consider third-party software.

Ease of use

Some systems' software is extremely versatile, but consequently lacks simplicity. If you will have multiple equipment operators and don't want to make huge investments in training, you should consider a simple user interface that may not necessarily have all the bells and whistles. If third-party software will be used, investigate how easy it will be to hand off data between the software programs.

Reliability and service

Does the instrument have good reliability when used in similar applications as yours? What is the availability of service and support in your geographical region? Does the supplier maintain a reasonable inventory of spare parts and any special accessories you may need?

Commonality

If you will be comparing your results with suppliers, customers, or other laboratories, you may need to consider purchasing the same equipment that they use. This is especially true for very precise surfaces or when special parameters or analysis techniques are used.

Instrument uncertainty

An uncertainty study is an essential part of the equipment runoff and final acceptance. Acceptable uncertainty limits and uncertainty testing procedures should be agreed upon prior to placing an order, and included in the purchase order. This study should be performed both at the supplier and in the intended final installation. One example of an uncertainty procedure is described on page **85**. Note: "repeatability" and/or "reproducibility" are not good measures of surface texture instrument performance—see the section, "**Can I do a gage R&R study on a surface texture instrument?**" on page **84**.

Price

The total cost of acquisition includes not only the instrument itself but also the installation, any necessary facility changes, electromagnetic and mechanical isolation, fixtures, material handling, operator training, and periodic calibration / maintenance after installation.

Additional information can be found in chapter 2 of ASME B46.1.

Should I buy a used instrument?

Used surface texture instruments can be a viable and affordable option, but there are some important considerations:

Instrument hardware

- The datum (if skidless) must be adequately straight for the application.
- The motors and motion must run smoothly at measurement speeds without adding excessive noise to the measurement and without diminishing the overall frequency response.

Hardware performance

- The noise level must be low enough for the application.
- The frequency response of the probe must be adequate for the application.
- The data point spacing must be in accordance with standards.

Software

- When buying used equipment it is important to realize that older software versions may incorporate parameter definitions that are based on a different generation of standards than you are currently using. For example, the ISO maximum peak height of the roughness profile (Rp) parameter prior to 1997 is different than the ISO Rp parameter definitions in ISO4287-1997.

When possible, it is recommended that used equipment be purchased from a reputable equipment supplier with a "certification" process for used instruments. If this isn't an option, it is advisable to have a reputable metrology company perform an evaluation or uncertainty assessment of your system. One example of an uncertainty assessment procedure is given on page **85**.

Review the section, "**What should I consider when buying a surface measurement system?**" on page **20** for more information.

Which measurement stylus should I use?

When measuring 2D surface texture with a stylus instrument, the results are completely dependent on the contact between the measurement stylus and that workpiece. The size, shape, and condition of the stylus are therefore critical factors in obtaining an accurate assessment of that surface. This section provides guidance on the proper selection of those factors, with links to other sections that provide additional information. Note that several other characteristics of your measurement system also affect the raw data: these include the sampling scheme, analog-to-digital conversion, electronic filters, sensitivity to vibration, etc. However, many of these are typically not under the control of the operator so will not be discussed.

Probe geometry

The structure to which the stylus is attached must be of the appropriate geometry to reach the features of interest. For example, the distance of the stylus from the end must be small if measuring near a step or fillet, or the vertical height must be sufficient to reach the bottom of a groove, or small enough to fit inside a hole. Probe length will impact aspects such as vertical resolution (see page **359**) or arcuate error on a curved surface (see page **230**). It should also be noted that the length and "slenderness" of the probe can be a considerable source of noise that can potentially give very misleading results. Note that some manufacturers refer to the "probe" as the sensor and stylus together.

Skidless vs skidded

When using a skidless probe, only the measurement stylus tip contacts the surface. A skidded probe contacts the workpiece surface with both the stylus tip and a much larger radius skid (see **Figure 10**). Skidded probes are commonly used in handheld instruments, noisy environments, or in situations where there is a high risk of damaging the stylus. However, a skidded instrument can only measure roughness. Waviness and form are not generally measurable with a skidded system. Also, data from a skidded instrument may contain artifacts not seen with a skidless instrument. See pages **29**, **31**, and **35** for more information.

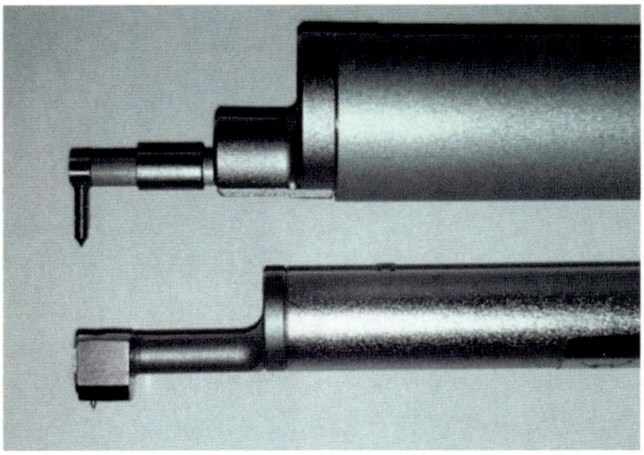

Figure 10: Examples of skidless and skidded styli

Stylus shape

The most common surface texture stylus geometry is a cone with a spherical tip. Other shapes may sometimes be desirable. For example, a knife-edge stylus may be useful for measuring along a small-diameter cylindrical surface to avoid the curvature imparted by the measurement axis being slightly different from the cylinder axis. Historically, pyramidal styli were rather common. However, the vast majority of today's styli are conical. Special geometries are rather rare. If one is required, this often requires a conversation with your instrument provider.

Stylus tip radius

The spherical radius of the stylus tip must be small enough to reach the bottom of the smallest (shortest wavelength) features of interest, and consistent with the specified short filter wavelength (lambda-s, or λs). Common values are 2, 5, and 10 µm. **Table 5**, taken from ISO 3274, provides recommended radii for λs. This table roughly translates the principle that the stylus radius should be no more than 80% of the smallest wavelength. Section 7 in ASME B46.1-2009 provides additional information about the interaction between the stylus and the workpiece.

Table 5: Typical stylus tip radii

λs (µm)	Tip radius (µm)
2.5	2
8	5
25	10

Smaller radii can cause more damage to the workpiece given the same contact force due to the resultant higher contact pressure (see **Table 7** and the equations in a footnote on page **40**). Under some very rare conditions, a smaller radius stylus may be more prone to damage, such as when measuring a very hard, very rough surface like a grinding wheel.

Stylus conical angle

Common styli have included angles of 40, 60, or 90° (see **Figure 11**). Tips with a 90° included angle have been most common, but there is a trend in industry in which most 2 μm tip radius styli are being introduced with a 60° included angle. In fact, some instrument manufacturers now only supply 60° styli. Smaller angles are less likely to experience flanking[4], so they may be desirable for some types of surfaces. However, smaller angles are more likely to break when they catch or "hook" on the flank angle. A picture of a chipped stylus is shown in **Figure 23** on page **43**. 40° styli are very prone to this form of damage. 90° styli should therefore be used when measuring over edges or in other situations where the stylus flank contacts the surface.

Many people have stated that a 5 μm tip radius doesn't break as often as a 2 μm tip radius. This is typically due to comparing a 90°, 5 μm stylus to a 60°, 2 μm stylus. Since the stylus is conical and breakage occurs along the sides of the cone, it is the angle that typically causes breakage, not the tip radius. A 90° conical angle provides the most durable stylus. The durability of a 2 μm tip is similar to the durability of a 5 μm tip—provided that they both have the same angle.

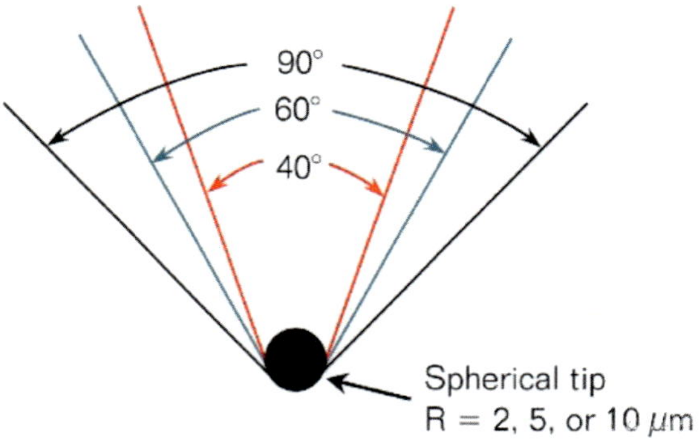

Figure 11: Common conical stylus dimensions

Stylus force

The stylus force must be adequate to follow the surface at the chosen measurement speed, but small enough to avoid damaging the surface. Both ASME B46.1 and ISO 3274 provide recommendations for the minimum force—see footnote **9** on page **40**. Most metallic workpieces are relatively insensitive to damage from the stylus, and any damage that does occur is so superficial that the impact on the measurement is insignificant—see page **40**. This may not be the case with

[4] Flanking is the condition when the surface contacts a cone angle (flank angle) rather than the spherical tip of the stylus.

non-metallic materials (plastics, glass, paper, films, etc.), and some may be fragile enough to need non-contact measurements.

Stylus condition

A dirty or damaged stylus will distort the data, and the impact on the data will be unpredictable. In fact, the result may be a profile graph that still looks like the expected shape. Styli must therefore be inspected frequently, especially when used in demanding applications. Visual inspection of the stylus at a minimum of 20X magnification is an essential first step. It should be performed on a routine basis, whenever something happens that may have caused damage (e.g., an equipment "crash"), or when the measurement data is suspect. Standardized procedures such as tracing across a sharp radius (e.g., a razor blade) may be useful—see page **45**. Routine measurement system uncertainty studies should be part of your normal laboratory procedure—see page **85** for one example.

What do skidded and skidless mean? Which should I use?

Contact surface texture instruments can be skidded or skidless. A skidded instrument uses a probe with a radiused skid and a sharp stylus, both of which contact the surface (see **Figure 12**). During a measurement the skid follows the general shape of the surface while the stylus detects the short wavelength roughness. This arrangement filters out the longer wavelength form and waviness because the movement of the stylus is always relative to the skid. It also protects the stylus from damage. The skidded system is commonly used with shop-floor instruments because of the protection it provides for the stylus, and because it is less sensitive to vibration and misalignment. It is also common on lower-priced devices.

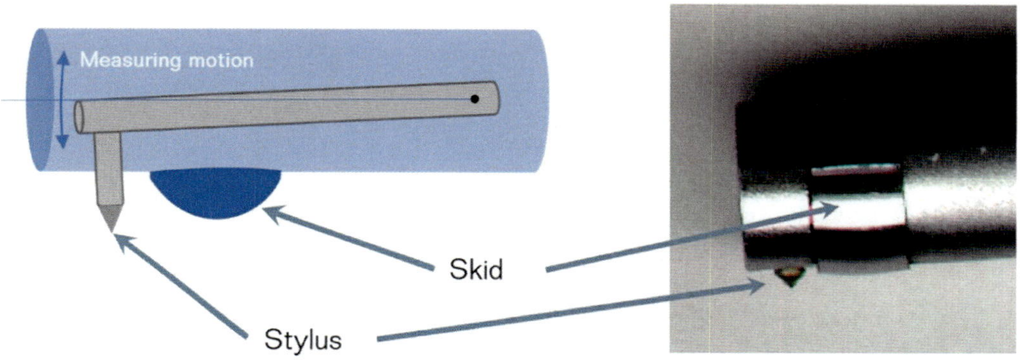

Figure 12: Skidded surface texture measurement probe

A skidless probe only contacts the workpiece with the sharp stylus (see **Figure 13**). An internal precision datum surface provides the reference for the stylus motion, rather than the skid (see **Figure 4** on page 11). Therefore a skidless measurement captures all of the wavelengths of the surface—form, waviness, and roughness—and gives a truer representation of the surface. Skidless systems typically leave the stylus exposed and are more prone to experiencing damage to the stylus tip. Thus, they tend to be more appropriate for laboratory conditions. Skidless systems are typically more sensitive to vibration and workpiece alignment, and also tend to be more expensive.

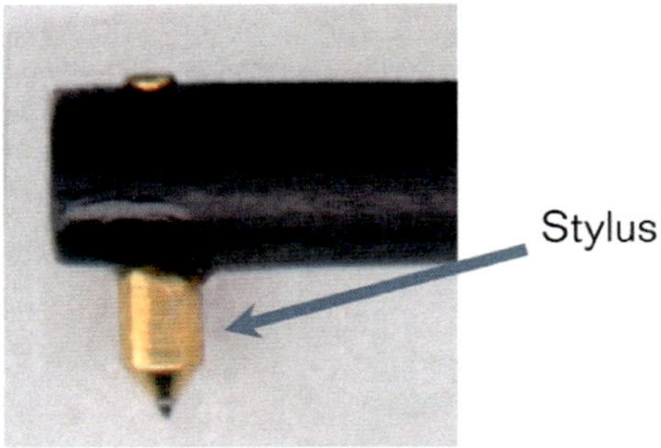

Figure 13: Probe and stylus of a skidless system

The choice of which system to use must primarily be driven by the characteristic being measured. If form or waviness are of interest, the skidless probe is the only option. If roughness data is the only requirement, then a skidded instrument may work. The environment in which the system is to be used is also a consideration—the rigors of shop floor measurements make skidded instruments very desirable—but this must be secondary to the needs of the product or process.

Both skidded and skidless systems can create artifacts in the data. Examples of these are shown in the section, "**I see unexpected things in my data. Are they real?**" starting on page **69**.

Which skid configuration should I use? Skid radius? Skid position?

Skidded instruments use a radiused skid that rests on the workpiece surface, providing a datum relative to which the stylus moves (see pages **27** and **29**). It is important to understand the effect on your measurements of the various skid designs.

The skid can be located to precede the stylus during a measurement, follow it, or surround it. In some applications, multiple skids have been employed. The goal is to allow the probe to travel parallel to the surface while minimizing any lateral forces or motion. The stylus location relative to the skid may be dictated by the geometry of the surface you are measuring, such as when you are measuring next to a shoulder or an edge.

Per the guidelines in ASME B46.1, if a single skid is used it should have a radius of curvature in the direction of the trace at least fifty times the roughness filter cutoff wavelength. If two skids are used that lie transverse to the probe, the radius should be no less than nine times the cutoff. Radii smaller than this will not fully separate the roughness from the sample waviness.

The distance between the skid and stylus should be chosen to not exaggerate any of the surface wavelengths. For example, if the skid-to-stylus distance is on the order of one-half the wavelength of surface features, the skid will be moving down while the stylus is moving up (and vice versa), effectively doubling the reported feature height—see **Figure 14** (a). If the skid-to-stylus distance equals the wavelength, or an integer multiple of the wavelength, the skid and stylus will move together (i.e., with no relative motion between them), resulting in a zero measured feature height—see **Figure 14** (b).

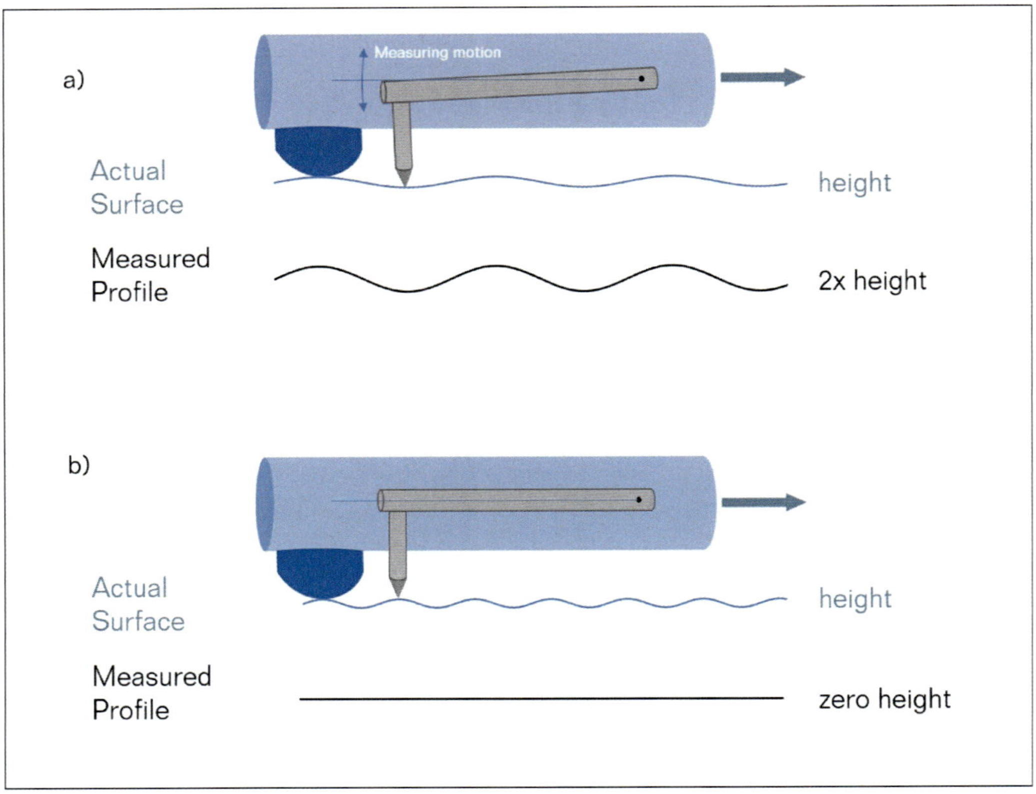

Figure 14: Illustration of the effect of an inappropriate spacing between skid and stylus

Debris on a surface can also lead to problems in a skidded measurement. In this case, the skid position will be reflected in the profile data. When the skid encounters debris, the stylus must extend farther down to maintain contact with the surface. This added extension results in an apparent "dip" in the measured profile—see **Figure 15**. The distance between the dip and the debris peak is the same as the distance between the skid and the stylus.

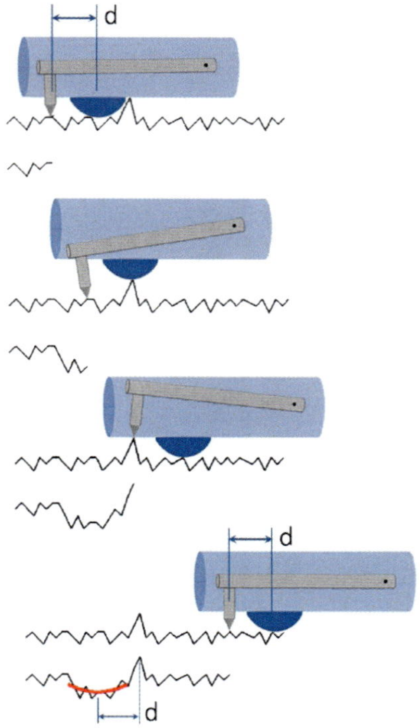

Schematic of an artificial dip being created by a skid traversing over a peak (possibly a piece of debris)

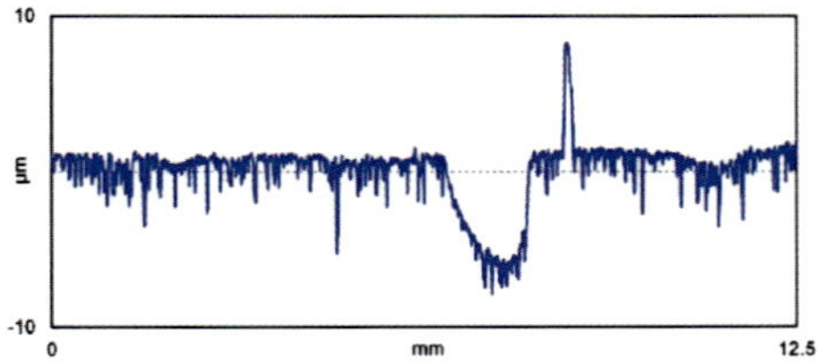

Actual profile from a skidded instrument

Figure 15: False dip in a skidded trace

It is important to note that the "dip" in the profile data can occur without a corresponding peak—see **Figure 16**. This skid-induced "dip" without an upward peak is very common, due to the fact that the skid is typically much wider than the stylus tip. Thus, the skid can encounter debris that the stylus misses.

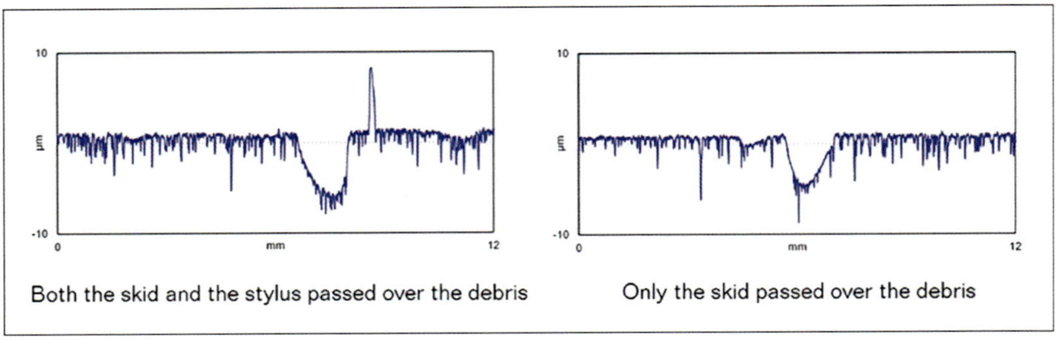

Figure 16: Skid passing over debris with and without a corresponding peak

Why are my skidded roughness results higher than I expect?

A skidded measurement system is typically rather robust against vibration. Thus, it may seem unusual to find that skidded parameter results are sometimes higher than those from skidless instruments. This problem can occur when there is debris on the surface which raises the skid, and the subsequent waviness filter does not adequately remove the effect (see **Figure 15**).

Figure 17 shows a measurement with a "dip" from the skid's encounter with debris. The debris is not contained in the profile graph because the stylus tip passed beside rather than over the debris. The "dip" or artificial valley is not completely removed from the roughness profile using the longer roughness cutoff shown on the left; this results in a higher roughness value. The artificial peaks on either side of the valley also increase the roughness value. The shorter roughness cutoff on the right removes all of the artificial valley, providing a more representative roughness value.

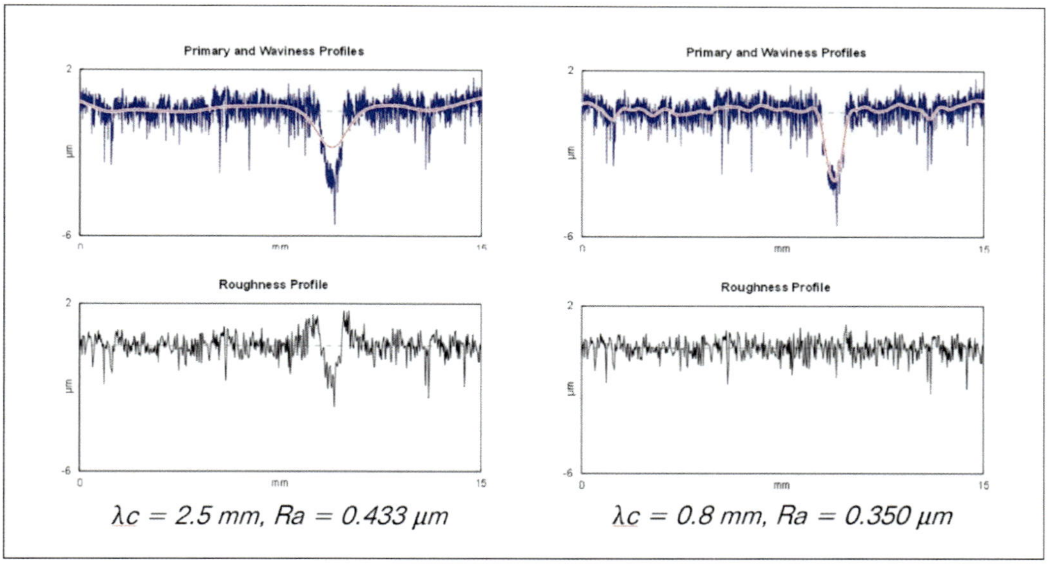

Figure 17: Effect of cutoff on roughness value from a skidded trace over debris

How does the stylus reach the bottom of deep valleys?

When viewing a surface texture trace the peaks and valleys can have very steep slopes which appear to be approaching 90°, as shown in the top image in **Figure 18**. This is an illusion resulting from the need to display micron-sized features on the z-axis of the graph and millimeter-sized features on the x-axis. These graphs therefore have an aspect ratio on the order of 500:1 or 1000:1. If a 1:1 aspect ratio is used (see bottom images in **Figure 18**) it can be seen that the sides of a feature have relatively shallow angles, so a stylus is usually able to stay in contact with the entire feature.

For some very deep and narrow valleys a large stylus tip radius may interfere with the sides and not reach the bottom, underestimating the depth of the feature. This is illustrated in **Figure 19** on page 38.

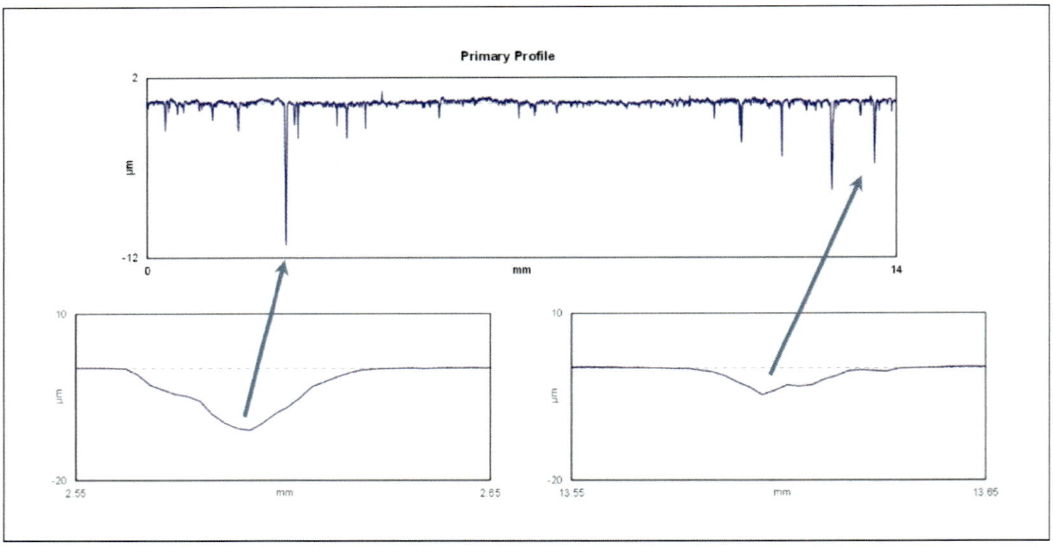

Figure 18: 1:1 aspect ratio views of deep valleys

Does the stylus tip radius make a difference?

Styli come in several shapes and sizes, including spherical, truncated pyramid, chisel point, chisel knife edge, and conical. The following will assume a conical stylus, but the principles also apply to other shapes.

Conical styli are available with different tip radii. Two, five, and ten micron radii are the most commonly available tips, and for the vast majority of important surfaces the 2 µm tip radius is most common. Included angles of 40, 60, and 90° are available. 90° styli have historically been the most common, but the current trend is toward tips with a 60° angle for the 2 µm radius. Some manufacturers are only supplying 60° included angles with the 2 µm stylus.[5] The smaller radii may be required to reach the bottom of deep narrow features, as shown in **Figure 19**. In this example a simulated 90° stylus with a 10 µm tip radius interferes with the sides of the feature and underestimates its depth (top right). A 2 µm radius stylus, however, is able to reach the bottom[6] and captures the full depth (bottom right). Note that the angles and radii are distorted due to the aspect ratio of the plots.

Any stylus will mechanically filter fine features, so ISO 3274 and ASME B46.1 recommend maximum tip radii based on the short filter cutoff wavelength setting[7]. These are listed in **Table 6**. If these recommendations are followed, it is likely that the mechanical filtering will be outside the specified transmission band.

[5] While a 90° stylus is very common, and most robust, 60° styli are commonly used for measuring surfaces with curves and angles. The 60° included angle reduces the risk of flanking.
[6] Note that this measurement was likely made with a slow-moving 2 µm stylus; hence the "perfect" fit with the simulated stylus in **Figure 19**.
[7] Section 7-4.2.4 of ASME B46.1-2009 provides the mathematical equation for calculating the maximum allowable tip radius for a surface texture modelled as a sine wave of wavelength λ and amplitude A. However, it is common to simply follow the guideline based on the λs value.

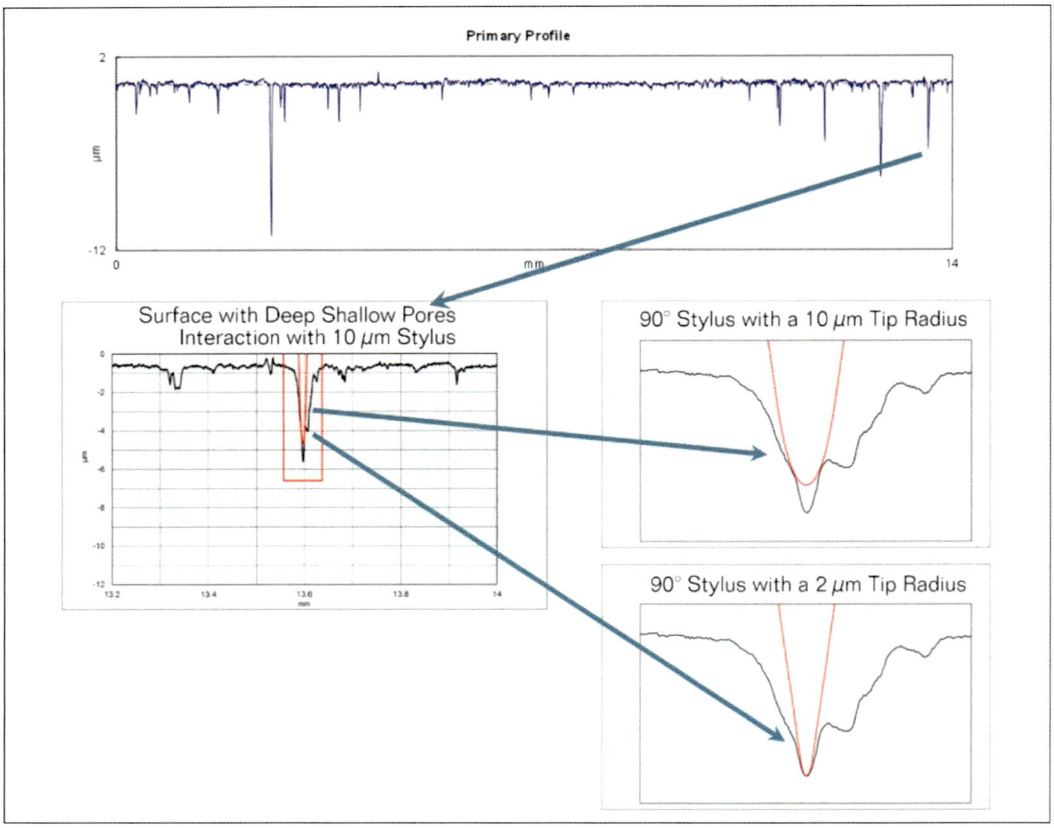

Figure 19: Large tip radius stylus interfering with deep narrow features

Table 6: Recommended maximum stylus tip radius based on lambda-s (λs)

λs (µm)	Tip radius (µm)
2.5	2
8	5
25	10

A smaller radius makes the stylus slightly more fragile and likely to be damaged. However, the included angle is a more important consideration for diamond tip durability. The crystalline structure of diamond is such that a 60° included angle is much more likely to chip or break than a 90° included angle.

Diamond breakage can be expensive. Worse yet, it often goes undetected. The instrument will still measure with a damaged stylus tip, and it can create data that is not representative of the surface. For more information, see the section on stylus damage starting on page 43.

Some stylus suppliers only provide 60° included angles for their 2 µm radius styli. If an instrument is being used in situations where tip breakage is more likely, consider switching to a 5 µm tip radius (if allowed[8]) with a 90° included angle.

Some people feel that the 5 µm tip is much more durable than a 2 µm tip. This isn't necessarily true. The relative difference between a 2 µm and a 5 µm radius is very small when considering that most stylus damage occurs well above the tip—see **Figure 20**. Furthermore, many styli break off at the attachment, not in the actual cone of the diamond.

Additional information on the tip radius can be found in section 7 of ASME B46.1-2009.

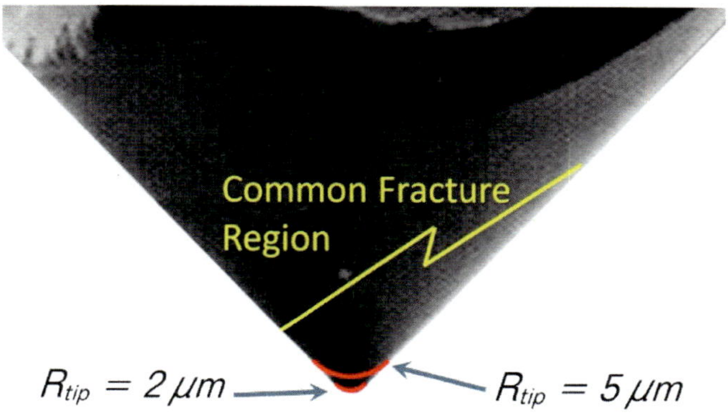

Figure 20: Tip radius as compared to typical breakage for a 90° stylus

[8] As a general rule, if the Ra of the surface is greater than 0.5 µm and if the roughness cutoff is 0.8 mm or larger, then a 5 µm radius can be used instead of a 2 µm without significantly impacting the final result.

Will the stylus scratch my surface?

If the stylus scratches my surface, can I trust the results?

The vertical force on the stylus must be sufficient to maintain contact with the surface. While these stylus loads are very small[9], when combined with the small stylus tip radii they create large contact stresses. **Table 7** contains the calculated[10] contact stress values for some common materials, assuming elastic behavior and static loading. These stress values are significantly higher than the materials' elastic limits[11] so, technically, the stylus is changing ("damaging") the surface. We often see evidence of this in the form of scratches. However, does this mean the measurement is invalid, and is the surface truly damaged?

Given the actual curvature of the surface texture on most surfaces, the contact conditions will be much the same over the full evaluation length. The deformation will essentially move the surface downward uniformly and the peak-to-valley distances will remain unchanged. All of the calculated parameters will therefore be valid. Some scanning electron microscope photos have indicated that the stylus damage may be different on rising features as compared to falling features. This can be related to the measuring speed and the slightly higher lateral force when pulling against a peak rather than sliding down a peak. The difference is typically insignificant.

Regarding surface damage, consider the radii of the contact areas shown in **Table 7**. The width of any "scratch" is significantly less than 1 µm, even on a soft lead surface, and much less than the grain size of most crystalline materials. The scratch depth is on the order of tens of nanometers. Any damage is therefore very superficial and of no practical consequence on typical engineered components[12]. If this amount of damage is of concern, non-contact measurement methods should be considered. For example, surfaces used in optics cannot tolerate stylus damage and thus alternative methods are used. Very soft surfaces may also have to be measured without a stylus.

---

[9] ISO 3274 recommends a nominal static stylus force of 0.75 mN. ASME B46.1 provides guidelines for a maximum force based on stylus tip radius: 0.7 mN for a 2 µm stylus, 4 mN for 5 µm, and 16 mN for 10 µm. Skidded instruments also have guidelines for the skid force, with ISO 3274 specifying a maximum force of 0.5 N. For reference, the diamond elastic constants used here are E = 1220 GPa and $\nu = 0.2$.

[10] Contact stress calculated using Hertzian stress equations and assuming a diamond sphere contacting a homogeneous elastic half-space. E_1, E_2, ν_1, ν_2 are elastic constants, R = sphere radius, F = contact force, a = contact radius, p_0 = maximum contact stress, and δ = displacement depth.

$$a = \sqrt[3]{\frac{3FR}{4E'}},\quad p_0 = \frac{1}{\pi}\left(\sqrt[3]{\frac{6FE'^2}{R^2}}\right),\ \text{and}\ \delta = \sqrt[3]{\frac{9F^2}{16RE'^2}},\ \text{where}\ \frac{1}{E'} = \frac{1-\nu_1{}^2}{E_1} + \frac{1-\nu_2{}^2}{E_2}$$

[11] For example, on a macroscopic scale, a steel surface hardened to 60 HRC will plastically deform under a contact stresses of approximately 3 GPa.

[12] It is acknowledged that the plastic deformation experienced at these stresses will create a larger scratch than that predicted by the elastic equations. However, it should still be small relative to the metallurgical and topographical features on typical components.

Further discussion of these surface effects can be found in Thomas's *Rough Surfaces*[13]. This in turn references several studies that attempted to measure surface changes from stylus measurements. No significant changes were found.

Note that the calculations in **Table 7** assume static loading. The dynamic loading during an actual measurement, especially at high traverse speeds, can be significantly greater. This must be kept in mind when measuring sensitive surfaces.

Table 7: Contact conditions on typical materials with a 2 μm diamond stylus and a 0.75 mN load

Material	Elastic Modulus (GPa)	Poisson's ratio	Stress (GPa)	Contact radius (μm)	Depth (μm)
Steel	207	0.3	11.1	0.18	0.02
Aluminum	69	0.33	5.8	0.25	0.03
Glass	65	0.25	5.4	0.26	0.03
Lead	16	0.4	2.3	0.39	0.08

[13] TR Thomas, *Rough Surfaces* (London: Imperial College Press,1999), 23-24

While stylus-induced surface damage is of no practical consequence for most engineered surfaces, the damage from a skid can sometimes be more significant. **Figure 21** shows the surface of a calibration specimen[14] which has been traced with both skidded and non-skidded styli. Within the skid track the surface peaks have been plastically flattened, and the deformed material has filled in the valleys. Any subsequent measurements on the skid-damaged area of this patch would have roughness parameter values different from any certified value.

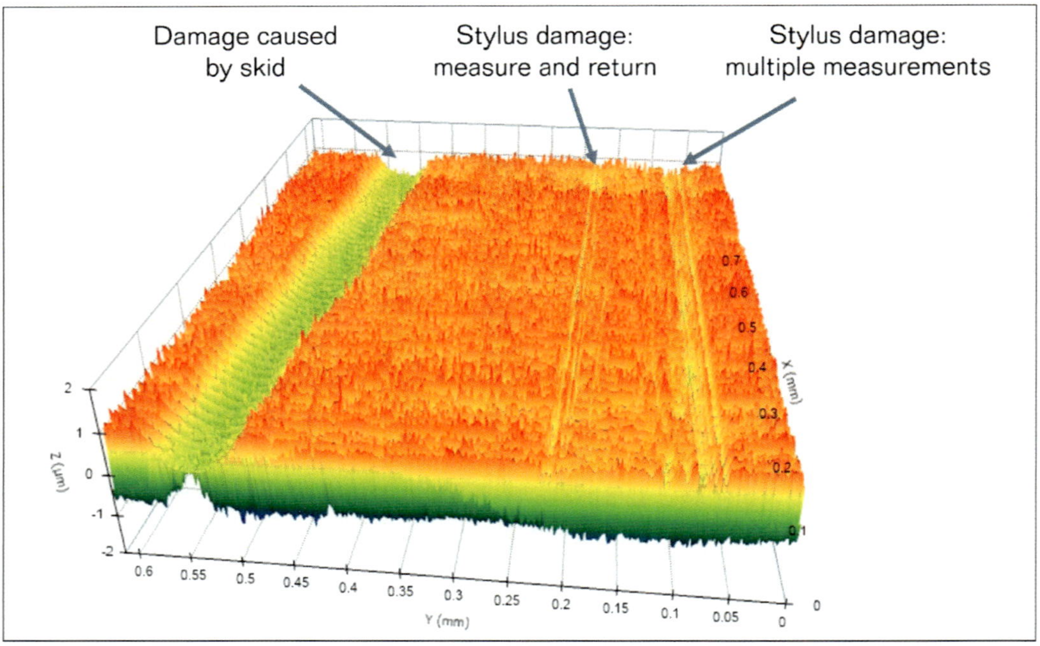

Figure 21: Damage from a the skid of a stylus measurement instrument.

[14] The mechanical properties of this electroformed nicked surface are not known, but one source of electroformed nickel claims a yield strength on the order of 480 MPa (70 ksi). See **http://www.electroformednickel.com/**.

Can a diamond stylus be damaged? How can I tell?

Diamond is the preferred stylus tip material because of its high hardness and low friction coefficient. While diamond's hardness makes it very resistant to wear, it must still be treated with care. Without proper precautions it can be chipped or may even become dislodged from the probe. **Figure 22** compares a new stylus with one where the diamond tip has been broken off. **Figure 23** shows a new and a damaged stylus where the diamond is still attached, but it is chipped and less able to follow the fine surface details.

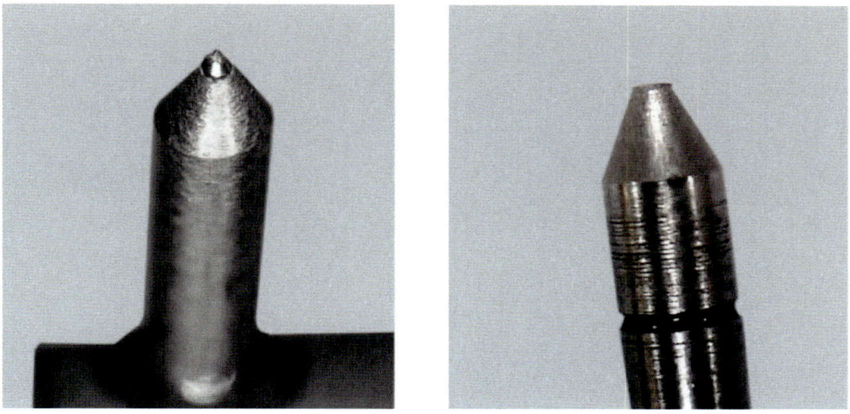

Figure 22: New stylus (left) and one with the diamond tip totally missing (right)

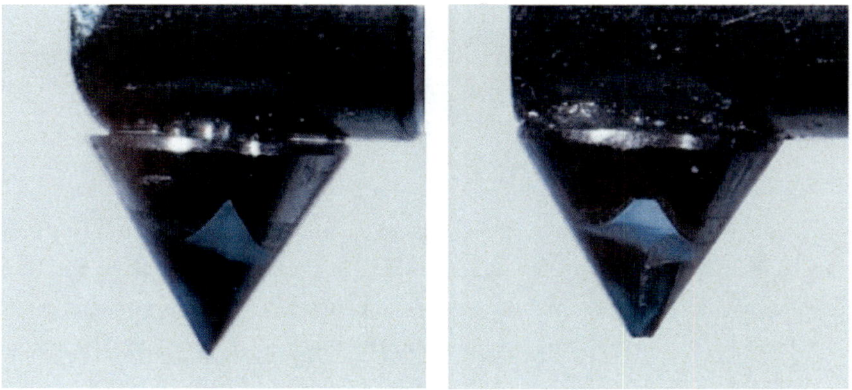

Figure 23: New (left) and chipped (right) diamond styli

Styli should be periodically inspected using, if available, a stereomicroscope with sufficient magnification to clearly see the tip (usually a minimum of 20X). An inexpensive USB microscope can also work. Periodic uncertainty studies (see page **85**) will help find stylus problems, as will traces over a series of narrow grooves (ISO 5436, type B2).

Figure 24 shows data gathered with the two styli from **Figure 23**, tracing the same workpiece at approximately the same location. Sample parameter values are included. The traces on the left were made with the new stylus while the right-hand traces were made with the chipped stylus. The bottom traces are enlarged views of the top ones. This example illustrates not only that stylus damage can change the measured profile but, more importantly, that the change is unpredictable, sometimes counterintuitive, and will impact various parameters differently.

The same parameter trends noted here applied to multiple traces made under the same conditions.

Note: This trend is not typical of all chipped or broken tips. Sometimes the roughness is considerably reduced rather than increased.

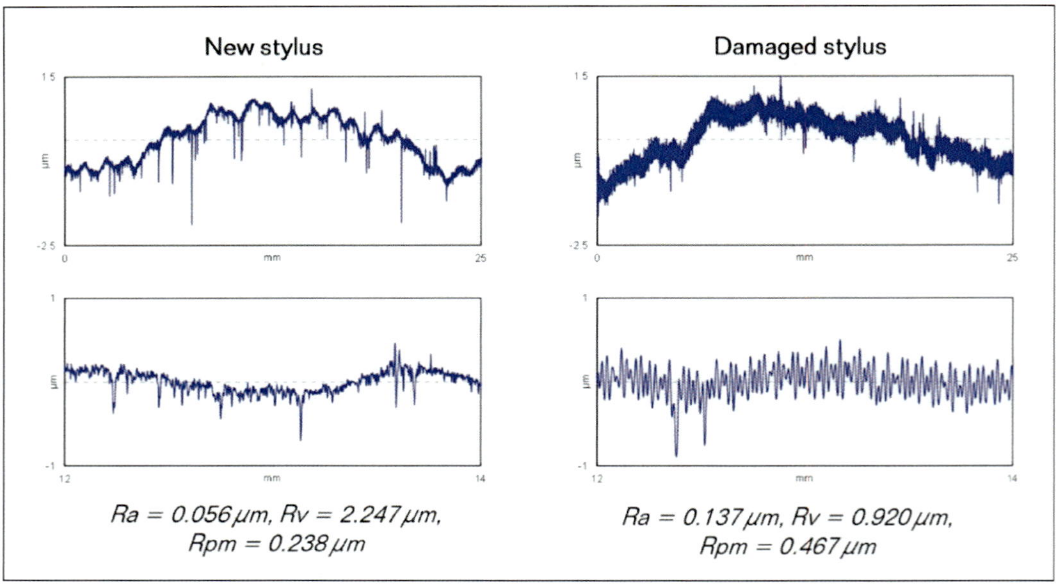

Figure 24: Comparison of data taken with new and damaged styli

If visual inspection of the stylus is not possible or is inconclusive, the tip radius can be measured more directly by tracing over a sharp edge such as an uncoated razor blade. If the radius of the edge of the blade is sufficiently small relative to the stylus tip radius[15], the output profile represents the tip radius. ISO 5436 provides the details for this evaluation technique (type B3). Use a curve-fitting algorithm[16] to derive the actual tip radius from the measured data. **Figure 25** shows a trace over the edge of a razor blade using a stylus with a nominal 2 µm radius and a 90° included angle (top). An

[15] A new uncoated razor blade has an edge radius of approximately 0.1 µm. The output radius equals the sum of the razor blade and stylus radii, so ignoring the blade radius creates errors of 5% or less for typical styli.

[16] For example: Vorburger et. al., "An iterative algorithm for calculating stylus radius unambiguously," *Journal of Physics*: Conference Series, vol 311.

enlarged view of the trace is also shown. This image can be used to calculate the actual stylus tip radius (1.98 μm) and the actual included angle (87°).

Further discussion on stylus damage can be found in the section on tip radii starting on page 37.

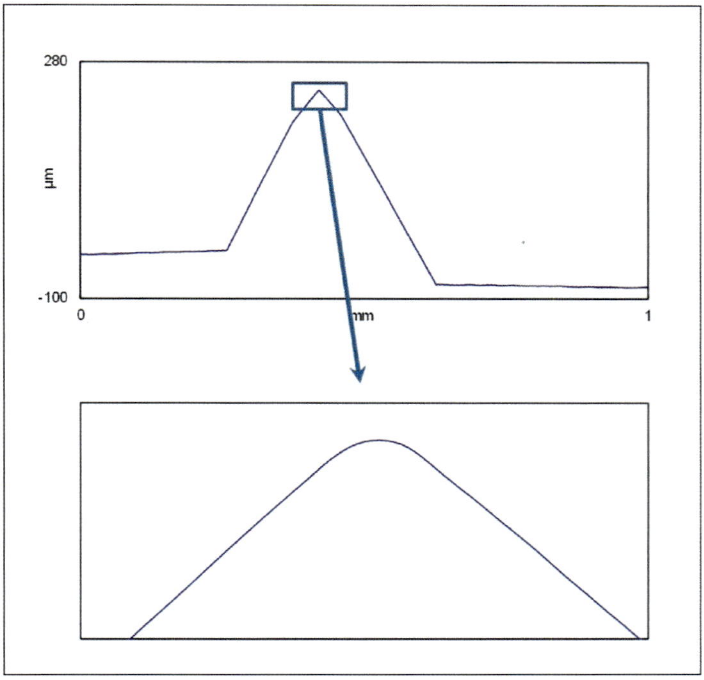

Figure 25: Measurement of the stylus tip radius using a razor blade

What are data spacing and data density?

Digital surface texture instruments sample the height of the surface at specific intervals, based on time or distance traveled. This x-axis spacing between data points—known as data spacing, sampling interval, or data density—can vary from about 0.25 µm to 5 µm or more, depending on the type of instrument and the measurement conditions[17]. The horizontal resolution can affect measurement results in several ways.

In extreme situations, undersampling can create an aliasing problem and a different apparent wavelength, as shown in **Figure 26**. The more common spacings will usually not significantly impact form or waviness because those wavelengths are much larger than the spacing. However, they will impact the data analysis and some roughness parameters, as discussed below.

- The application of the short filter requires adequate data density. A minimum of five data points is usually required[18] to define a given wavelength. So, for example, if the cutoff wavelength (lambda-s, or λs) is set to 2.5 µm, the spacing can be no larger than 0.5 µm. A larger spacing is compatible with a larger λs. This is reflected in the equipment requirements documented in ISO 3274 (see Table 1 in ISO 3274:1996, see also **Table 4** on page **21** of this document). If the actual spacing is too large for the specified λs, the short filter will not be applied and the shorter wavelengths will be included in the analysis results[19].

- Some roughness parameters are sensitive to the data spacing because that spacing is included in the calculation. For example,

- **Table 8** lists the analysis results for two parameters from the same measurement (see **Figure 27**) at three different data spacings. While an averaging parameter like Ra shows essentially no change, the change in others such as Rdq can be significant. The application, or not, of the short filter can further increase these differences, and may, in some situations, be the primary source of those differences. Some of these sensitive parameters are used in comparative studies in, for example, tribological simulations. Care must be taken that these comparisons incorporate data taken with the same spacing.

- A lower data density reduces the likelihood of detecting the maximum peak heights and valley depths. **Figure 28** shows an example of this, where a valley disappears when the data

---

[17] While many instruments may not allow direct adjustment of the data density, the choice of measurement conditions can affect the density. For example, some systems impose a limit on the number of data points. A longer traverse length will therefore result in a larger data spacing. This may be a gradual increase in spacing or a step increase at a certain length.

[18] Note that some software only requires 3 points per λs.

[19] Some instrument's software may automatically increase the λs for the larger spacing. It is often not clear which approach is used by a given measurement system.

spacing is increased. While the 10X change in spacing used in **Figure 28** may seem extreme, such differences have in fact been seen in two different instruments from the same supplier.

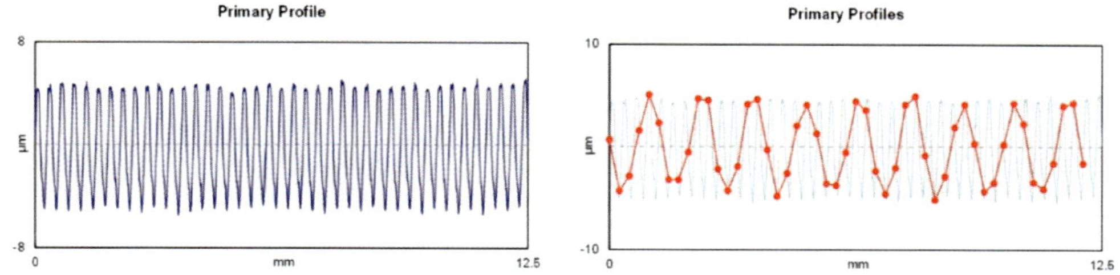

Two traces of the same surface. The data in blue has a data spacing of 0.58 µm. The data in red has a 250 µm spacing. The extreme undersampling in the red profile creates an artificially long dominant wavelength, an effect known as "aliasing."

Figure 26: Effect of extreme undersampling

Table 8: Influence of data spacing on values of selected parameters

Data spacing	0.5 µm	1.0 µm	1.5 µm
Ra (µm)	0.472	0.471	0.471
Rdq (°)	16.192	11.874	9.257

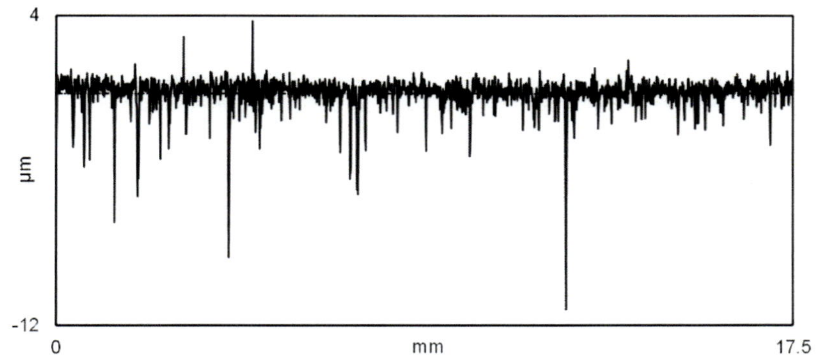

Figure 27: Profile used to calculate the values in **Table 8**

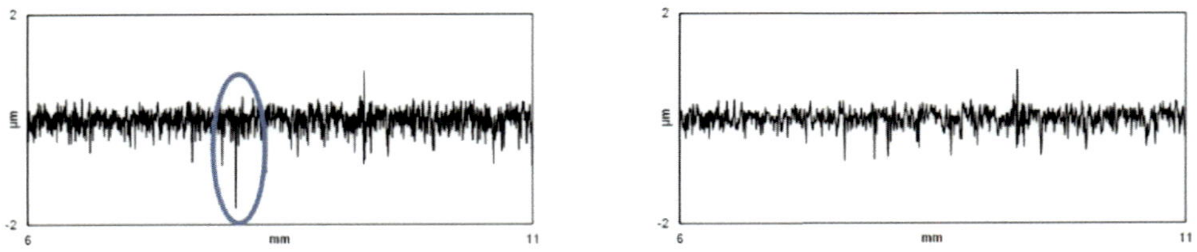

A simulation of the potential effect of data point spacing on detecting high peaks and deep valleys. The *roughness* profile on the left has a spacing of 0.5 µm. The one on the right is the same trace desampled to a spacing of 5 µm. The larger spacing totally misses the deep valleys seen just left of center in the left trace.

Figure 28: The effect of data density on detecting peaks and valleys

How fast can I measure?

While lab productivity often emphasizes making measurements as quickly as possible, the ability of the system to mechanically respond to the surface features and electronically capture the data will limit the traverse speeds that can be used. Mechanical noise may also be more significant at certain traverse speeds. A measurement uncertainty study (e.g., page **85**) should always be performed at the proposed measuring conditions before using those conditions.

The system's mechanical response is the ability of the stylus to stay in contact with the surface with a relatively uniform contact force. Factors that affect this include the stylus tip radius, the spring force, vibration, gravity, and damping due to the workpiece material or the friction at the pivots.

The electronic response is the ability of the system electronics to sense and capture the movement of the stylus. This is influenced by the analog-to-digital sampling rate, anti-aliasing filtering, electronic filtering, and electronic noise.

Figure 29 is an example of the design of the electronics affecting the data density in an older (1992) instrument. For this instrument, traverse speeds of 2.5, 0.25, and 0.025 mm/sec result in data spacings of 6.25, 1.25, and 0.125 µm, respectively. As discussed on page **46**, these data density differences can affect some of the roughness parameter values. The data density also impacts the application, or not, of the short wavelength filter (lambda-s, or λs); see page **54**.

For most instruments, a speed of 0.5 mm/sec is a good starting point. If a faster speed is desired, the surface should first be measured with the slower speed to determine "typical" results. Next, the surface can be re-measured at the faster speed and the results can be compared. To avoid the influences of surface variation, it is often a good idea to mark a small area of surface and then measure at least ten times (uniformly covering the area) with each setting. The mean and standard deviation of these measurement can then be compared.

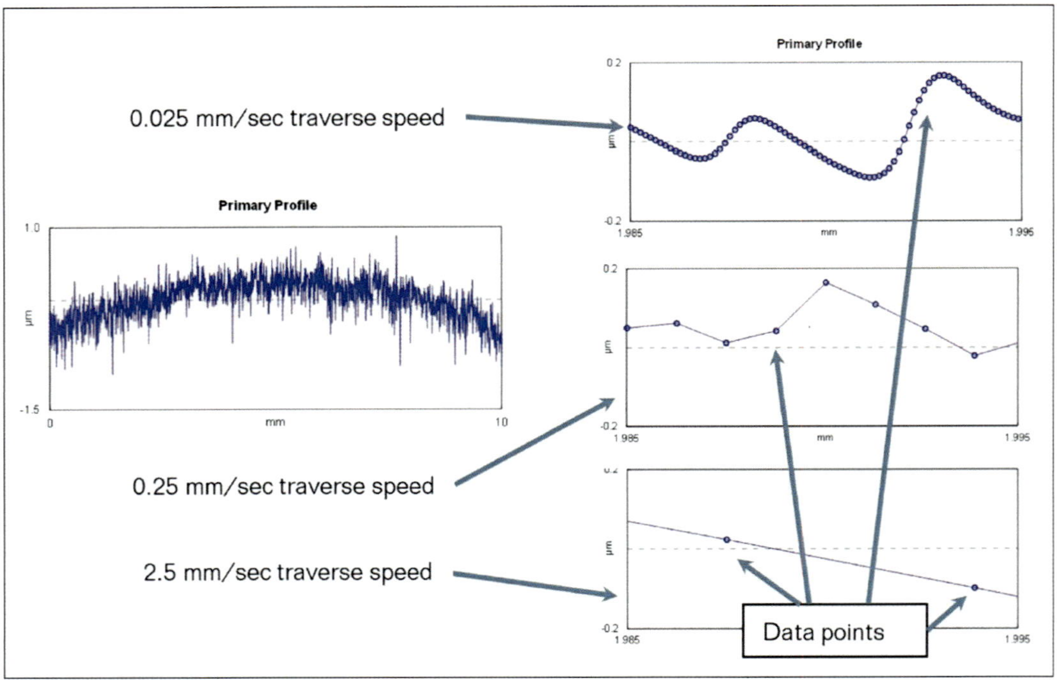

Figure 29: Effect of traverse speed on data spacing with one older (1992) instrument

How long should a measurement be?

Is the length different for roughness and waviness?

The "correct" measurement length must balance the desire for as much information as possible with practical considerations such as the instrument characteristics and equipment productivity. The measurement must be long enough to include the longest wavelength(s) of interest and should contain an adequate number of any characteristic features to properly represent their sizes. The lengths for roughness and waviness can be different, but they don't have to be. Some technical drawings may specify the length, either as a number of sampling lengths or as an actual length (see pages **194** and **207**); this requirement must be followed, and if the measurement system is incapable of doing so, this must be discussed with the customer. If the evaluation length isn't defined, or if you are designing the component, consider the following factors in choosing the measurement length.

Note that when discussing lengths there are several terms that must be understood, and properly used: traverse or tracing length, evaluation or assessment length, and sampling length. If any of these are confusing, see **Figure 30** (below) or read their descriptions in the ***Definitions*** section starting on page **292**. For communicating lengths on engineering drawings, the "evaluation length" is preferred over "tracing length," since the evaluation length is the length that is used for calculating surface texture parameters. Various instruments discard different lengths of data from the end regions, so different instruments may require longer or shorter tracing lengths to achieve the same evaluation length.

Roughness evaluation length considerations

The relevant specifications and most measurement instruments use a default evaluation length of five sampling lengths (a sampling length is equal to the filter cutoff wavelength). Thus, a default evaluation length for a 0.8 mm filter cutoff is 4 mm of data. This may or may not be adequate or possible for your specific needs. Several factors must be considered when selecting an evaluation length:

- How long is the feature? If it is 100 mm long, you may feel that evaluating a 4 mm sample (5 x 0.8 mm sampling length) is inadequate. This would be an even larger concern if you were using a 0.25 mm roughness cutoff (default evaluation length = 1.25 mm). Note that if your feature is very short you should not assume that you are required to have five sampling lengths. It is not advisable to reduce your cutoff value to achieve that length, since doing so changes the definition of roughness and waviness. It is acceptable to have fewer than five sampling lengths; refer to the short surface discussions on pages **62** and **207** for more information.

- How long is the traverse range of the instrument that will be used? Be careful specifying long lengths (> 50 mm) unless you know your instrument or your supplier's instrument is capable of making the measurement or is willing to upgrade and become capable.
- Will longer traces yield a lower data density? Some instruments have limited memory with room for a limited number of data points; thus, for longer traces the data point spacing is increased. If so, will this still be adequate for the subsequent analysis? See page 46 for more information on traverse length and data density; see page 54 for how this can affect the application, or not, of the short wavelength filter (lambda-s, or λs).
- Will the roughness values from locally "bad" areas (e.g., high roughness) be averaged out by taking a longer trace? Or, will a short trace miss any high roughness regions? Special considerations are required if a surface is suspected of having an inconsistent roughness. For example, you may choose to take a long trace and then visually examine the *roughness* profile to identify inconsistent regions. If such regions are found, short traces at those locations can then be taken. Alternatively, some software includes parameters that report the highest roughness in a long trace.
- If your specification includes some of the "count" parameters—for example, valley count (Rvc) or high spot count (Rhsc)—the evaluation length directly affects the reported value. If the drawing doesn't specify the length, the customer must be consulted. For these parameters the evaluation length must be clearly stated in the lab report.

Waviness evaluation length considerations

There is no default evaluation length for waviness. As stated above, most instruments use a default evaluation length of five sampling lengths where a sampling length is equal to the roughness filter cutoff wavelength. Thus, a default evaluation length for a 0.8 mm filter cutoff is 4 mm of data. This is often inadequate for the measurement of waviness. In some situations, people have selected a longer filter cutoff for waviness in order to get a longer trace. This is not recommended as it results in a gap between the specified wavelength band for roughness and that for waviness, potentially missing important features.

- The evaluation length should be long enough to capture enough of the wavelength(s) of interest to adequately characterize them.
- If the measurement is on a surface with a nominal form or geometry, consider using bandpass waviness to separate the features, rather than shortening the trace to minimize the influence of the form (see page 328 for more on bandpass analysis).
- Since waviness features are by definition longer than roughness ones, a waviness evaluation length will likely be longer than the roughness measurement. If you wish to make only one measurement, the length must be driven by the waviness parameters. If this results in inadequate data density, separate roughness and waviness measurements must be taken. See

the section, "**Can I evaluate roughness, waviness, and shape with one long trace?**" on page **54** for more on this topic.

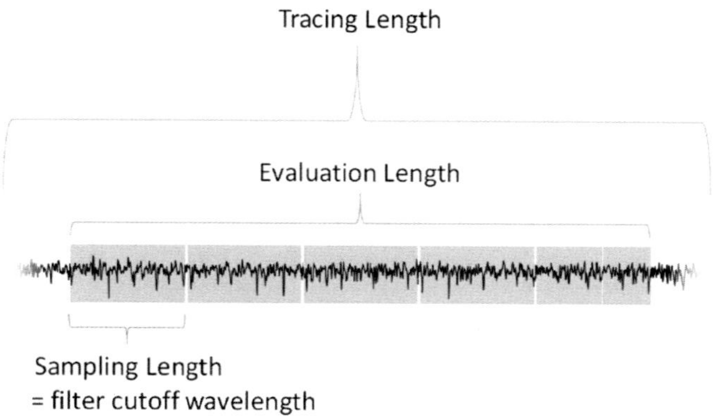

The Evaluation Length is used to determine texture parameters. All instruments discard data from both ends of the Tracing Length in order to achieve the Evaluation Length—but the discards vary greatly between instruments (fading regions in the image above).

Figure 30: The relationship between tracing, evaluation and sampling lengths

This section speaks about traverse length and evaluation length and the fact that there is no consistent relationship between them (see the text in **Figure 30**). How, then, does an inspector know the appropriate traverse length when he or she is asked to measure a surface with a specified evaluation length? The answer is that the inspector must determine the traverse length based on the characteristics of their measurement system; no more specific guidance can be given.

Can I evaluate roughness, waviness, and shape with one long trace?

The correct answer to this question depends on the surface, the surface texture specification, and the measurement system.

In theory, using one long trace to evaluate all three domains—roughness, waviness, and shape—is a reasonable approach. Making one long measurement is usually[20] a more efficient use of laboratory resources than taking multiple traces of different lengths. In addition, with a long trace the roughness is evaluated over a larger sample of the surface and you are more likely to find poorer quality regions than with a short trace[21]. However, if the goal is to measure shape, waviness, and roughness with only one trace, issues may arise with some measurement systems.

Consider this surface texture callout which uses a short filter cutoff wavelength (lambda-s, or λs) of 0.0025 mm and the evaluation length of 45 mm.

0.0025–0.8 / ln=45 / Ra 0.25
0.8– / ln=45 / Wt 6

The above callout seems reasonable. However it may not be measurable on some measuring systems. This is due to the fact that some measuring systems can only handle a limited number of data points in a profile. When a longer trace is required, these systems increase the spacing between the points.

The 2.5 µm λs specified above requires a minimum data spacing of 0.5 µm. If the spacing is larger, this λs filter cannot be applied and your measurement of the 45 mm evaluation length would be out of compliance with the callout.

What should you do in a situation such as this?

Unfortunately, other than switching to a different measurement system, there is no perfect answer to this question. The following are possible responses, each of which is problematic in some way, with the extent of the problem being dependent on the nature of the surface. Whatever you choose

[20] It should be noted that some instruments can perform waviness and form measurements at a faster traversing speed. Thus, if roughness is not required over the full length, it might actually be faster to perform the waviness/form trace at a higher speed and then go back for a shorter roughness trace at the slower speed.

[21] While a long trace has a higher chance of sampling any poor quality regions than a randomly-placed short trace, that does not guarantee that the poor quality regions will be identified. For example, with an averaging parameter such as Ra or Rz, the average of the high and low roughness values may still be within specification limits, effectively "hiding" the poor-quality regions. A visual examination of the surface profile is always recommended with, if appropriate, a selective analysis of any rougher (or smoother) regions.

to do must be discussed with your customer and documented in your report. The response options are listed here and explained more fully below.

- Take one 45 mm trace and ignore the λs issue
- Take multiple short traces to evaluate the roughness and one long trace to evaluate waviness
- Take multiple short traces, each of which complies with the λs requirement, with a combined length of 45 mm
- A fourth option, changing λs, is a design question that is outside the scope of this book. However, some comments will be provided at the end of this section.

Take one 45 mm trace and ignore the λs issue

How big of a deal is it if λs is not applied? This depends on the nature of the surface and the characteristics (parameters) being evaluated. **Table 9** contains examples of various types of surfaces[22], showing the change in value of some common averaging and extreme value parameters with different data spacings. The original data sets have a spacing of 0.25 or 0.5 μm (dx in the tables). Data sets with a 1 μm spacing were created by mathematically desampling the original data. A 2.5 μm λs was applied to the 0.25 and 0.5 μm spaced data, but not to the 1 μm data. All profiles were analyzed using a least squares line reference geometry, a 0.8 mm λc, and a form-following spline roughness filter, except for the plateau-honed surface (d), where a robust filter was used. All values in **Table 9** are in micrometers. The * indicates those data sets to which the short filter was not applied.

[22] Note that the graphics in **Table 9** do not have consistent axis scales, and some are only a portion of the full trace. This shows the nature of the surface texture; the surfaces' overall shapes are of less importance for this discussion.

Table 9: Effect of data spacing on selected surface texture parameters

a)

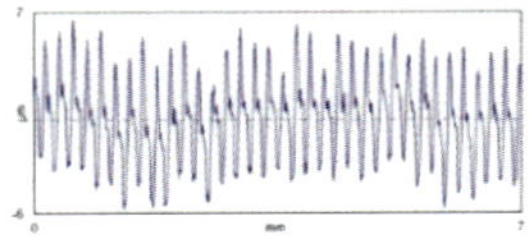

dx	Ra	Rz	Rpm	Rp	Rv	Rt	Wt
0.25	2.080	9.713	5.255	6.338	5.177	11.52	16.15
1*	2.081	9.734	5.264	6.337	5.189	11.53	16.14
Δ%	0.03%	0.21%	0.16%	0.01%	0.22%	0.10%	0.06%

b)

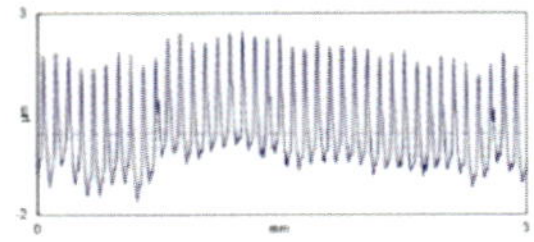

dx	Ra	Rz	Rpm	Rp	Rv	Rt	Wt
0.25	0.719	3.330	2.227	2.464	1.523	3.987	2.039
1*	0.720	3.376	2.260	2.518	1.553	4.071	2.038
Δ%	0.14%	1.40%	1.45%	2.18%	1.95%	2.09%	0.03%

c)

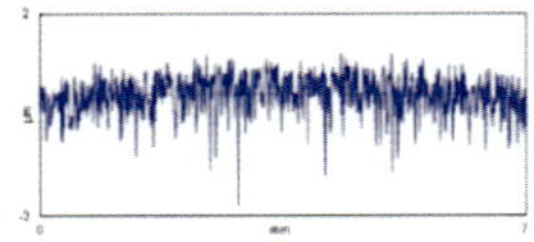

dx	Ra	Rz	Rpm	Rp	Rv	Rt	Wt
0.25	0.353	2.949	1.040	1.558	2.974	4.531	4.492
1*	0.364	3.035	1.096	1.559	3.017	4.577	4.492
Δ%	2.92%	2.91%	5.39%	0.09%	1.47%	1.00%	0.01%

d)

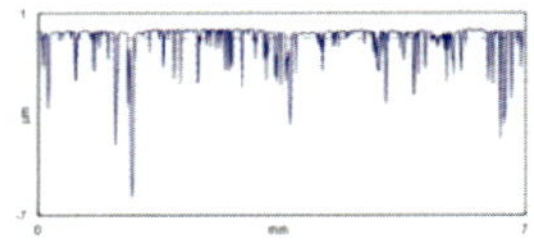

dx	Ra	Rz	Rpm	Rp	Rv	Rt	Wt
0.25	0.340	3.910	0.051	0.091	12.95	13.04	0.440
1*	0.341	3.937	0.066	0.162	12.96	13.13	0.439
Δ%	0.29%	0.69%	29.4%	78.0%	0.11%	0.66%	0.23%

e)

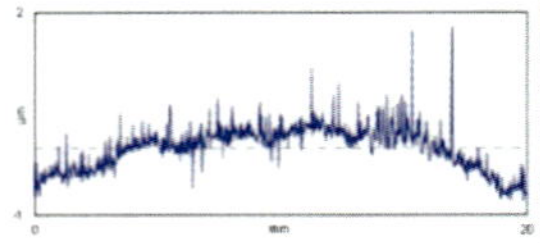

dx	Ra	Rz	Rpm	Rp	Rv	Rt	Wt
0.5	0.098	1.039	0.660	1.952	1.168	3.119	3.497
1*	0.099	1.048	0.663	1.951	1.177	3.128	3.497
Δ%	1.10%	0.94%	0.42%	0.01%	0.80%	0.29%	0.00%

f)

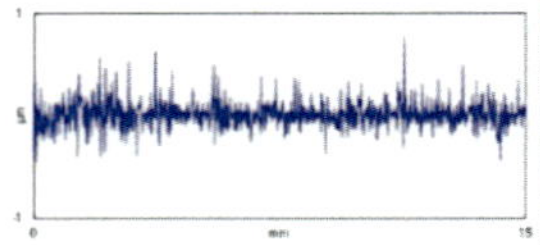

dx	Ra	Rz	Rpm	Rp	Rv	Rt	Wt
0.5	0.088	0.989	0.583	0.802	0.517	1.319	0.157
1*	0.127	1.147	0.630	0.860	0.674	1.534	0.157
Δ%	44.0%	16.0%	8.13%	7.28%	30.3%	16.3%	0.04%

g)

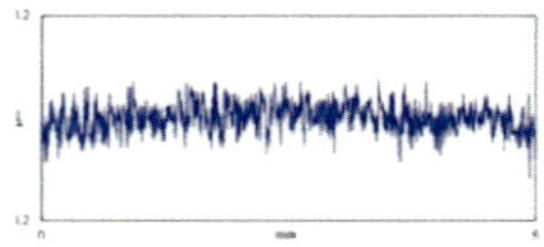

dx	Ra	Rz	Rpm	Rp	Rv	Rt	Wt
0.25	0.022	0.234	0.091	0.214	0.352	0.566	1.535
1*	0.026	0.265	0.107	0.215	0.380	0.595	1.535
Δ%	18.4%	13.3%	17.8%	0.19%	8.07%	5.08%	0.01%

The Ra values in **Table 9** cover a range of 2 to 0.02 μm. As would be expected, the effect on Ra of the application, or not, of the λs filter is generally larger for the smoother surfaces. This is because the amplitudes of the short wavelength "noise" which λs removes will make a larger contribution to the overall roughness of a smooth surface. This trend is not perfect, however, as shown by profiles (e), (f), and (g), where the Ra of (f) is impacted much more by the λs filter than would be expected from the "trend" implied by (e) and (g).

The effect of λs on the extreme value parameters, and even on the "averaged extreme value" parameters Rz and Rpm, is less predictable, showing almost random changes as the surface becomes smoother.

The effect on the waviness (Wt) is essentially zero. This is expected, since waviness is evaluated for wavelengths longer than 0.8 mm (in these examples) and the short features affected by λs are not considered. The small changes noted in **Table 9** are most likely an artifact of filtering.

This information supports the need to meet the data spacing requirement of the specified λs filter. The validity of the surface texture parameter values is impacted in two different ways if this doesn't occur:

1. Different instruments may handle the increased spacing differently. Some software will not apply the short filter. Other packages may increase the λs cutoff. It may be difficult to determine the approach taken by your instrument's software.
2. We cannot easily predict the results due to two opposing factors:
   a) Removing the short filter can increase roughness values by letting in more noise or high frequency content (see **Table 9**).
   b) Larger spacings can decrease roughness values by missing the most extreme points (see **Table 10** and the discussion about changing λs at the end of this section).

If it is not possible to generate data with a sufficiently small spacing, this lack of compliance must be noted in the data documentation.

Take multiple short traces to evaluate roughness and one long trace to evaluate waviness

This is usually the best answer. Unfortunately, it impacts laboratory productivity, which may be unacceptable in a busy production environment. Also, it is not in compliance with the evaluation length specified for roughness. If extreme value roughness parameters are specified, the results have to be manually sorted to find the true worst case.

Take multiple short traces, each of which complies with the λs requirement, with a combined length of 45 mm

This is acceptable for roughness, but it is the worst possible solution regarding waviness, especially the total waviness in a single-sided waviness specification such as the one given above. The reason is obvious, and is illustrated in **Figure 31**: when a surface contains long-wavelength waviness, assessing that surface with short segments will reduce the total waviness value. The problem is less severe with a straight surface, but the severity of the effect can only be determined with a full-length measurement. If you have time to make this full-length measurement, you can follow the first recommendation given above.

This approach may be less problematic if the waviness specification is bandpass rather than single-sided, especially if the traverse lengths are longer than λf. The λf filter would remove the longer-wavelength form, an effect not too dissimilar from evaluating within a short "window" (i.e., a short traverse length). **Figure 32** illustrates this. However, this option should still be avoided if at all possible, since the errors in the example in **Figure 32** can still exceed 30%. Other profiles could be worse.

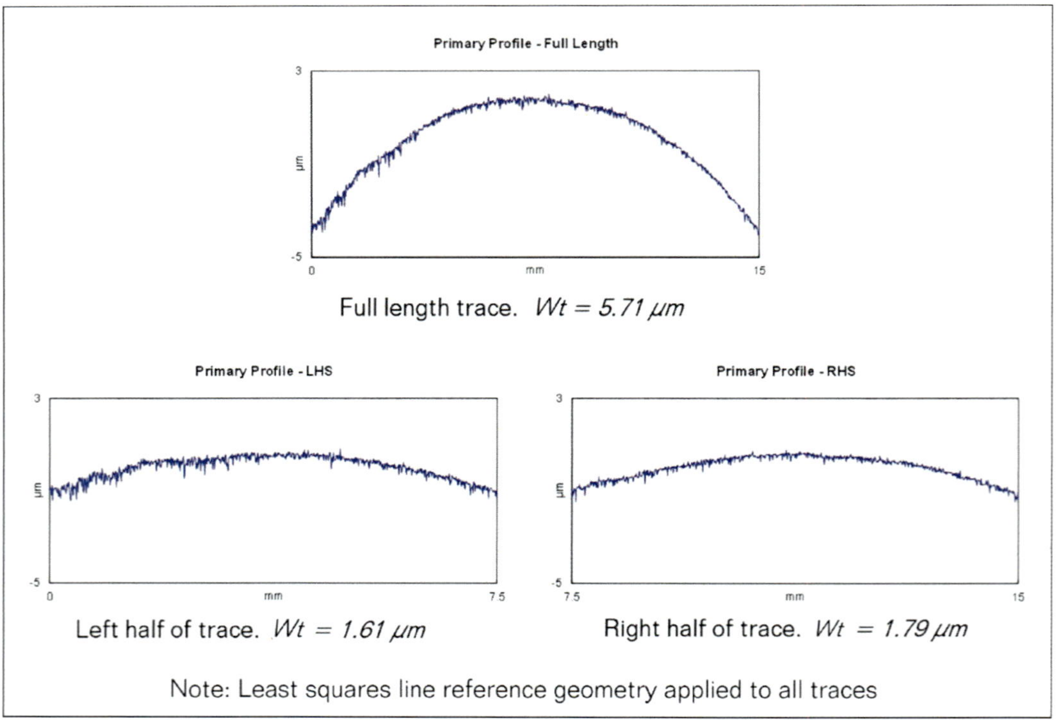

Figure 31: Effect of short evaluation lengths on Wt

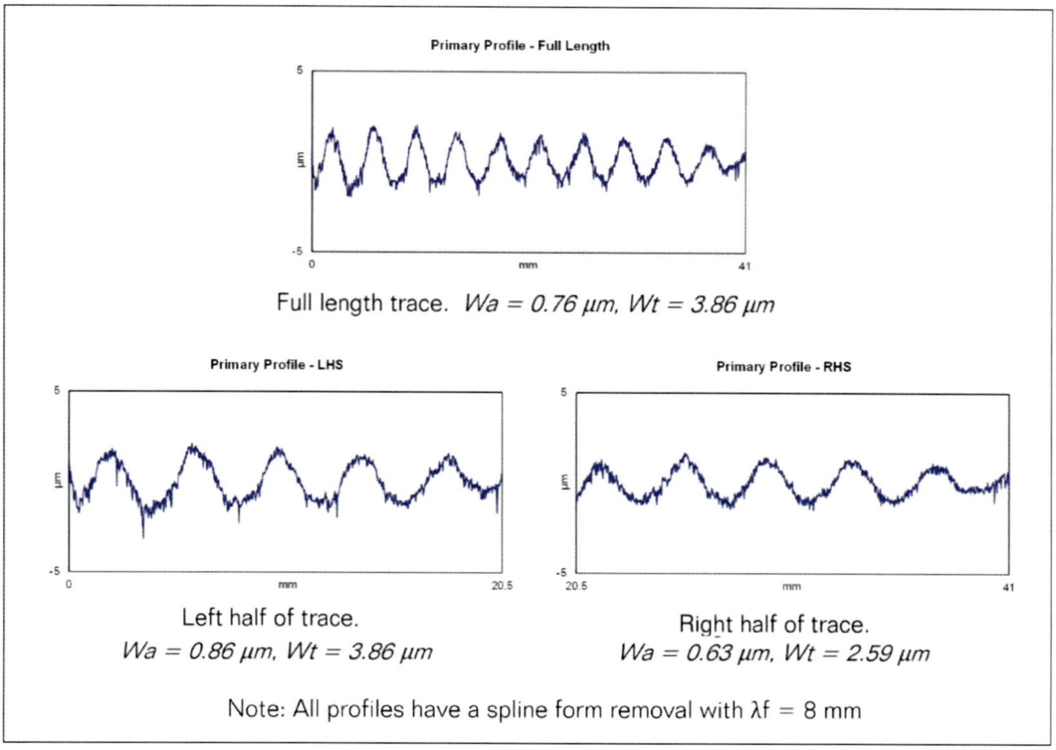

Figure 32: Effect of short evaluation lengths on bandpass waviness

Change λs

Changing any aspect of the specification is a design or engineering function and is outside the scope of this book. The following comments are for information only and are not to be considered as recommendations.

Effects of a λs filter change

The examples in **Table 9** show the changes in parameter values resulting from changes in data point spacing, changes occurring due to surface wavelengths in the region of the short filter cutoff. **Table 10** shows the effect of changing the short filter cutoff from 2.5 µm to 8.0 µm for these surfaces.

Table 10: Effect of data spacing on selected parameters with λs = 8 μm

	dx	Ra	Rz	Rpm	Rp	Rv	Rt	Wt
a)	0.25	2.077	9.599	5.211	6.313	5.081	11.394	16.153
	1	2.076	9.595	5.209	6.314	5.080	11.394	16.144
	Δ%	-0.01%	-0.03%	-0.04%	0.01%	-0.03%	-0.01%	-0.06%

	dx	Ra	Rz	Rpm	Rp	Rv	Rt	Wt
b)	0.25	0.712	3.186	2.126	2.377	1.367	3.743	2.039
	1	0.712	3.179	2.121	2.374	1.363	3.737	2.038
	Δ%	-0.05%	-0.21%	-0.25%	-0.10%	-0.32%	-0.18%	-0.03%

	dx	Ra	Rz	Rpm	Rp	Rv	Rt	Wt
c)	0.25	0.314	2.517	0.839	1.161	2.769	3.931	4.492
	1	0.314	2.504	0.835	1.127	2.756	3.883	4.492
	Δ%	-0.15%	-0.51%	-0.44%	-2.93%	-0.48%	-1.20%	0.01%

	dx	Ra	Rz	Rpm	Rp	Rv	Rt	Wt
d)	0.25	0.339	3.772	0.044	0.080	12.892	12.972	0.441
	1	0.339	3.765	0.045	0.080	12.891	12.971	0.442
	Δ%	0.00%	-0.19%	2.27%	0.00%	-0.01%	-0.01%	0.23%

	dx	Ra	Rz	Rpm	Rp	Rv	Rt	Wt
e)	0.5	0.093	0.970	0.630	1.920	1.122	3.042	3.497
	1	0.093	0.967	0.628	1.916	1.120	3.036	3.497
	Δ%	-0.09%	-0.32%	-0.31%	-0.24%	-0.18%	-0.22%	0.00%

	dx	Ra	Rz	Rpm	Rp	Rv	Rt	Wt
f)	0.5	0.043	0.498	0.284	0.442	0.372	0.815	0.157
	1	0.044	0.500	0.285	0.432	0.365	0.798	0.157
	Δ%	1.07%	0.33%	0.24%	-2.24%	-1.86%	-2.07%	-0.04%

	dx	Ra	Rz	Rpm	Rp	Rv	Rt	Wt
g)	0.25	0.017	0.154	0.058	0.149	0.246	0.395	1.535
	1	0.017	0.156	0.060	0.154	0.245	0.400	1.535
	Δ%	1.45%	1.17%	3.49%	3.69%	-0.32%	1.19%	0.01%

Note that many of the parameters in **Table 10** become smaller with the 1 μm spacing (the **Δ%** is negative). This is most likely an example of observation 2b (above)—the larger data spacing results in the measurement not capturing some of the peak and valley extremes. For these data sets the change tends to be small.

While increasing λs reduces the sensitivity to data point spacing, changing the specified λs must not be taken lightly. **Table 11** compares the parameter values resulting from short filters of 2.5 and 8 μm, applied to the smaller spacing data in **Table 9**. Roughness parameter values decrease with the longer λs because more data (longer wavelengths) are being excluded from the parameter calculations. The change is often not trivial; see, for example, surface (f), where Ra decreases by half. The impact of a λs change on perceived component quality, and on component function, must therefore be carefully evaluated. In some situations, a change in parameter limits might be considered.

Table 11: Effect of λs on parameter values

a)

λs	Ra	Rz	Rpm	Rp	Rv	Rt	Wt
2.5	2.080	9.713	5.255	6.338	5.177	11.515	16.153
8	2.077	9.599	5.211	6.313	5.081	11.394	16.153
Δ%	-0.16%	-1.18%	-0.85%	-0.38%	-1.86%	-1.04%	0.00%

b)

λs	Ra	Rz	Rpm	Rp	Rv	Rt	Wt
2.5	0.719	3.330	2.227	2.464	1.523	3.987	2.039
8	0.712	3.186	2.126	2.377	1.367	3.743	2.039
Δ%	-0.89%	-4.32%	-4.53%	-3.55%	-10.28%	-6.12%	0.00%

c)

λs	Ra	Rz	Rpm	Rp	Rv	Rt	Wt
2.5	0.353	2.949	1.040	1.558	2.974	4.531	4.492
8	0.314	2.517	0.839	1.161	2.769	3.931	4.492
Δ%	-11.04%	-14.66%	-19.34%	-25.46%	-6.87%	-13.26%	0.00%

d)

λs	Ra	Rz	Rpm	Rp	Rv	Rt	Wt
2.5	0.340	3.910	0.051	0.091	12.951	13.041	0.440
8	0.339	3.772	0.044	0.080	12.892	12.972	0.441
Δ%	-0.29%	-3.53%	-13.73%	-12.09%	-0.46%	-0.53%	0.23%

e)

λs	Ra	Rz	Rpm	Rp	Rv	Rt	Wt
2.5	0.098	1.039	0.660	1.952	1.168	3.119	3.497
8	0.093	0.970	0.630	1.920	1.122	3.042	3.497
Δ%	-5.06%	-6.55%	-4.49%	-1.61%	-3.91%	-2.47%	0.00%

f)

λs	Ra	Rz	Rpm	Rp	Rv	Rt	Wt
2.5	0.088	0.989	0.583	0.802	0.517	1.319	0.157
8	0.043	0.498	0.284	0.442	0.372	0.815	0.157
Δ%	-51.17%	-49.63%	-51.19%	-44.85%	-28.00%	-38.24%	-0.01%

g)

λs	Ra	Rz	Rpm	Rp	Rv	Rt	Wt
2.5	0.022	0.234	0.091	0.214	0.352	0.566	1.535
8	0.017	0.154	0.058	0.149	0.246	0.395	1.535
Δ%	-25.69%	-34.13%	-36.56%	-30.59%	-29.99%	-30.22%	0.00%

I have a short feature. Do I still need five sampling lengths?

It is a common misperception that a roughness measurement must contain five sampling lengths. These five sampling lengths make up what is called the "evaluation length." Some measurement system software even automatically adjusts the filter settings so the evaluation length contains five sampling lengths. This is poor practice. The recommendation of a number of sampling lengths is only meant to provide a reasonable sampling of the workpiece surface and is not fundamental to the analysis.

The sampling length partitions the profile for some parameters, such as peak-to-valley roughness (Rz). The sampling length is numerically equivalent to the roughness filter cutoff wavelength (lambda-c, or λc) and is established when λc is selected (see the section, "**Are "cutoff" and "sampling length" the same thing?**" on page **327**). λc should be determined by the nature of the surface, and not by its length, because it represents the transition wavelength between roughness and waviness. This roughness/waviness transition is determined either by the details of the manufacturing process or, better, by the functional requirements of the surface. Depending on the nature of the surface, changing λc may change the value of the roughness and waviness parameters, and may hide features that need to be controlled, especially if waviness is not specified. **Figure 33** shows a short feature where one might be tempted to change from a 0.8 to a 0.25 mm cutoff. The resultant changes in, for example, Ra and Wt demonstrate why this may not be wise. **Figure 73** on page **118** provides another example of how changing the cutoff changes the *roughness* and *waviness* profiles.

If a feature's length is less than one sampling length, the *primary* parameter should be used instead of the *roughness* parameter (e.g., Pa instead of Ra).

Page **207** presents the ISO 1302 methodology for specifying a smaller (or larger) number of sampling lengths, either by directly stating the number or indirectly by specifying the evaluation length. This approach should be used to avoid the ambiguity inherent in measuring short features.

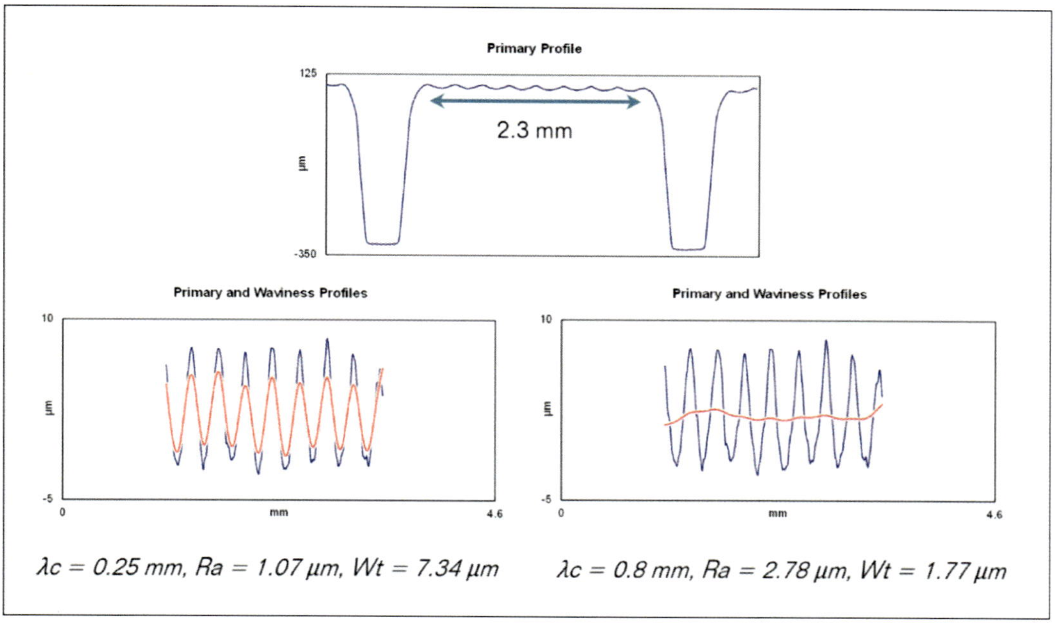

Figure 33: Effect of λc selection on roughness and waviness parameters on a short surface

Why do I have to measure 5.6 mm to get a 4 mm evaluation length?

Standard Gaussian filters can't be applied all the way to the ends of the data, so some amount of data must be discarded from each end of a trace. This is illustrated in **Figure 34**, showing that one half of one cutoff length has been discarded at each end, where the averaging with the Gaussian weighting function begins and ends. With some instruments the discarded amount is one-half of a cutoff length; on others, it's a full cutoff length[23]. With a 0.8 mm roughness filter cutoff, one cutoff length from each end of a 5.6 mm trace leaves 4 mm of usable data.

Many people have stated that we need to discard data due to the acceleration/deceleration of the stylus motion. This is not generally true today. Most modern instruments account for the acceleration/deceleration, and the entire primary profile is trustworthy; thus, the discarding is based on the filtering, not the motion.

Spline filters can be used to provide a filtered profile without edge discarding. A visual comparison of this behavior is shown in **Figure 35**.

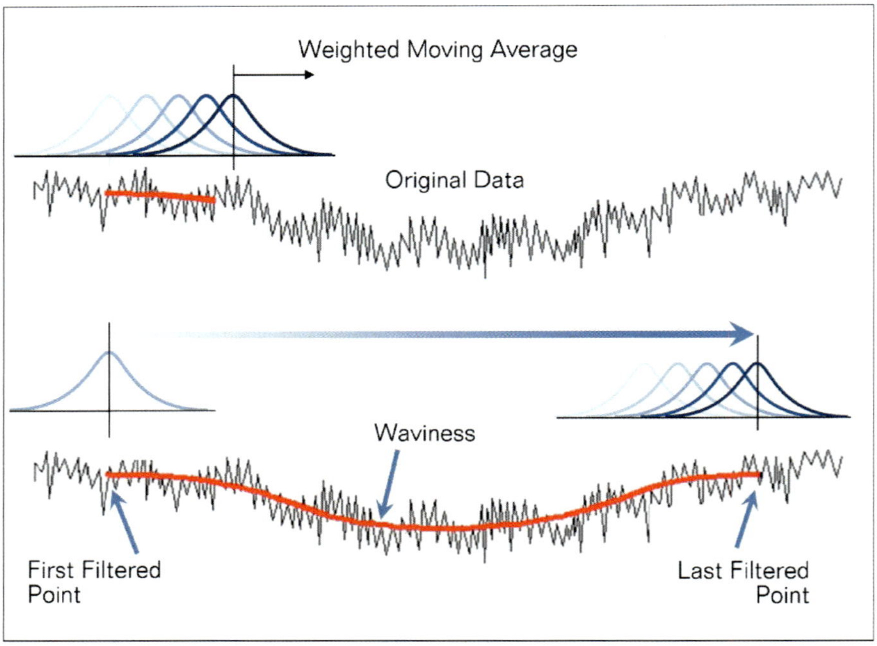

Figure 34: Gaussian filter with one-half cutoff length discarded at each end

[23] The amount of discard is determined by how the Gaussian weighting function is truncated. A true Gaussian curve has infinitely long tails, so in a practical application those tails must be truncated. Instruments vary in whether this is at one-half or at one cutoff length.

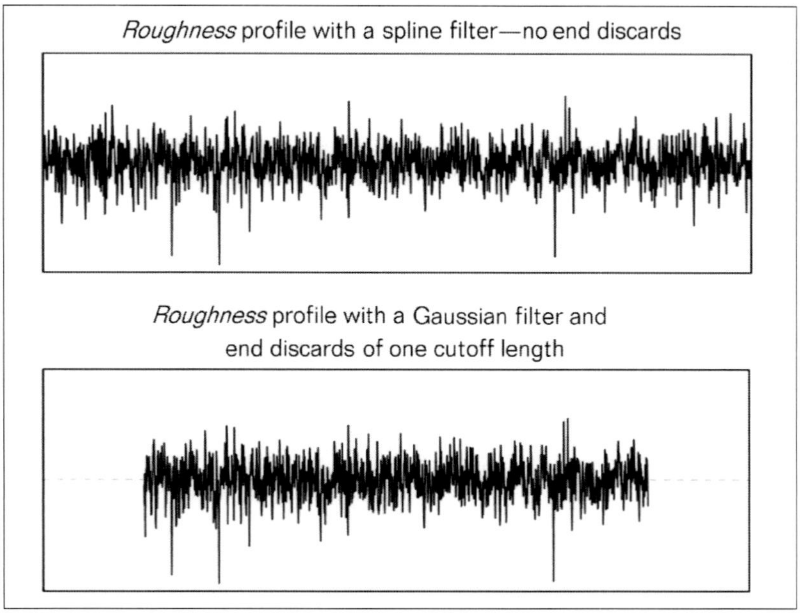

Figure 35: Spline and Gaussian filter effects on end discards

How many times should I measure my part?

Considerations such as employee efficiency and laboratory productivity make it tempting to take only one measurement per part. Sometimes that is sufficient. However, that one measurement may not provide an accurate assessment of the surface, especially if the workpiece has a complex geometry, variable material properties, or some other characteristic that affects its response to manufacturing processes. Customer expectations, quality control procedures, or process changes may also drive a decision to make a larger number of measurements.

The following are four scenarios, each having a different "correct" answer to the question of how many measurements should be made. Additional scenarios can be envisioned, but the point is, there is no one "right" number of traces; each situation must be evaluated based on the factors listed below.

Scenario 1

You've been manufacturing this same component using the same manufacturing process for several years. You make periodic measurements to audit this process, taking the measurements from a region on the component which you know to be most sensitive to process changes. One measurement will be adequate if its frequency is tied to relevant processing steps; e.g., tool change or grinding wheel dress. This is an example of a process audit.

Scenario 2

Your customer requests data demonstrating that an individual part conforms to their specification. You take multiple measurements at various locations and report the worst reading. This is an example of a conformity assessment.

Scenario 3

Your process unexpectedly drifts and some measurements are now out of specification. The specification incorporates the 16%-rule (see the section, "**Max-rule or 16%-rule for the quality control decision**" on page **67**). On parts with an out-of-specification reading you make five additional measurements and only reject those parts with two or more readings outside the designated limits.

Scenario 4

Your colleague asks you to reverse engineer a competitor's product. You take multiple measurements from multiple locations to obtain some understanding of the functional aspects of the texture, and to explore the nature and limits of their manufacturing processes.

Factors that influence the number of measurements include the following:

The reason for the measurement

Is the measurement intended to satisfy a customer's quality control requirements? to monitor a known process? to develop a new process? as part of a failure investigation? Each reason drives a different number.

The uniformity of the surface texture

If the texture is consistent over the whole surface, then a small number of measurements (maybe just one) is adequate. If the texture varies due to workpiece geometry, material properties, process details, or other factors, more measurements will be necessary. Note that if the texture varies in a predictable manner, identification and measurement of the worst area may allow fewer measurements because each one is "worst case."

The uniformity of the surface texture across a batch of parts

If the texture varies in a systematic manner across a production batch, then careful selection of the sample parts may minimize the number that need to be measured. For example, if a ground surface is roughest after the wheel is dressed, then samples from immediately before and immediately after dressing will bracket the range of textures achieved. However, if texture variation is random, a larger sample size is appropriate.

Presence of surface flaws

If surface flaws are controlled by a visual or functional specification they should not be included in a surface texture measurement. If a flaw is accidentally included, that measurement result should be discarded and another measurement taken in a different area. If flaws are not controlled separately, then their inclusion in a surface texture measurement is valid.

Max-rule or 16%-rule for the quality control decision

If max-rule is specified, only one measurement needs to be taken, with part acceptability dependent on that one measurement (as long as due consideration is given to the other factors listed here). If you are using the 16%-rule, the number of measurements will depend on the results of the initial data, since the 16%-rule allows 16% of the data from a surface to be out of specification. More information on implementing the max-rule and 16%-rule is provided on page **188**.

The specified evaluation length

Short evaluation lengths, such as the default five sampling lengths for roughness, may provide an inadequate "look" at a surface, especially with short sampling lengths on a long feature. Multiple measurements in multiple locations may therefore be appropriate, For example, if a 100 mm long feature is specified with a λc of 0.25 mm and the default number of roughness cutoffs, only 1.25 mm of that long feature would be evaluated in any single measurement.

All of these considerations relate to the issue of uncertainty, a measure of how wrong your measured value might be. Measurement uncertainty includes a contribution from your instrument (see page 85) and a contribution from the surface. In general, the surface variation's contribution to uncertainty can be reduced by increasing the number of measurements, as $\sqrt{(1/n)}$. The section, "***How do I know if my data is accurate?***" on page 159 provides more detail. Your tolerance for uncertainty, whether quantified or not, is therefore the ultimate guide to how often you need to measure.

I see unexpected things in my data. Are they real?

Artifacts in your data can be confusing and can generate inaccurate parameter values and inappropriate responses by the recipients of the data. Looking at the measurement traces—especially the *primary* profiles—instead of just at the numbers is an important step in avoiding this. Asking yourself the question, "Can my manufacturing process really create that feature?" may also be useful. Following are some examples of measurement artifacts and what you can look for to identify them.

- Dirt (**Figure 36**)
- Vibration from the drive, from bumping the instrument table or a door closing, etc. (**Figure 37**)
- Backlash or pre-travel problems in the instrument's drive mechanism (**Figure 38**)
- A worn or dirty datum (**Figure 39**)
- The skid (**Figure 40**)
- Inadequate vertical resolution (**Figure 41**)
- Steps in the profile due to the instrument losing its absolute height or "zero" reference (**Figure 42**)
- Deep grooves producing non-existent "raised metal" (**Figure 43**)
- Regions of non-cleanup producing non-existent "raised metal" (**Figure 44**)

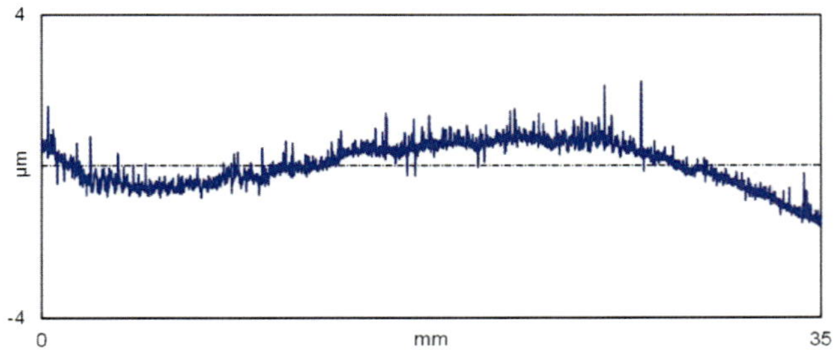

The higher peaks on this lapped surface are due to inadequate cleaning prior to the measurement. A lapping process would not be expected to leave random high peaks such as these.

Figure 36: Debris on a lapped surface

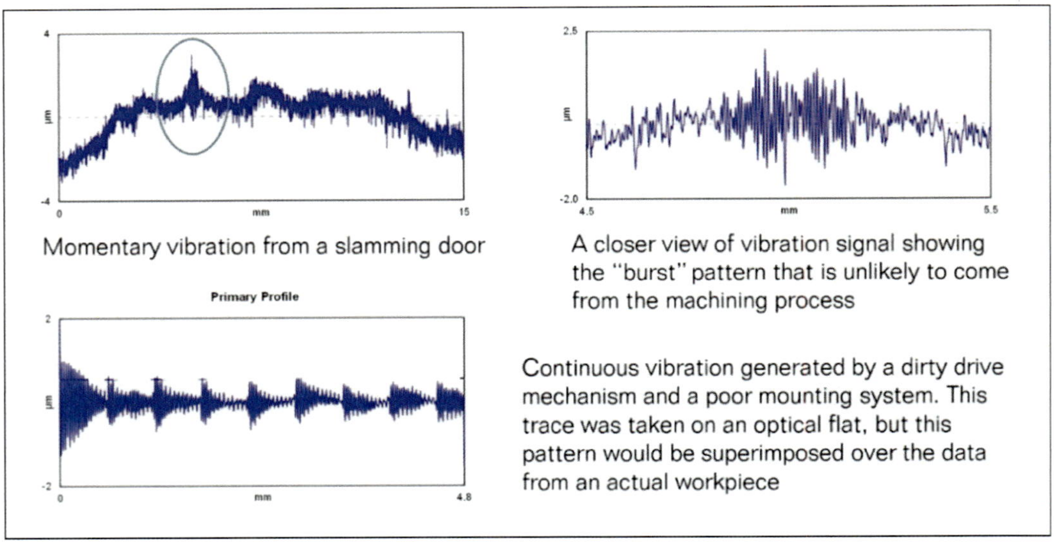

Figure 37: Vibration from the environment or the instrument drive mechanism

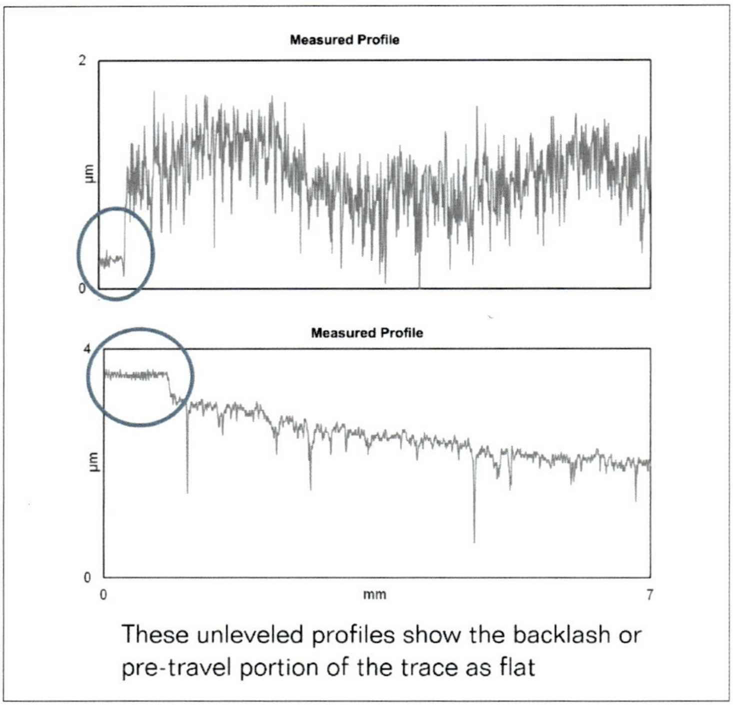

Figure 38: Backlash/pre-travel problems at the beginning of a measurement

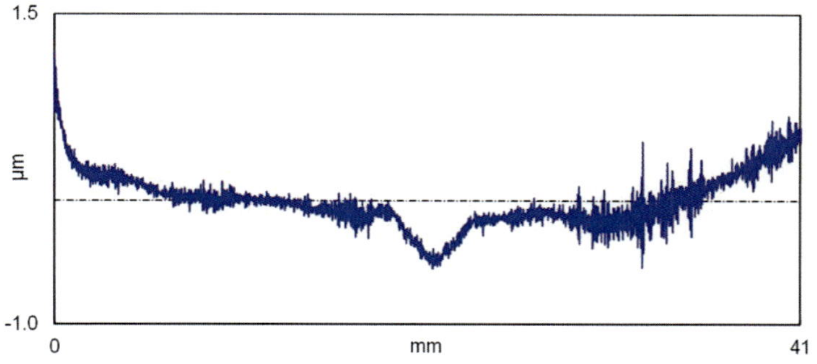

This trace illustrates three zones with different issues, all of which can be found by tracing an optical flat. At the far left we see where the datum has worn, causing the carriage to drop and the stylus to rise. This wear is in the area where the measurements normally start. In the middle, we see a dip where the carriage rose (perhaps over debris), causing the stylus to fall. Most measurements on this instrument were 17.5 mm long, so the carriage could have caused dust/debris to accumulate at that location. On the right, we see a noisy signal which could be due to friction characteristics, the drive signal, etc. This portion of the datum was seldom used so it may have oxidized.

Figure 39: Artifacts from an instrument datum with wear and built-up debris

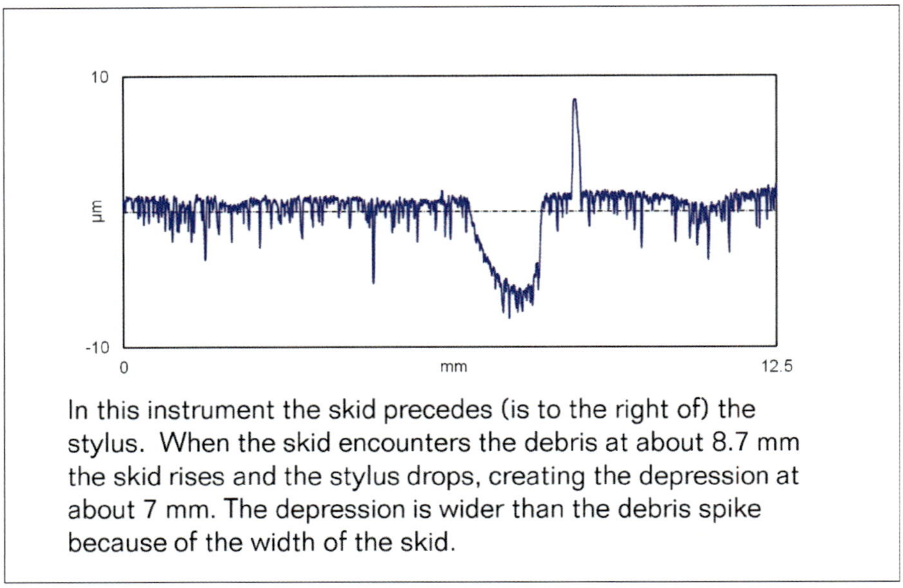

Figure 40: Artifact created by a skid traversing debris

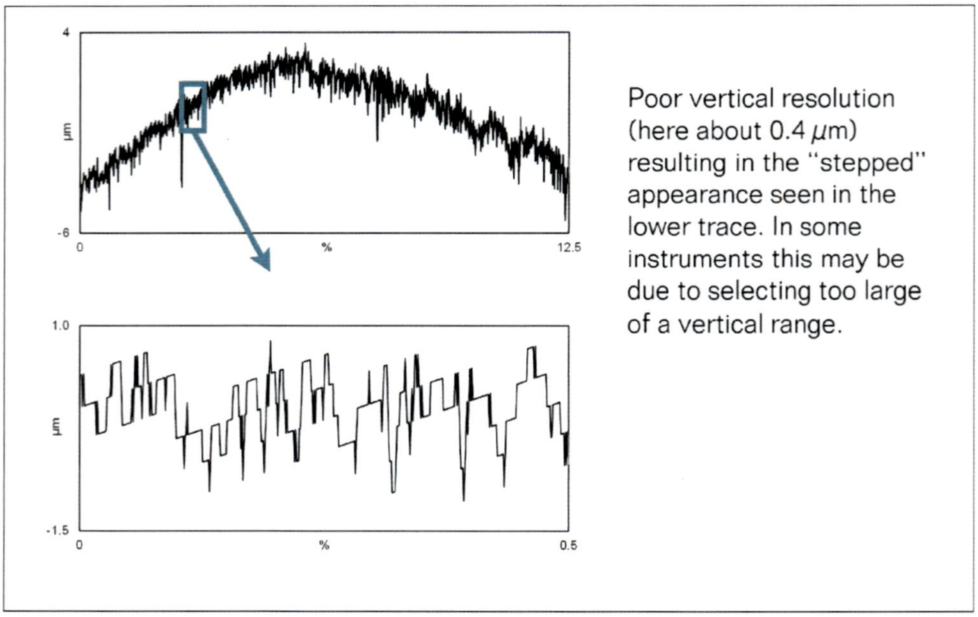

Figure 41: Trace from an instrument with poor vertical resolution

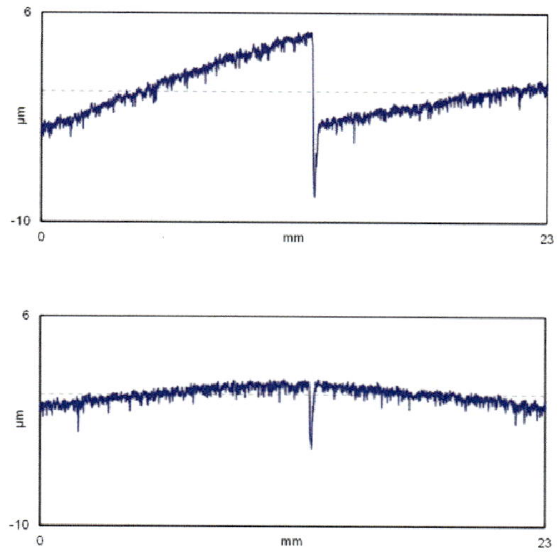

The top profile shows a step due to the instrument losing its absolute height or "zero" reference after a sudden drop into the deep scratch. The bottom profile taken adjacent to the top profile shows the actual shape of the surface.

Ideally, the top profile should be traced in the opposite direction to prove if the step is real.

Figure 42: Step in profile caused by instrument's loss of absolute height or "zero" reference

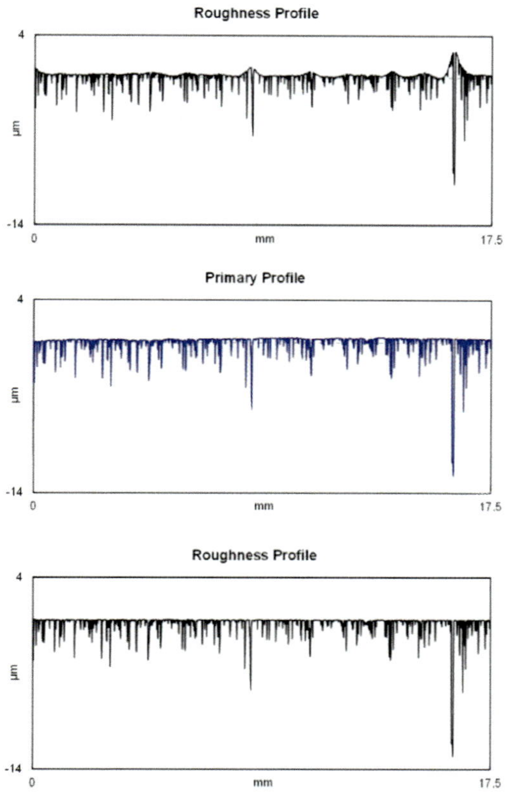

Roughness profile showing "raised metal" beside deep grooves

Primary profile, a true representation of the surface showing that the raised metal isn't real

Artifacts eliminated from *roughness* profile by using a robust filter

Figure 43: Raised metal analysis artifacts occurring beside deep grooves

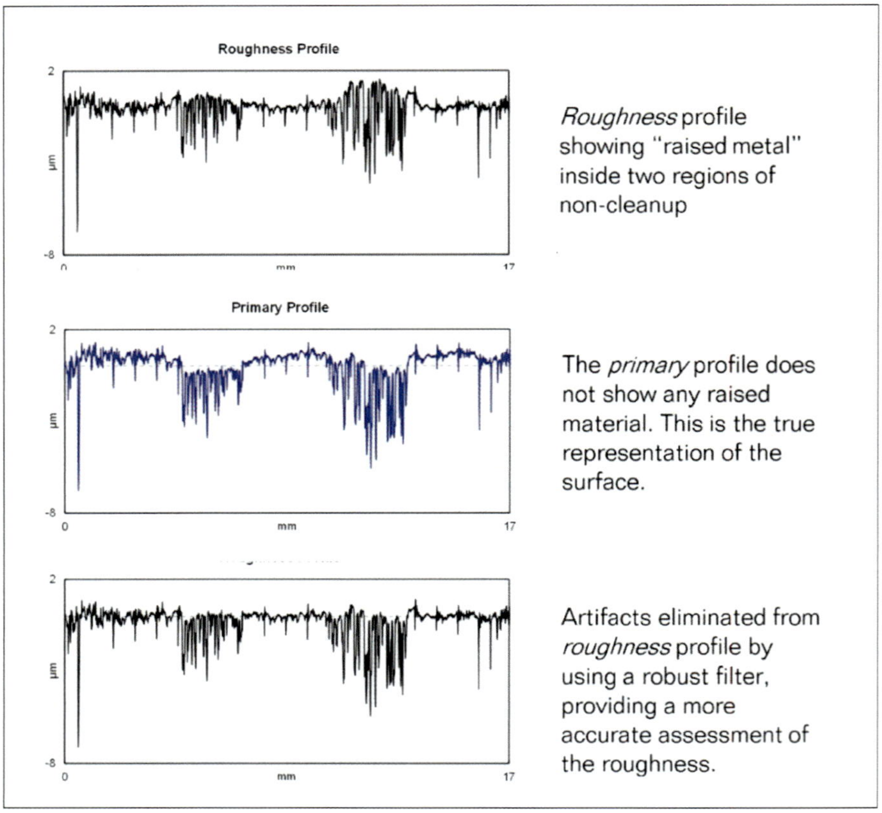

Figure 44: Raised metal analysis artifacts from regions of non-cleanup

How much noise can I tolerate in my measurement system?

Ideally, noise in your measurement system will be eliminated by servicing or by isolating the equipment from its environment. However, this is not always practical, so a decision must be made about how much noise is too much. Your tolerance level will depend upon the type of surface being measured. An acceptable amount of noise when measuring a rough surface might be totally unacceptable when measuring, say, a ground and polished surface. **Figure 45** shows simulated examples of this affect, with a noisy trace on an optical flat combined with a finely polished surface and with a turned surface. The impact on Ra differs dramatically. **Figure 46** plots the percentage of error in Ra caused by relative amounts of noise; this can be used to help determine when system improvements need to be made.

Depending on the surface and the parameters that are being reported, the impact of noise can vary widely. For example, there are cases where noise can increase parameters such as roughness peak height (Rpk) or average maximum peak height (Rpm) by 100% or more while having only a marginal influence on Ra.

It should be noted that most of this error due to noise cannot be subtracted from the measurement. High frequency noise, with a wavelength shorter than lambda-s (λs), will be attenuated by the short wavelength filter. However, noise at longer wavelengths will combine with the actual surface features and cannot be removed. Since feature amplitudes typically increase with wavelength, the major portion of the noise amplitude cannot therefore be "filtered out" and will remain as part of your perceived surface texture.

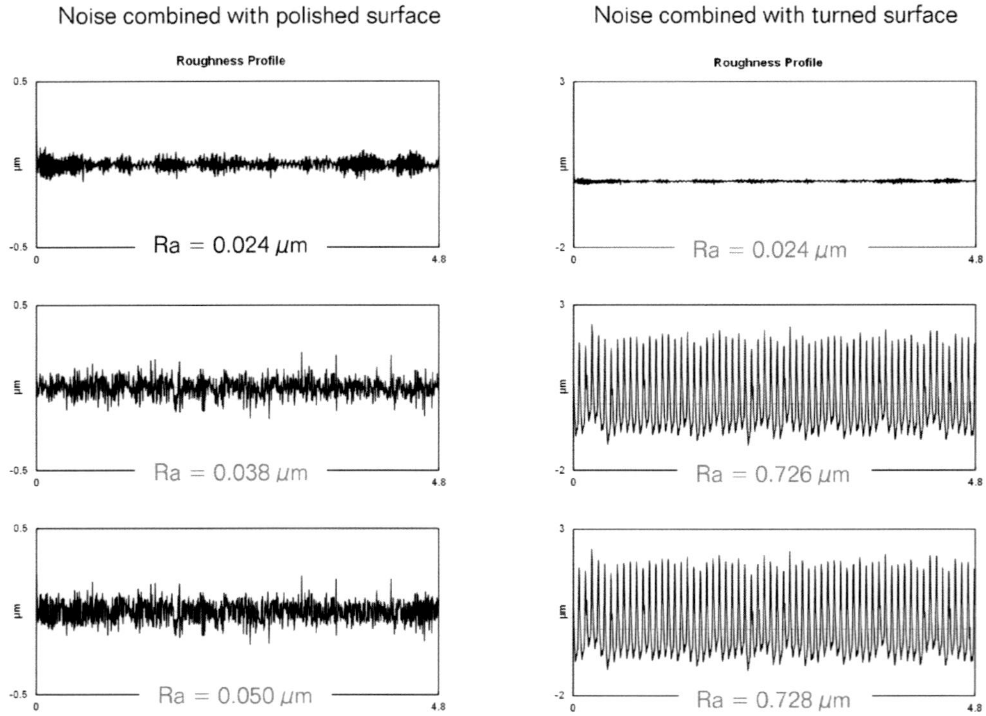

Top traces are noise on an optical flat. Middle traces are the original surfaces - a polished surface on the left and a turned surface on the right. Bottom traces are the combinations of the upper two. The noise increased the Ra of the surface on the left by 32%, but the surface on the right by only 0.3%.

Note: Left and right sets of traces are at different scales.

Figure 45: Simulated influence of noise on two levels of roughness

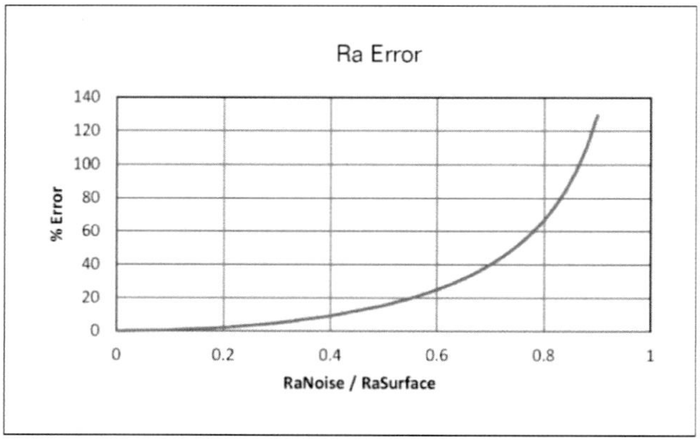

Figure 46: Percent error in Ra versus the amount of noise (as a fraction of the surface roughness)

How do I get noise out of my measurement system?

Surface texture instruments are susceptible to mechanical noise or vibration because of the small stylus forces and the sensitivity to small deflections. Electromagnetic noise can also interfere with accurate measurements. Vibration from internal or external sources will add to the actual roughness of the workpiece, and may even overwhelm the true roughness of very smooth surfaces. Vibration may be less critical for a skidded instrument, but it should still be monitored and controlled as much as possible.

Since vibration and noise often fall in the same frequency range as the surface texture, these errors cannot be easily "filtered out." The filtering process will also remove real surface features. Even modern "notch" filters (for example those used in audio signal processing) can cause other unwanted effects such as "ripples" in surface texture data.

Several steps can be taken to identify the presence of vibration or noise:

1. Watch for unusual patterns in your measurements. Depending on the source, vibration may look like a superimposed sine wave or a "burst" of increased amplitude at random positions along the trace. **Figure 37** on page **70** contains examples of both of these.
2. Measure an optical flat. Ra values greater than the certified value (usually about 0.01 to 0.02 μm) should be investigated further. Optical flat measurements are important even if you do not see a noise signature in your measurement (see item 1) because the noise may be continuous (e.g., electronic noise or a constant vibration) rather than random (e.g., a slamming door or a passing forklift)—see **Figure 47**.
3. Some instruments can gather data without moving the stylus. This will identify whether the source of any noise is internal or external to the drive mechanism.

If vibration is a problem, try the following:

1. Review your measurement fixtures to be sure they are stable and the workpiece is well supported.
2. Perform the recommended maintenance to keep the instrument's internal datum surfaces and drive mechanism clean and properly lubricated.
3. Install isolation pads under the legs of the instrument table. Both passive and active pneumatic systems are available. Ideally, vibration isolation should occur at one point. Installing vibration isolation in two places (for example, below the table legs and also on the table top) can sometimes increase the amount of noise in the system.
4. Remove or modify obvious vibration sources such as slamming doors or heavy foot traffic. Your instrument may be telling you to look for a machine tool with bad bearings or a poorly isolated air compressor, so consider treating your "noisy" instrument as a diagnostic tool for some other piece of equipment.

5. Place the instrument on a granite table, whose large mass will reduce the sensitivity to higher frequency vibration.
6. Find a quieter location for the instrument. This may require a site survey by a person experienced in vibration detection and analysis.
7. Cut the floor around the instrument lab to isolate the installation from external vibration.

As with mechanical vibration, electromagnetic noise may come from the instrument itself or from the environment. For example, induction heating units approximately 30 meters from the measurement laboratory were found to affect one instrument's readings. Equipment shielding and electrical isolation should therefore be evaluated if a noise problems turns out to be electronic or electromagnetic and not mechanical.

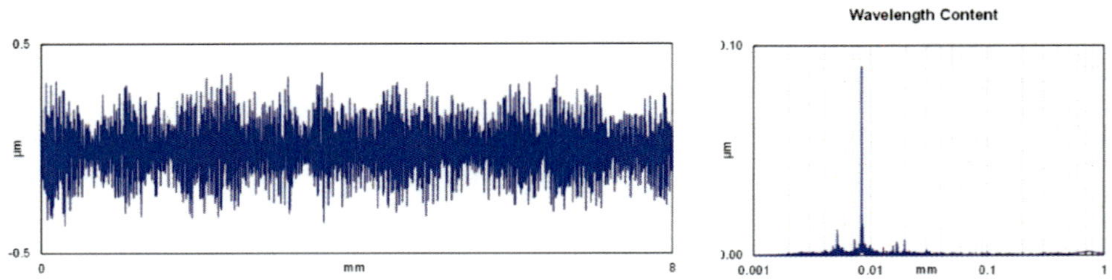

The trace on the left is from an optical flat with a measured Ra of 0.123 µm versus an expected value of about 0.01 µm. The wavelength content spectrum on the right shows a peak at 0.0082 mm. At the traverse speed of 0.5 mm/sec this represents a 60 Hz frequency which remained constant at varying traverse speeds. This turned out to be an external mechanical vibration problem but could have been electrical, since the electrical line frequency was also 60 Hz.

Figure 47: Example of continuous noise affecting a surface texture measurement

How important is sample alignment when I make a measurement?

To understand the role of sample alignment we will consider the alignment in all three directions:

- Tilt in the x–z plane, which creates a slope in the data
- Alignment of the stylus travel with the axis of the part
- Position along the y-axis which, for a curved component, relates to the crowning of the stylus on the sample

For a flat sample, the alignment that matters most is the tilt, which has to be small enough that the stylus stays in range over the whole traverse length. Optimizing the alignment allows the instrument to function within the smallest possible range and to provide the best possible resolution. Also, depending on the probe's performance, staying in the central part of the range may provide better accuracy.

For a curved sample, all three directions matter:

- The tilt matters for the same reason as for the flat sample.
- Alignment of the stylus travel with the part axis must be good enough to avoid adding significant curvature to the form from the cross-section of the curved sample. The smaller the cross-section radius of curvature, the more critical the alignment[24]. If a part with an axial profile is being measured, misalignment will add to, or subtract from, the actual curvature, depending on the type of profile and whether an OD or ID surface is being measured. **Figure 48** and **Figure 49** provide simulated examples of this phenomenon on a straight OD and an axially-profiled ID, respectively.
- Lack of proper crowning, if severe enough, will cause the stylus to contact the sample on the flank rather than on the tip, a problem known as flanking. Improper crowning will combine with axial misalignment to add slope to the trace, as shown in **Figure 48** and **Figure 49**, bottom profiles.

---

[24] The misalignment-induced change in profile height at point i (Δz_i) equals $K(R - \sqrt{R^2 - {y_i}^2})$, where R is the radius of curvature of the sample surface and y_i is the amount of offset at point i of the actual trace from a perfectly aligned trace. $y_i = x_i tan\theta$, where x_i is the distance from the misalignment center of rotation and θ is the misalignment angle. K is -1 if the surface is convex and +1 if concave, as viewed from the stylus.

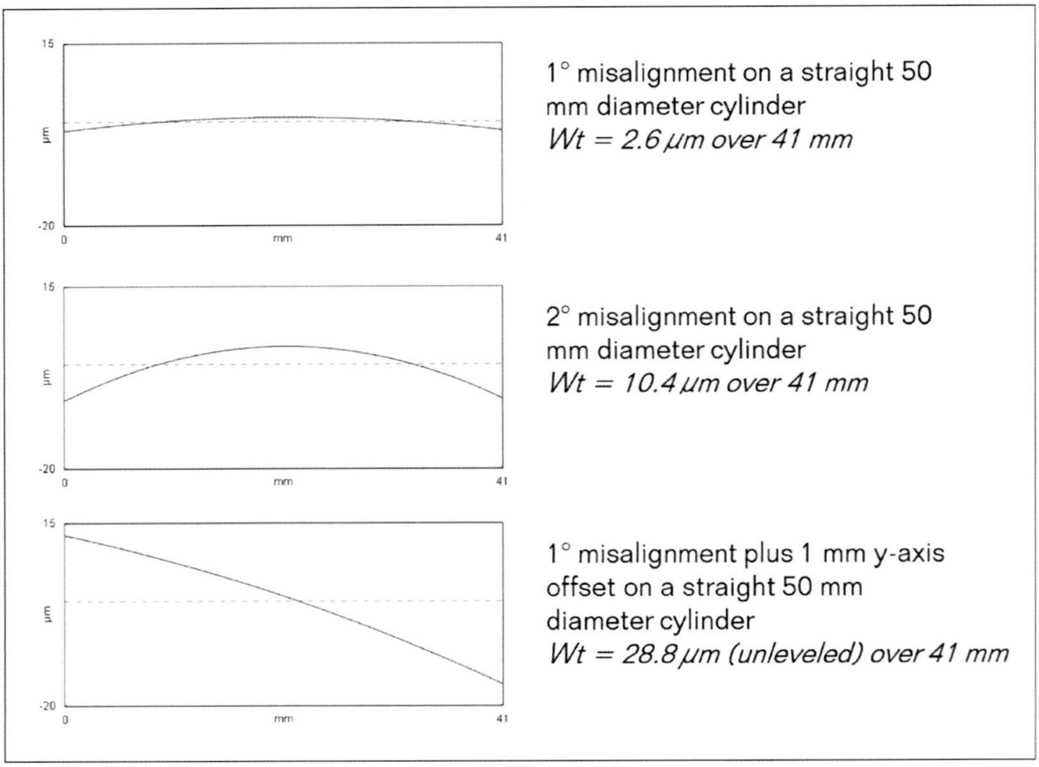

Figure 48: Simulated results of misalignment on a straight cylinder (OD surface)

The alignment of cylindrical samples can be made less critical by measuring with a knife-edge stylus which, on a flat sample, would produce a line contact perpendicular to the direction of stylus travel (**Figure 50**).

Note that it should not be assumed that the measurement instrument's supporting columns are always well-aligned with T-slots or other features in the base. This alignment can be checked by tracing a straight, small-diameter rod, such as a gage pin. Any curvature in the trace is indicative of misalignment.

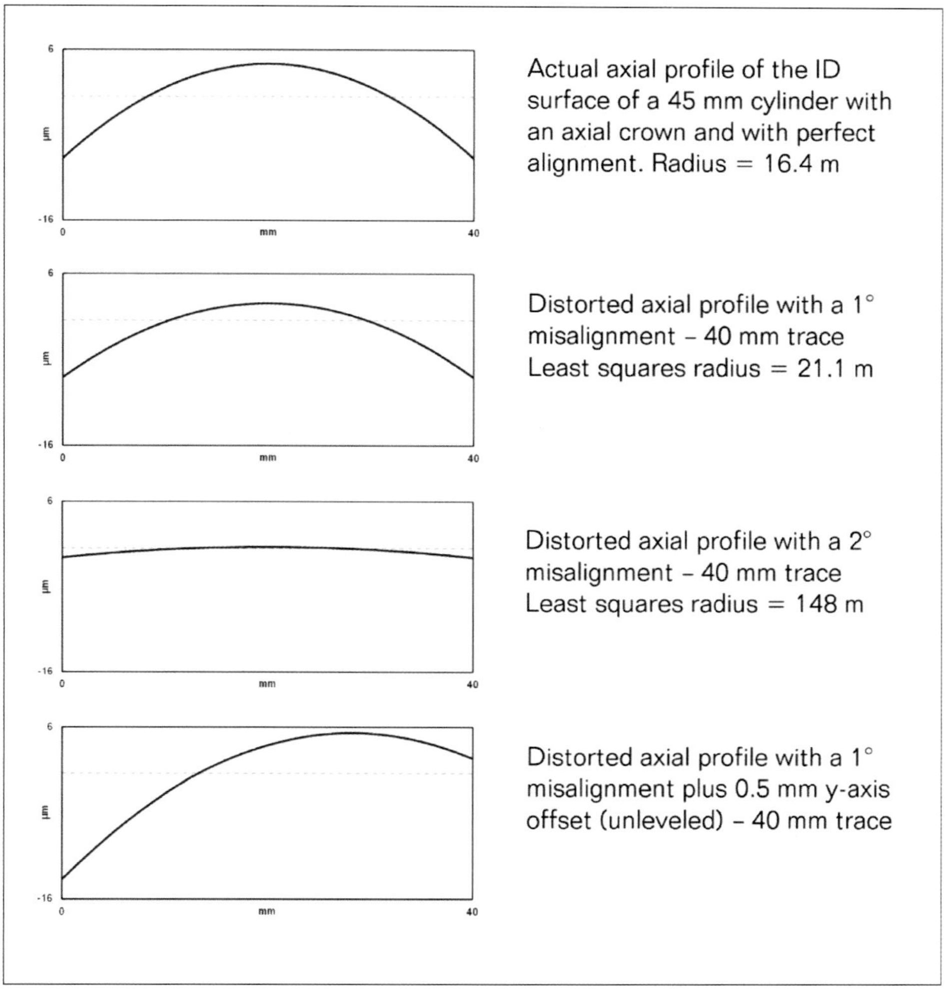

Figure 49: Simulated results of misalignment on ID of a 45 mm diameter bore with axial crown

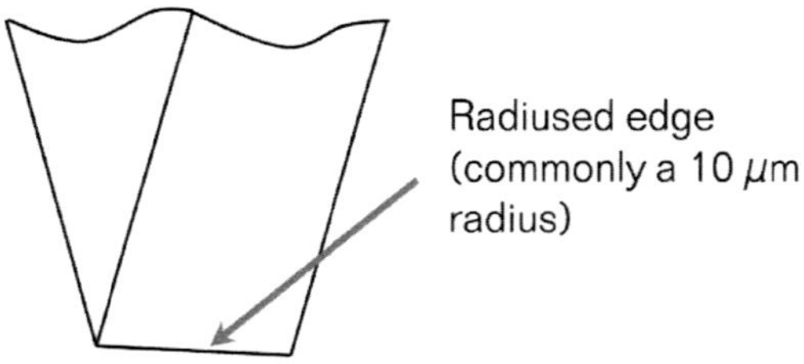

Figure 50: Knife edge stylus

In addition to these effects on the measured form, sample alignment can also influence roughness and waviness parameters. Surface texture specifications assume measurements are made perpendicular to the surface lay, the dominant machining direction (see page **219** for more information about lay). Misaligned measurements will distort spacing measures such as mean width (Rsm). In severe cases, it can move features from one domain to another (e.g., from roughness to waviness). **Figure 51** illustrates this relationship of measurement direction to the surface lay.

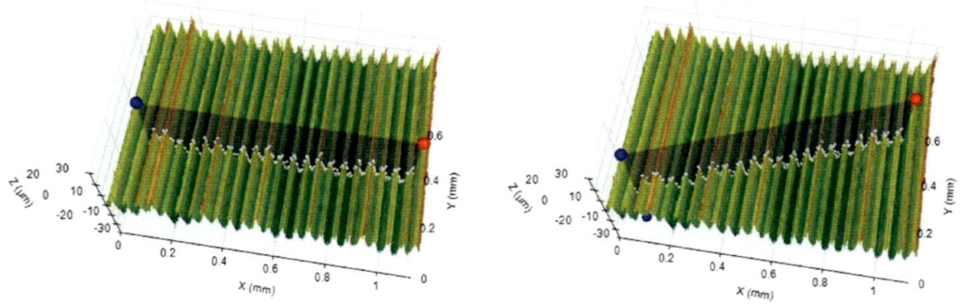

The measurement on the left (shown by the gray cutting plane on this 3D representation of a turned surface) is aligned perpendicular to the lay. It provides a more accurate assessment of the surface than the misaligned measurement on the right. The misaligned profile will have a wider spacing of features, which could cause some roughness features to be moved into the waviness domain.

Figure 51: Sample alignment and surface lay

Why does a cheap instrument repeat better than an expensive one?

Buying a roughness instrument is like buying a fine audio system. If you want to listen to music with the best possible quality, you want stereo equipment that properly transmits all the high and low frequencies. Similarly, if you want the best possible roughness measurement system, you want one that properly transmits all of the surface's short and long wavelengths.

The challenge comes in the measurement of real surfaces.

An instrument that can detect all of the surface's finest details can also detect the fact that traces aren't always made in the exact same place on the surface. At the microscopic level, a slight change in the measuring position can sometimes reveal an entirely different surface. Inexpensive, low accuracy measuring systems that are not sensitive to these fine details may not see a change in the surface. Therefore, these inexpensive, low accuracy systems often appear to be more "repeatable." If repeatability is the primary consideration, you will almost always choose the "worse" instrument.

Ultimately, the more "dead" (i.e., less responsive) a surface measuring system is, the more repeatable it will be. A large tip radius, a high measuring force, and heavily filtered electronics can all lead to more repeatable measurements. However, all of these factors will hide fine surface details.

A surface measuring system is used to measure surface texture. The exact placement of the measurement location isn't the primary concern. Thus, the placement of the trace shouldn't penalize the instrument in terms of testing its performance. Repeatability (and reproducibility) should not be used to assess surface texture instruments. A "measurement uncertainty estimation" approach is preferred. One possible method for assessing this uncertainty is described on page **85**.

Can I do a gage R&R study on a surface texture instrument?

Surface texture R&R studies are not a good idea. While repeatability and reproducibility (R&R) studies are appropriate for the evaluation of many measurement systems, this is not the case for surface finish measurements. This is true for the following reasons, all of which violate the basic premise behind an R&R assessment:

- It is impractical to place the stylus at the exact same location multiple times, and there is the potential of high variability from one location to another.
- During each trace the stylus contact force may alter the surface, leading to a different measurement on the next trace.
- Measurement results may be more dependent on the fixture than on the instrument.
- The system's accuracy can't be set by adjusting the zero point because, while one surface may read too low, another surface may read the opposite.

For a dynamic measurement system, the most repeatable system will be one which is insensitive to these changes. Therefore, a relatively "dead" instrument may perform the best, and one that is unplugged will be the most repeatable of all! However, the measurement will bear no relation to the actual surface. Additional comments on dead instruments are given on page **83**.

The preferred method for assessing a surface texture instrument's performance is an uncertainty study, which answers the question, "How wrong can my results be?" While there are no standardized procedures for conducting an uncertainty study, one sample method is given on page **85**. Note that such a study only assesses the uncertainty of the measurement system and does not take into account the uncertainty present in the workpiece surface. Further discussion on this topic can be found in the section, "**How do I know if my data is accurate?**" starting on page **159**.

How do I perform a surface texture measurement uncertainty study?

The amount of confidence we have in a measurement or the "width of the plus-minus 2-sigma confidence interval" is called measurement uncertainty. This estimation of uncertainty must include the errors of the measuring instrument and variation in the surface itself. For the sake of clarity, we will speak of "instrument uncertainty" and "surface variation." This section only addresses the instrument's contribution. Surface variation will be addressed in the section, "**How does surface variation impact my measurement results?**" on page **93**.

For the purposes of this discussion we will assume that the "average value" of a surface texture parameter is of interest. This is called the "measurand."

Many measurement systems are evaluated using repeatability and reproducibility (R&R) studies. An uncertainty study is the preferred assessment method for surface texture instruments because of the reasons given on page **84**. Instrument uncertainty estimates how wrong an instrument might be by looking at the factors that can influence that instrument when applied in a specific measurement scenario. This is similar to FMEA—Failure Mode and Effect Analysis—a common approach used in design and manufacturing.

To assess instrument uncertainty, error sources or "uncertainty contributors" must be identified for your specific measurement task. These contributors to surface texture instrument uncertainty can include the following:

Traceability

Is the instrument actually measuring true height values?

Linearity

Does the answer depend on probe position?

Noise

Is my "roughness" result real or is it from mechanical vibration or electronic noise?

Frequency response

Can the instrument detect all of the wavelengths? Is my probe moving too fast?

Datum straightness

Is that waviness really in my surface or is it in the instrument?

Software and math

Does 2 + 2 really equal 4.00001?

Stylus condition

Is the tip radius correct and intact?

Measurement setup errors

Is the stylus perpendicular to the surface?

There is no agreed upon method for conducting an uncertainty study, but any study will likely use artifacts traceable to a national laboratory (e.g., NIST, PTB, NPL)[25]. For illustration purposes, one method is described here. It uses the artifacts shown in **Figure 52**. A complete study using this procedure can take a few hours, and consists of the following:

- Trace an optical flat to measure noise (a 4.8 mm trace—see **Figure 53**) and datum straightness (as long a trace as possible). The optical flat should be at least as large in diameter as your intended longest traverse length so you can check your datum over that length.
- Trace roughness specimens, such as Type D standards, to assess frequency response. The different specimens have repeating patterns with different levels of roughness (see **Figure 54**); use the data most representative of the roughness of your manufactured component(s).
- Trace the roughness specimens with the probe at the extremes and center of its vertical range to test the linearity.
- Trace reference specimens to check the calibration of the instrument. These reference specimens can have various patterns. Some have a sinusoidal pattern with a certified Ra value, such as shown in **Figure 55**. Others have grooves of certified depths, as shown in **Figure 56**. Note: This test should use a specimen of a type that is different than the one used to set the instrument's gain/calibration.

The results of the above measurements are statistically analyzed to yield an overall uncertainty value. The ISO Guide to the Expression of Uncertainty in Measurement[26] (also called the "GUM") gives guidance on the combination of these uncertainty contributors.

To those accustomed to other types of gaging, the uncertainties associated with even good quality surface texture instruments and installations may seem high[27]. This can be especially true when measuring smoother surfaces where factors such as noise have a relatively large influence.

---

[25] NIST = National Institute of Standards and Technology (US); PTB = Physikalisch-Technische Bundesanstalt = National Institute for Science and Technology (Germany); NPL = National Physical Laboratory (UK)
[26] ISO/IEC Guide 98-3:2008, Uncertainty of measurement—Part 3: Guide to the expression of uncertainty in measurement

If you supply customers with manufactured components or measurement data, you will need to reach agreement on:

- The acceptable level of instrument uncertainty
- The frequency of instrument uncertainty assessments
- The strategy for incorporating instrument uncertainty into quality control decisions

One possible set of guidelines is given in **Table 12**. Note: these are suggestions by the authors and are not derived from any national or international standard.

High instrument uncertainties may be reduced by the following actions. More detailed recommendations are given on page **91**.

- Maintaining equipment and workpiece cleanliness
- Performing routine servicing and calibration
- Monitoring the condition of the stylus
- Isolating the equipment mechanically and electromagnetically
- Keeping software up-to-date.

Instrument uncertainty studies typically focus on Ra; the masters may only be certified for Ra, and occasionally Rz. The rationale is that the only difference between the various parameters is the math applied to the raw data, so focusing on one or two parameters is adequate. If there is interest in the uncertainty or correlation of other parameters, you may need to develop you own procedures and samples.

Note that when evaluating an actual surface, the measurement instrument uncertainty is only part of the total. The variability of the workpiece surface is the other half, and this may in fact be larger than the influence of the instrument. Further discussion on this can be found on page **159**.

---

[27] To illustrate the difference in expectation, national laboratories will normally claim an Ra measurement uncertainty no better than approximately 5%. In contrast, the length of a gage block will have an uncertainty on the order of 0.001%.

Table 12: A possible set of instrument uncertainty guidelines

Instrument uncertainty as % of tolerance	Assessment	Action required
< 10%	Excellent	None
10–20%	Good	None
20–30%	Marginal	Management signoff required. Explore improvements.
> 30%	Poor	Customer signoff required. Improvements mandatory.

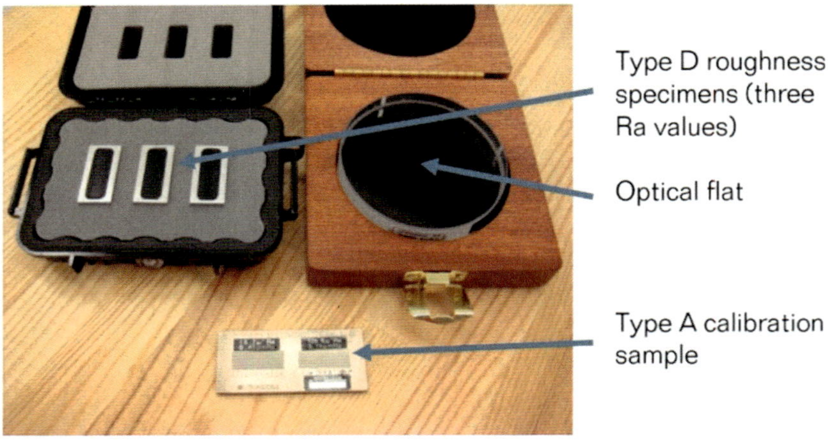

Figure 52: Typical set of masters for an uncertainty study

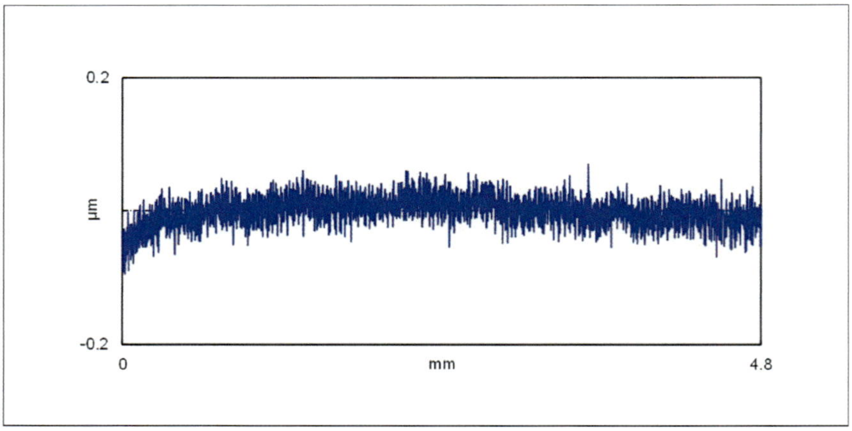

Figure 53: A trace on an optical flat. Ra = 0.017 µm

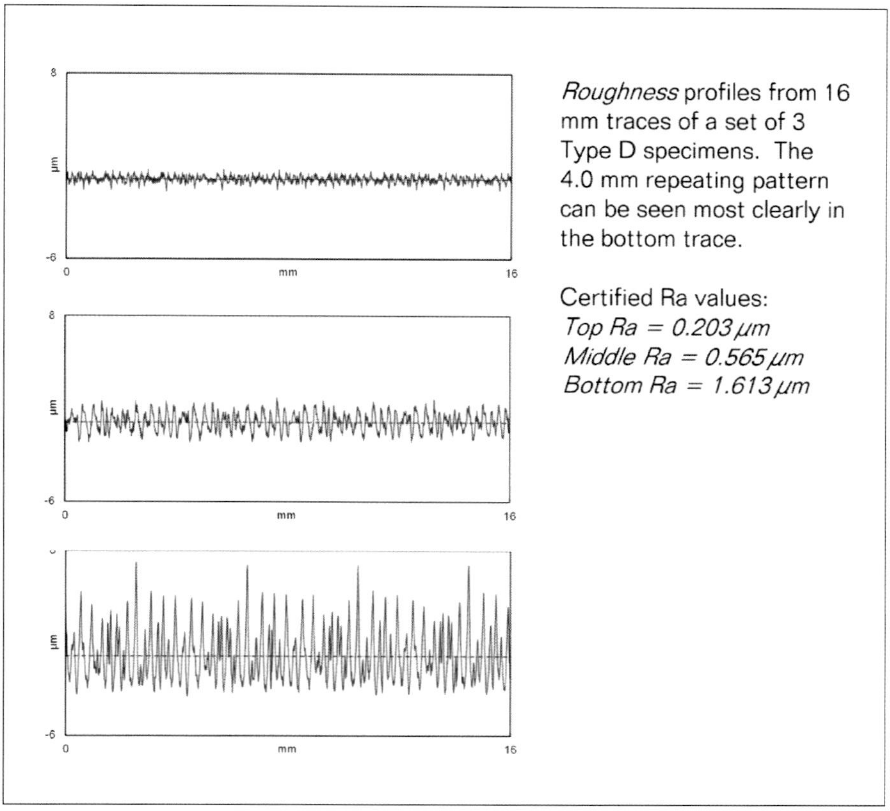

Figure 54: Sample traces from a set of Type D specimens

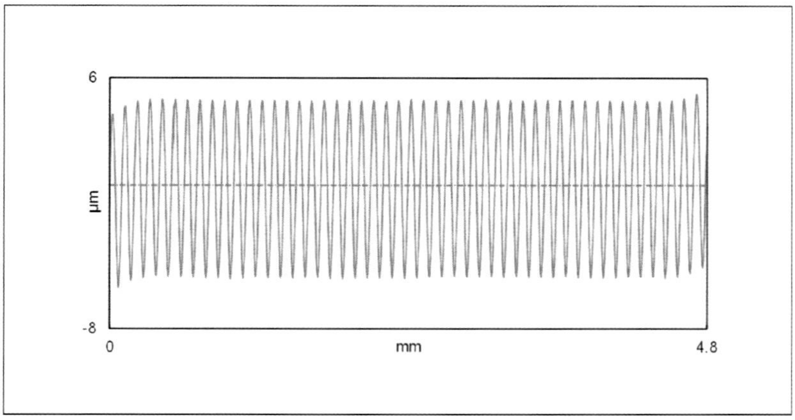

Figure 55: Trace from a sinusoidal calibration standard with a certified Ra = 3 µm

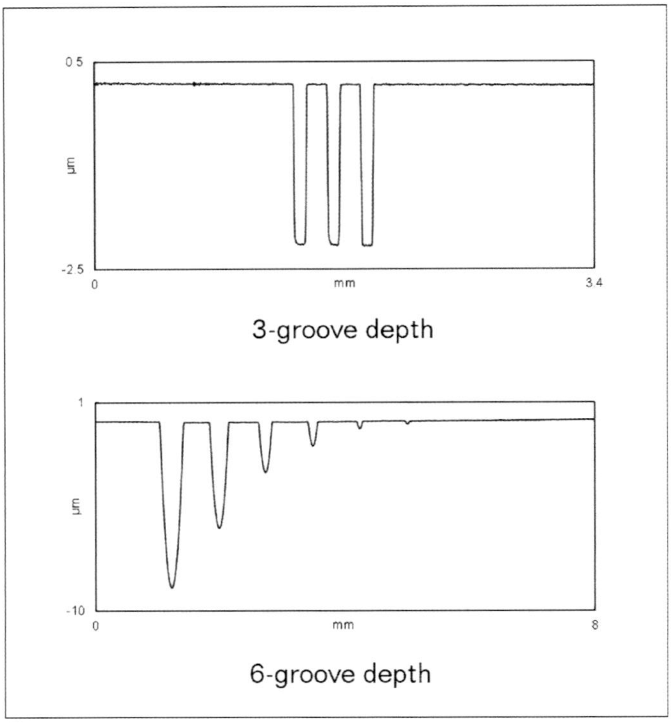

Figure 56: Traces from ISO 5436-1 type A depth measurement standards

How can I lower my instrument's measurement uncertainty?

A high instrument uncertainty can be caused by characteristics of the measurement system or of the environment. The several outputs of an uncertainty study (e.g., see page **85**) can help pinpoint the main sources of error and how they can be improved.

An optical flat reveals vibration, either from the environment or from the instrument itself. Some instruments support the gathering of data while the stylus is stationary, allowing the separation of these two potential sources. If this option is not available, measurements can be made when the environment is quieter (e.g., during an off-shift at a factory) to assess the external effect. If the noise is from the environment, it can be reduced by:

- Ensuring that the workpiece is stable and well supported
- Installing rubber pads under the legs of the equipment table
- Installing active or passive pneumatic isolation pads
- Supporting the instrument on a granite table or other high-mass surface
- Moving the measurement system to a quieter location
- Isolating the lab floor from the rest of the facility

If the noise is from the instrument itself, it can be reduced by:

- Using a different traverse speed
- Cleaning and lubricating the datum bar and the carriage pads
- Ensuring the instrument is properly assembled with attachment points torqued correctly
- Servicing or replacing the drive mechanism

Examination of an optical flat trace will reveal straightness or cleanliness issues with the datum bar (see **Figure 39** on page 71). Some instruments will map the datum bar and numerically adjust the raw data to remove the effects of datum wear.

Issues identified on a calibration specimen are addressed by following the appropriate calibration procedures. These procedures are typically instrument-specific, but may include:

- Mechanically adjusting the probe length. For example, a high Ra reading or large step height value may be indicating that the probe is too long.
- Measuring a gage ball or other precision artifact and adjusting the instrument gain.

Roughness specimen data reveals frequency response issues, stylus condition, calibration problems, and noise (especially on the smoothest specimen). Stylus condition may also be reflected in the calibration data. Appropriate responses to this data include:

- Stylus condition can be determined by visual examination under high magnification. Dirty styli can be cleaned in, for example, an ultrasonic cleaner with a mild soap solution. Damaged styli must be replaced.

- More rigorous stylus tip evaluations can be performed using techniques described in ISO 5436-1.
- Frequency response can be improved by reducing the traverse speed.

Roughness specimen or calibration sample measurements at the extremes of the instrument's vertical range will reveal linearity problems. These can often be corrected by the adjustment of internal system parameters (similar to the gain). This exercise may reveal linearity problems only at the extremes of the range, in which case measurements should attempt to avoid these range extremes.

Careful examination of the actual traces of the various artifacts, and not just the numerical data, will reveal additional problems such as backlash (see **Figure 38**) on page **70**), sticking or damaged pivot bearings, software or machine set-up errors, and other problems.

Note that a calibration procedure usually must be performed whenever the measurement probe is changed. Specialized styli may have their own calibration procedures. The equipment operator must be familiar with these requirements to ensure that he or she is supplying high quality data.

How does surface variation impact my measurement results?

When measuring a surface, we are typically assessing a very small area to estimate the value for the entire surface. Let's consider making one profile trace on a large surface in an attempt to estimate the average value for a roughness parameter over the entire surface. In this case, the single profile is not perfect—it has "instrument uncertainty" as described on page **85** in the section, "**How do I perform a surface texture measurement uncertainty study?**" The value calculated from this profile is also influenced by the variation in the surface itself.

In practice, it is very common to find that the surface variation is much, much larger than the instrument uncertainty.

To arrive at an overall "measurement uncertainty" we must consider the instrument *and* the surface. One simple method for doing this is to consider the standard deviation associated with the instrument uncertainty and the standard deviation of the surface being measured.

Ultimately, we can combine the instrument's standard deviation with the surface's standard deviation:

$$\sigma_{measurement} = \sqrt{\sigma_{instrument}^{2} + \sigma_{surface}^{2}}$$

Which leads to a total measurement system uncertainty of:

$$U_{measurement} = 2\sigma_{measurement}$$

(based on the common coverage factor k=2).

It is not uncommon to find an instrument standard deviation on the order of 4–8% while the surface standard deviation is on the order of 10–20%. Thus, the overall measurement system uncertainty is often much more influenced by the surface itself.

Note: The above discussion applies to the case of a single measurement as a prediction of the *mean* value. The influence of surface variation can be reduced by averaging multiple measurements.

How do I know if my instrument is capable of measuring my surfaces?

Several factors determine whether your measurement system can measure a given surface. The flowchart in **Figure 57** provides some guidance on this decision. Additional information can be found in the sections: "**What should I consider when buying a surface measurement system?**" on page **20**; "**How much noise can I tolerate in my measurement system?**" on page **75**; and, "**How do I perform a surface texture measurement uncertainty study?**" on page **85**. The standards ASME B46.1, ISO 3274, and ISO 4288 also contain relevant information.

There are factors regarding the integrity of the measurement system that also affect its suitability for a given measurement; these are in addition to the inherent instrument capabilities discussed above.

- Calibration
- Stylus tip condition
- Datum condition
- Frequency response
- Vertical and lateral resolution
- Vibration/noise
- Algorithm verification
- Workpiece fixturing and probe alignment/accessibility

These factors are all under the control of the operator and are addressed by routine servicing and inspections (including uncertainty assessments), software updates, being knowledgeable about the software and the user interface, and careful design and maintenance of the fixtures.

The initial installation of the system will also play a role. For example, the alignment of the stylus travel with workpiece alignment features (e.g., T-slots in the instrument base) will affect the validity of form and waviness analyses. The operator should verify these factors, and not assume that they are correct.

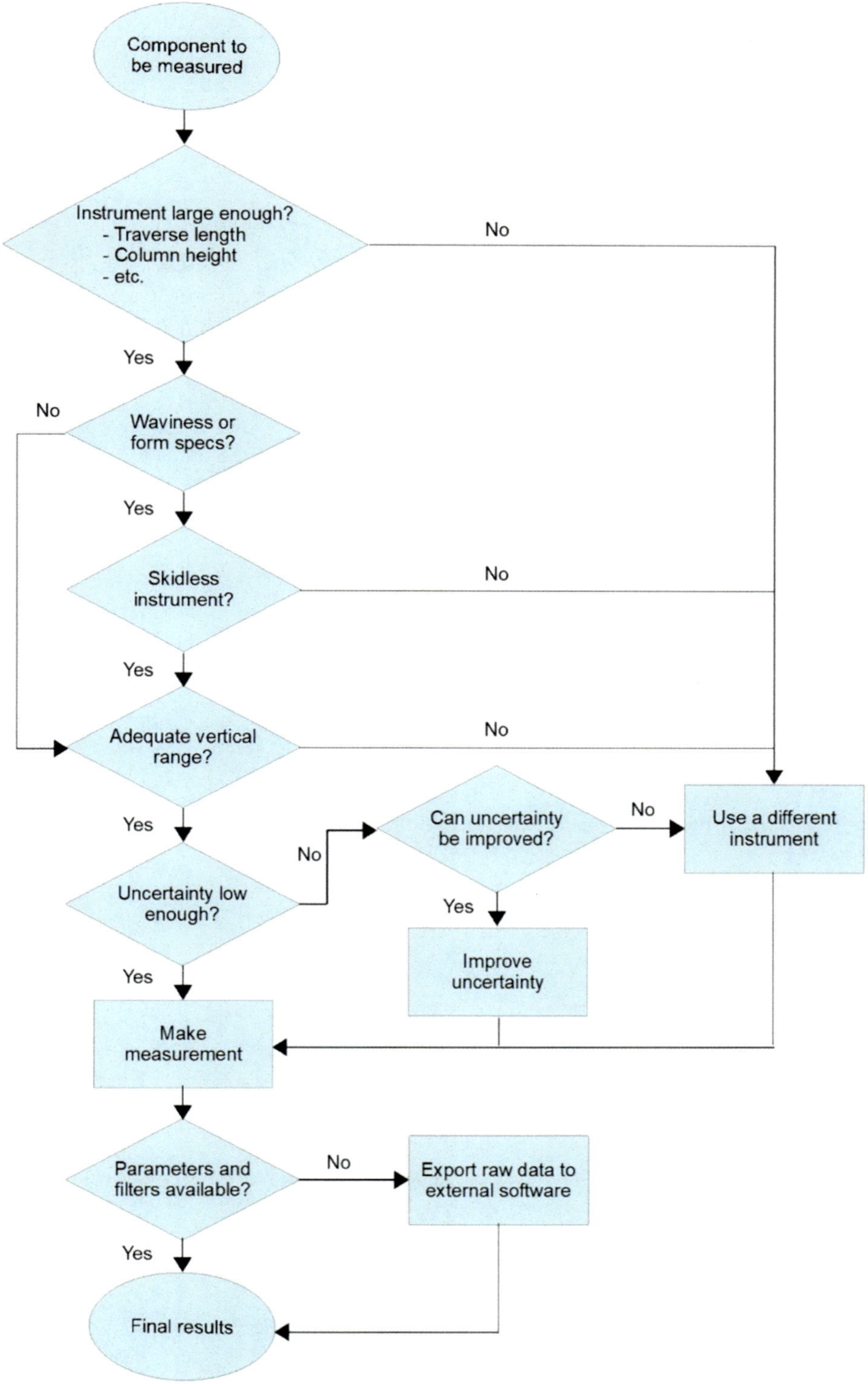

Figure 57: Flowchart for determining if a measurement system can measure a surface

What if my supplier's instrument doesn't correlate with mine?

The measurement of surface texture is very difficult to reproduce from one instrument to another—let alone from one company to another. The following are some of these potential causes for this disagreement:

A misunderstanding of what "acceptable" correlation is

Measurement equipment uncertainty (see page **85**), point-to-point and part-to-part variability (see page **93**), and the small size of the features being measured can all contribute to a poorer degree of correlation than quality personnel expect, especially if they are more used to dimensional rather than surface texture data.

Preparation of the surface being measured

If one laboratory doesn't clean the part as well as the other, and if the specified parameters are sensitive to peaks, results can differ significantly.

Measurement location

This is especially true if some areas of the workpiece surface are significantly different than others. For example, tape polishing a non-uniform shape such as a camshaft lobe can produce varying finishes around the circumference because the tape pressure, a key factor in the amount of material removal, is affected by the local radius of curvature.

Fixturing

The workpiece alignment and stability are critical in terms of achieving comparable results.

Equipment condition

Is the datum clean and unworn? Is the stylus the right size and undamaged? Is the data density adequate (in both x and z)?

Measurement system environment

Mechanical vibration and electronic noise must be minimized.

Analysis settings

Filter types, cutoff wavelengths, and evaluation lengths can all impact the reported results. If these are poorly documented, correlation issues can be expected.

Software versions and analysis algorithms

Parameter and algorithm definitions can change over time, so different vintage software can report different values. Software developers can also create different parameter values through their programming techniques or algorithm choices. In one study, analysis of the same raw data by different software programs revealed differences as large as 40% in some parameters[28].

The following steps can help resolve these issues:

Supplier and customer discussions

The supplier and customer should reach agreement on what is an acceptable level of correlation. They should also agree how to apply measurement uncertainties to the specifications they are measuring to. ISO 14253, part 1, provides one set of decision rules, but any rules acceptable to both parties can be used.

The measurement approach

A review of the condition of the measurement system, of the workpiece fixturing, and of the local environment is required.

Documentation of measurement and analysis settings

Check that the filter types, cutoff wavelengths, evaluation lengths, measurement locations, etc. are fully specified in the engineering documentation and are being followed. If not, both laboratories should document their own practices (with photos if possible), reach agreement on which factors matter, and agree to standardize on those factors. (Note that this last statement only applies to factors that are not explicitly defined in the engineering documentation (e.g., measurement location). Requirements that are spelled out in that documentation must be met. The laboratories must also conform to requirements such as filter settings that are not listed in these documents but for which default values are specified in the applicable national or international documents (e.g., ISO 1302).)

---

[28] Internal company report by Carl Musolff, 2008. Other studies have documented differences in the reference software of national labs such as NPL, NIST, and PTB; see, for example, L.D. Todhunter, R.K. Leach, S.D.A. Lawes, and F. Blateyron, "An analysis of Type F2 software measurement standards for profile surface texture parameters," *Measurement Science and Technology*, 28(6)(2017). These studies help identify how varying interpretations and implementations of the defining standards (e.g., ISO 4287) can produce different parameter values from the reference standards of national laboratories, and subsequently from the different instrument manufacturers' software that may be "calibrated" using their national lab's software. The studies can also point out ambiguities in the defining standards themselves, ambiguities which can lead to non-unique parameter values.

Evaluating measurement uncertainty

Perform instrument uncertainty studies using the same types of artifacts on both instruments. Determine the measurand uncertainty through repeat measurements on the surface(s) of interest. Compare the total uncertainty to the expected correlation.

Exchanging raw data

If both laboratories have access to software that can read the other's raw data files, exchange and analyze those files to help separate software issues from hardware issues. Also exchange information on the software settings using settings files, screen captures, or printouts.

Exchanging round robin parts

Perform a carefully documented series of measurements on the same physical parts. Exchange the parts and, if possible, the raw data files to help separate measurement variation from analysis variation. Note: the round robin study approach should be based on the comparison of multiple traces in a small location as discussed in the section, "**How do I compare measurements between two instruments?**"

How do I compare measurements between two instruments?

When comparing two surface texture measuring devices, individual traces are often misleading. This is due to the fact that the stylus can never be placed in the exact same position twice. A better approach is to mark an area on the test surface where at least ten closely spaced parallel measurements can be made. The area is typically around 5 mm wide, and the length of the area should be long enough to contain the tracing length plus any additional length for skids. In practice, it is easier if the marked area is a few millimeters longer than the trace (on each end) and a few millimeters wider than the measured zone (on each side). The measurements should be evenly distributed over the marked region—see **Figure 58**.

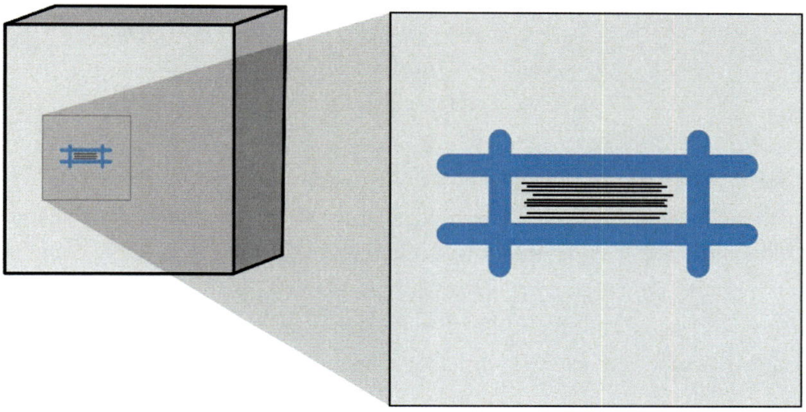

Figure 58: Schematic of designating a specific area for taking multiple traces

With multiple measurements with two instruments you can perform a statistical comparison of the distributions—*not* the individual traces (see **Figure 59**). It should be noted that a simple comparison of mean values could lead to a determination that the two systems do not agree. However, when looking at the distributions in **Figure 59**, we see that the two systems agree very well.

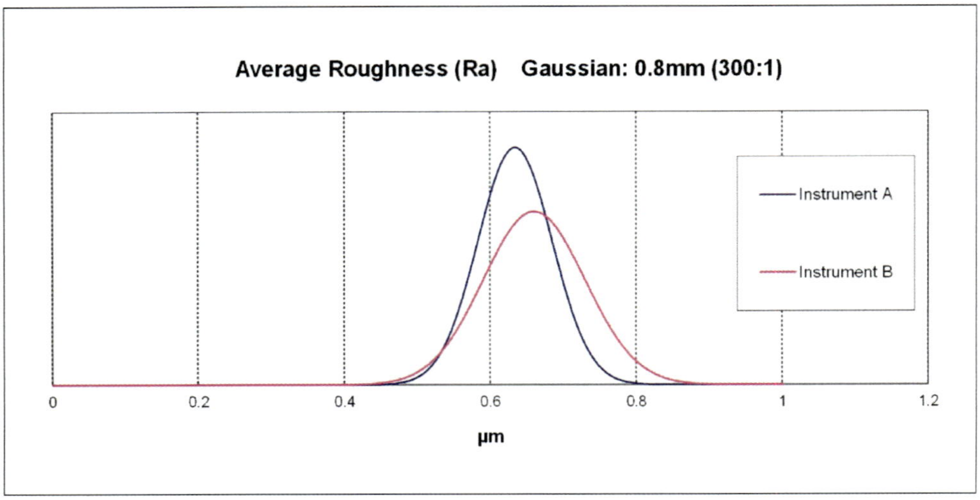

Figure 59: Comparison of multiple measurements from two instruments

Many practitioners attempt to do a correlation study based on a single profile trace on several workpieces with a range of roughness values. As stated above, this approach is highly influenced by surface variation and rarely leads to an understanding of the correlation between the instruments. In many cases, surface variation is many times larger than instrument differences. If a range of surface roughness values is important for the study, then the above multi-trace method should be conducted on two or more surfaces with differing roughness values.

Note: this comparison on a workpiece should always be accompanied with an understanding of each system's measurement uncertainty as described in the section, "**How do I perform a surface texture measurement uncertainty study?**"

My supplier is measuring higher values. Should I adjust my tolerances?

Many measurement devices can be "re-zeroed" or shifted to attain agreement with other similar systems. These devices are typically "one-dimensional" and include devices such as calipers and dial indicators.

Unfortunately, the measurement of surface texture is not a one-dimensional task. In fact, not only is the measurement of a surface profile a two-dimensional task, but it also includes aspects of time and dynamics as the profile is mechanically traced. The results from this kind of multi-dimensional, dynamic measurement cannot be scaled, shifted, or "adjusted" to achieve agreement with another measurement.

Let's consider the case where an instrument has a blunt stylus tip. That instrument may consistently measure lower roughness values because it cannot detect certain surface features. Scaling or shifting the reported Ra value from this instrument does not make those missing features appear. It simply corrupts the reported value.

Similarly, if an instrument consistently measures high due to noise, the correct response is to mitigate the source of the noise. Adjusting the results to account for the noise will also alter the surface data, again corrupting the result.

Can I just measure roughness with my fingernail?

Fingernails are the oldest "stylus-based" surface texture devices. They can work quite well in deciding whether surface A is smoother or rougher than surface B. In some industries "roughness comparison samples" are used for fingernail-based surface assessment (see **Figure 60**). If you design or manufacture components for which this level of accuracy is sufficient, then your fingernail is an acceptable choice.

Unfortunately, fingernails are subjective, non-quantifiable, and, when you cut them, non-repeatable. All of these aspects tend to disqualify them from most lists of acceptable instrument options.

Additional information on roughness comparison specimens can be found in section 12 of ASME B46.1 and in ISO 2632-1.

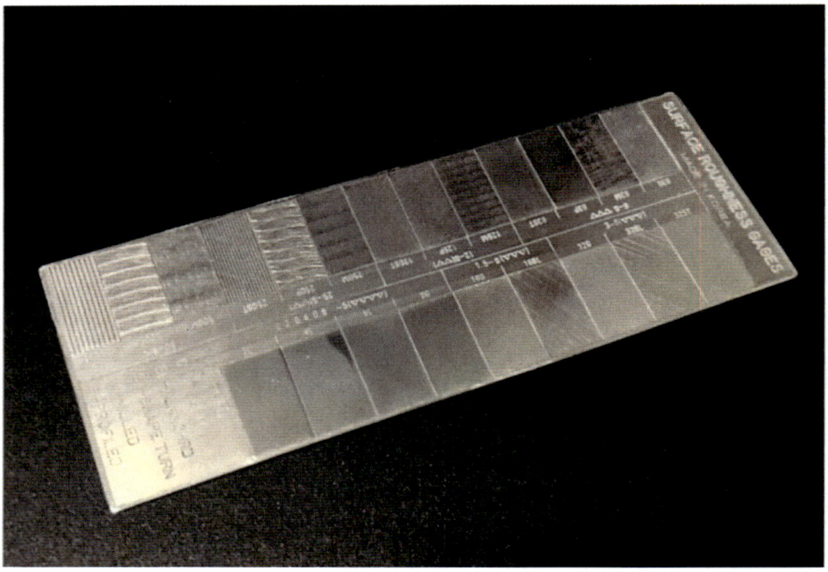

Figure 60: A common surface finish comparator incorporating various machining methods

It should be noted that some of the surfaces in these comparison samples have very wide spacings. These wide spacings can lead to very different results if the specimen is traced with a stylus and analyzed for roughness with typical filtering. For example, a surface like the 500 microinch "horizontal milling" sample in **Figure 60** may only give an Ra value on the order of 100 microinches when measured and properly filtered.

Can I measure surface texture on my CMM?

The answer here is, "yes and no."

Yes, it is possible to measure roughness with some of today's CMMs that are equipped with special roughness-measuring probes. For example, some modern CMMs allow for a small roughness drive unit to be mounted. The CMM positions the drive unit and then the drive unit internally measures roughness, in a manner consistent with a traditional roughness instrument. Another acceptable approach uses the CMM to move a skidded roughness probe along the surface. These attachments are gaining in popularity.

Also, since waviness is a part of "surface texture" we can say "yes" in that some scanning CMMs with small probe tips can measure some types of waviness.

It follows that this topic can only be addressed on a machine-specific basis. However, in the most general case, while a CMM is an excellent tool for measuring size and shape, it is typically not capable of measuring surface texture, for the following reasons:

Probe radius

A 2 µm or 5 µm stylus tip radius is required to resolve the roughness features on most machined component surfaces. Typical CMM probe radii are 100 to 1000 times larger. These will mechanically filter the surface features, significantly changing the *roughness* profile appearance and the roughness parameter values (see **Figure 61**)[29].

Lateral data spacing

Surface roughness data typically has a lateral spacing of 0.5 µm to 1.5 µm, which supports the application of standard noise filters and provides adequate definition of the surface features. A typical CMM's data spacing is much larger than this.

Vertical resolution

Surface texture instruments can have vertical resolutions of a nanometer or less, with common instruments having resolutions of ten or twenty nanometers. Typical CMM resolutions are much larger, on the order of microns.

---

[29] Section 7-4.2.4 of ASME B46.1-2009 provides the mathematical equation for calculating the maximum allowable tip radius for a surface texture modelled as a sine wave of wavelength λ and amplitude A.

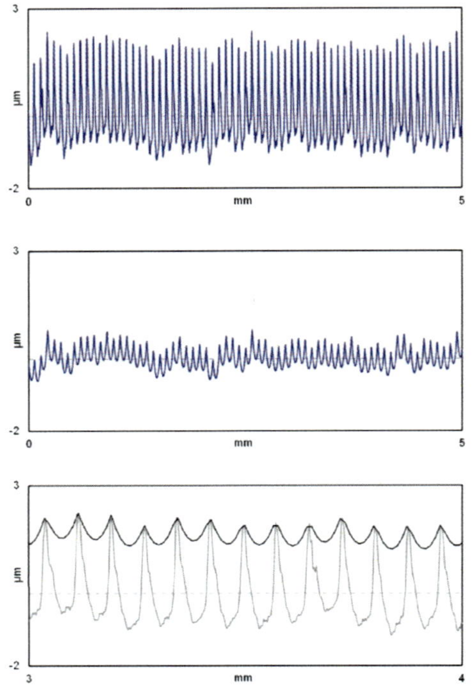

Primary profile of a turned surface, measured with a 2 µm tip radius stylus
Ra = 0.723 µm, Wt = 0.645 µm

Primary profile of the same turned surface as it would appear when measured with a 1 mm radius ball
Ra = 0.202 µm, Wt = 0.583 µm

Magnified view of the interaction of the 1 mm radius ball (black line) with the turned surface (gray line), showing how it is unable to reach into the valleys, thus reducing the apparent roughness. This black line produces the primary profile shown in the middle graph. There is less impact on the waviness than on the roughness.

Figure 61: Tracing surface texture with a 1 mm radius ball

DATA ANALYSIS

DATA ANALYSIS

Data analysis takes the raw measurement data and calculates the desired profiles, functions, and parameters. This involves the following steps and user-selected options (see **Figure 62**):

1. Suppression of a reference geometry. The underlying geometry is removed from the measured profile to yield the *form-suppressed* profile. This step may include the application of the form removal filter (lambda-f, or λf) to separate form from waviness.
2. Short wavelength filtering. The wavelengths shorter than the short filter wavelength (lambda-s, or λs) are assumed to represent noise and are discarded. The longer wavelengths are retained as the *primary* profile.
3. Roughness filtering. The roughness filter is applied to the *primary* profile. The long wavelengths are retained as the *waviness* profile. The *waviness* profile is subtracted from the *primary* profile to give the *roughness* profile.
4. Parameter computation. Using the profiles derived in the preceding steps, various analysis functions (material ratio curve, amplitude distribution function, etc.) are calculated. These analysis functions and the profiles are then used to calculate the parameters.

To get the correct results the data analyst must have a basic understanding of these steps and must be able to interpret what is on the technical drawing or in the product documentation in order to apply those requirements to the data. This se ction explores questions that can arise about this analysis.

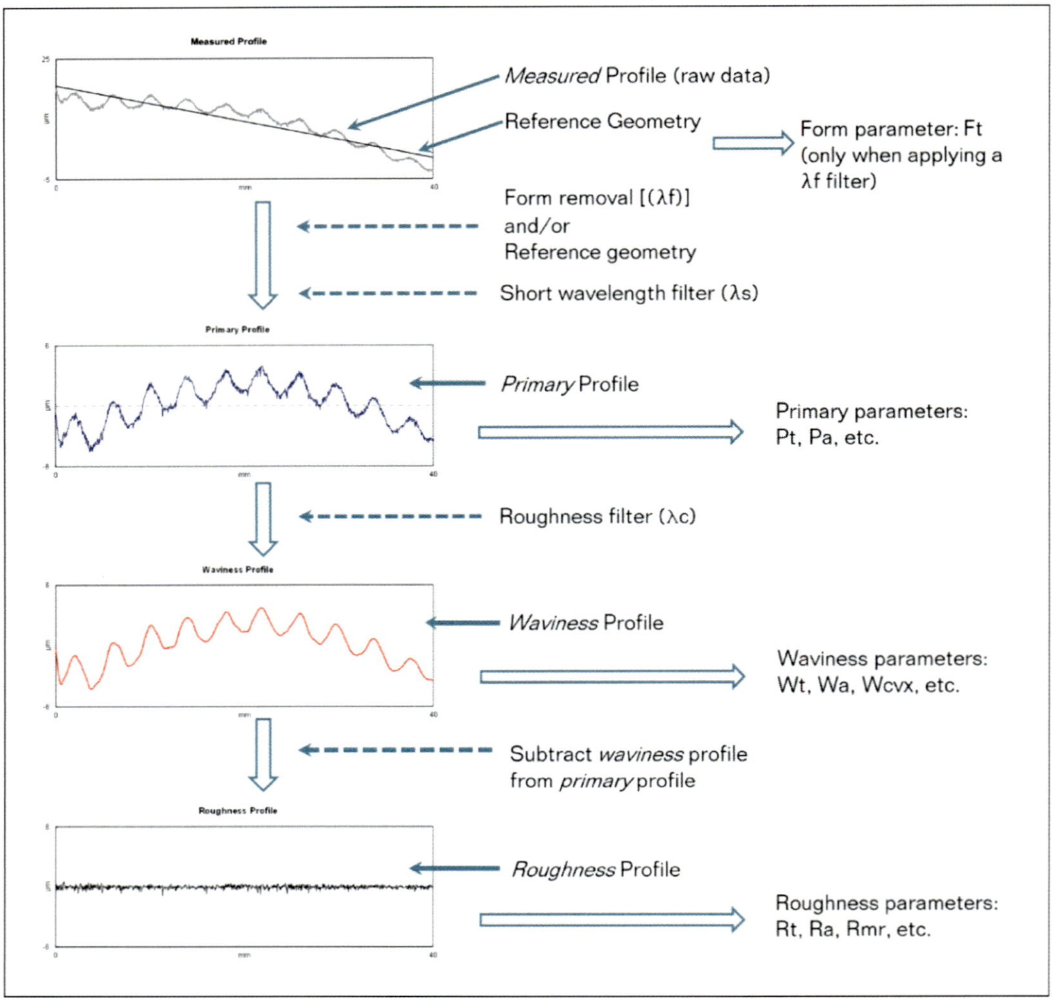

Figure 62: Simple schematic of data analysis

What is form removal?

"Form removal" is a term that can be confusing because of the difference between the general and the technical meanings of "form," The general meaning of form is synonymous with "shape," with options such as tilted, curved, convex, concave, etc. The technical meaning of form within the surface texture field is "the long-pass output of the form filter (lambda-f, or λf)." This type of *form* profile is typically encountered when using bandpass waviness.

In the context of bandpass waviness, the surface wavelength domains are named as shown in **Figure 63**. In other settings, we refer to form as geometric elements, such as a straight line, an arc, etc. Any discussion of "form removal" must differentiate between these two meanings: a filtered *form* profile vs. a nominal geometric form or shape.

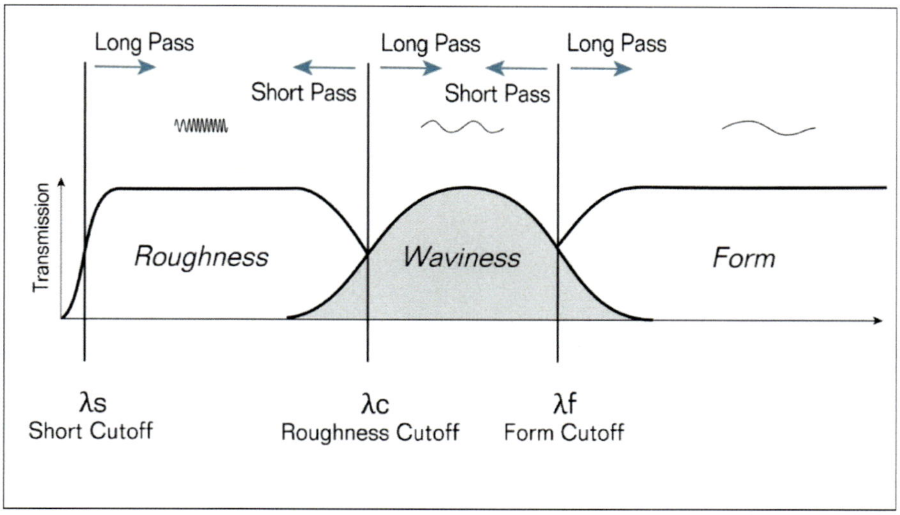

Figure 63: Cutoff wavelengths separate roughness, waviness and form domains.

In this book, the term "reference geometry" will be used to refer to the "shape" while "form" will be used in the "filtered form profile" sense[30]. A reference geometry (a least squares line, a least squares arc, a polynomial, etc.) will be used to remove tilt or a nominal curvature from a profile, and that reference geometry will be represented by a *reference geometry* profile. A *form* profile will only be created by the application of a λf filter, and it is only in this circumstance that form parameters (e.g., Ft) will be calculated.

[30] Unfortunately, "form" is so commonly used that you will likely see terms usch as "form removal" or "form-suppressed" applied to situations which are in fact reference geometry removal. This will occasionally occur in this book as well, where we feel it is most helpful to convey the intended meaning. The context will typically clarify which meaning is being used.

It should be noted that in recent ISO standards the notion of an "f operator" has been introduced. The "f" or "form" operator can be a fitting operation (e.g., least squares line) or a filtering operation (e.g., a Gaussian filter). Some of these operations (for example, the least squares line) lead to uninteresting profiles (e.g., a straight line). These are classified as geometry removal. On the other hand, filter-based "f operators" can extract interesting shapes from a profile, and as such their output is referred to as "form."

Given the potential ambiguity between reference geometry and form, it is suggested that the form/geometry removal process be clearly documented in measurement reports and product documentation.

With this ambiguity in mind, and in order to accurately address the original question, we will restate and answer that question in two versions:

What is reference geometry removal?

Measurement data may contain long wavelengths that are artifacts of the measurement process (e.g., a slope from tilting of the workpiece) or that are actual characteristics of the workpiece (e.g., curvature from tracing a curved surface). Shapes in a profile caused by alignment errors are typically of no interest, so these should be removed prior to roughness and waviness analysis. On actual curved components it is often desirable to analyze departures from the nominal shape, so removal of that shape is required. This is the purpose of reference geometry removal.

Reference geometry removal involves subtracting some shape from the *measured* profile to create the *form-suppressed* profile. That shape may be a predefined form, such as a fixed radius arc, or it may be derived from the measurement data, such as a least squares line or a polynomial. In most situations this will represent the nominal shape of the surface being measured; e.g., an arc for a nominally circular surface. Examples of some of these user-selectable options are given in **Figure 64**, showing the relationship between the *measured* profile and the *primary* profile[31].

What is form removal?

Form removal is most commonly used when performing bandpass waviness analysis, where it is desirable to exclude the longest wavelengths from the waviness. The data is filtered with a form cutoff wavelength (lambda-f, or λf), where λf is the long wavelength limit for waviness (see the discussion on transmission bands on page **324** for more information on λf). The long wavelengths passed by this filter constitute the *form* profile while the shorter wavelengths constitute the *form-suppressed* and *primary*[31] profiles—see **Figure 65**.

A form removal analysis may automatically include a reference geometry removal. For example, the *measured* profile in **Figure 65** has a tilt that is likely caused by inadequate leveling of the component. A least squares line reference geometry will commonly be applied to remove this tilt,

[31] The application of the short wavelength (lambda-s, or λs) filter to the *form-suppressed* profile yields the *primary* profile.

followed by the λf filter. The sequence is shown in **Figure 66**. This two-step procedure will most likely be invisible to the operator, except that form parameters will be calculated from the leveled *form* profile, as in **Figure 66** (b), rather than from the tilted profile, as in **Figure 66** (a).

Summary

To summarize the previous discussion, reference geometry removal is concerned with subtracting a nominal shape from the data. Form removal is about filtering to separate form and waviness.

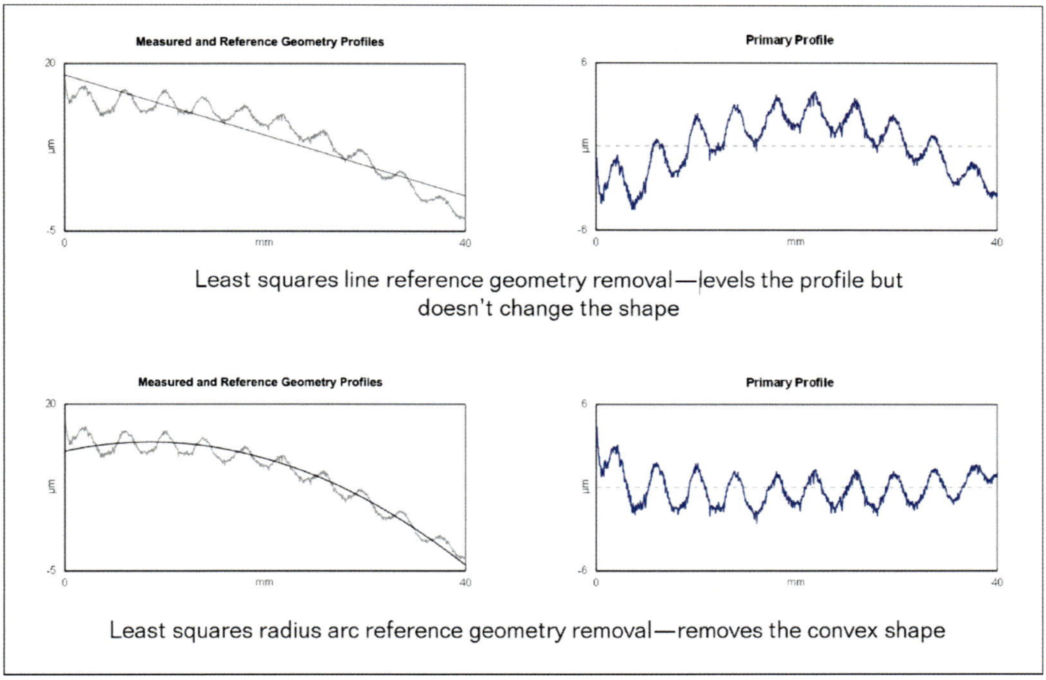

Figure 64: Examples of reference geometry removal and the effect on the *primary* profile

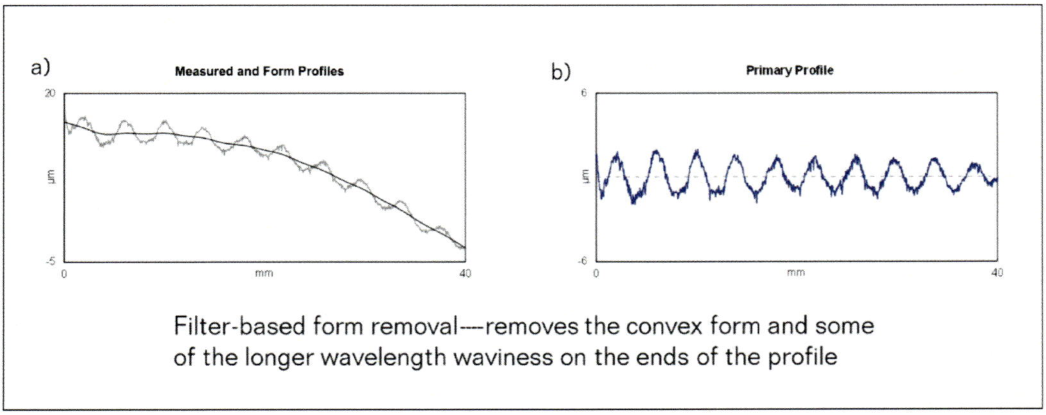

Figure 65: Example of form removal with the resultant *primary* profile

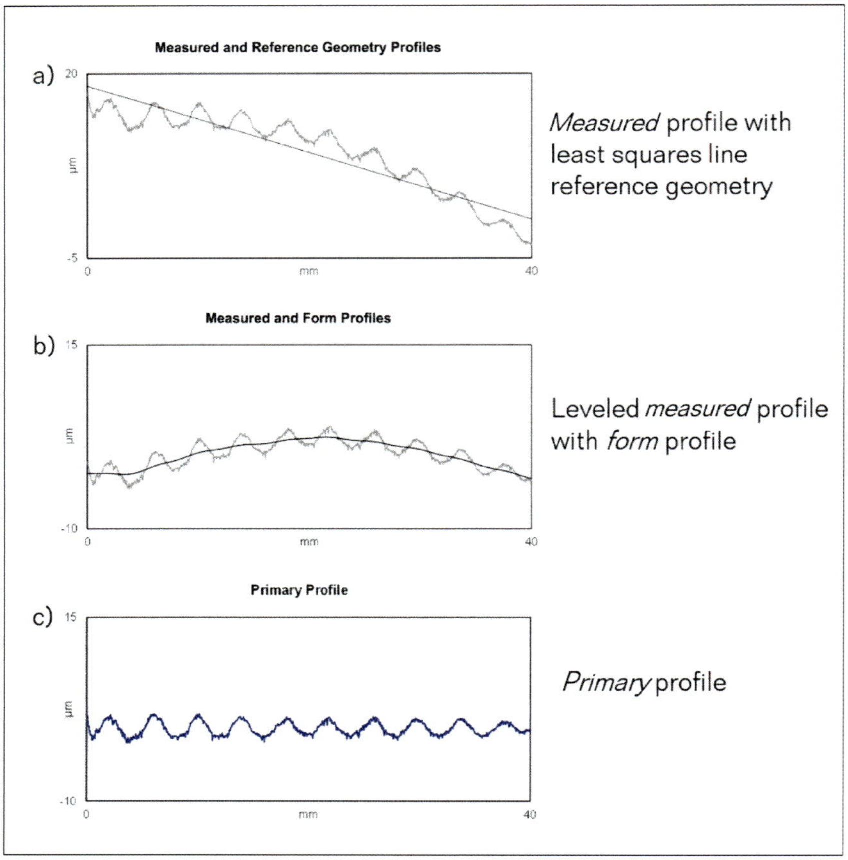

Figure 66: Typical sequence for calculating *form* and *form-suppressed* profile

The GD&T term "profile"

GD&T documentation uses a third term, "profile," in a manner similar to "shape" or "form." For example, with reference to a callout for the profile of a line or the profile or a surface, the "profile error" is the departure of the actual surface from the nominal shape. This departure is sometimes also called "form error," using the more generic meaning of "form." Further discussion of the relationship between GD&T and surface texture terminology is found in the section, **How do GD&T and surface texture specifications relate to each other?** on page **181**.

Residuals

When using a reference geometry, the differences between the *measured* profile and the *form-suppressed* profile are considered to be "residuals." These residuals can be calculated either perpendicular to the nominal geometry or vertically. Vertical residuals project the actual profile onto the x-axis and can result in noticeably-shortened features if the *measured* profile has significant slope or curvature. Perpendicular residuals "unroll" a curve or rotate a tilted feature, resulting in a *form-suppressed* or *primary* profile that will be longer than the *measured* profile. Individual feature heights can also be affected. **Figure 67** shows the difference between vertical and perpendicular residuals, and **Figure 68** shows a comparison of these effects for an actual measurement of a convex surface.

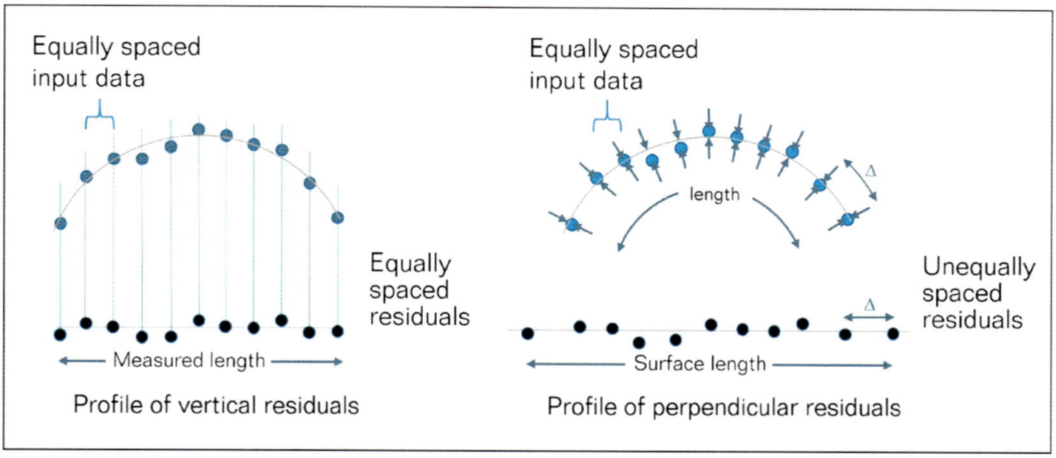

Figure 67: The length of the resulting profile depends upon the type of residuals

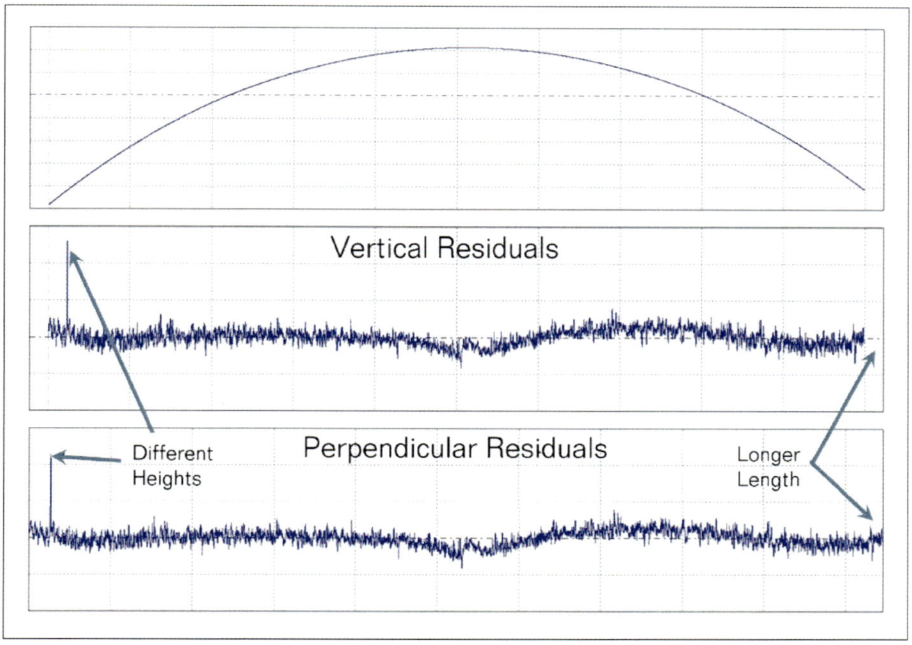

Figure 68: Comparison of vertical and perpendicular residuals

For many surface texture data sets, the type of residual has little effect on the results. It may, however, be important when, for example, the length of a feature is being calculated, as shown in **Figure 68**.

What is "filtering?"

Filtering is the process of separating the measurement data into various wavelength domains. For example, filters can separate short wavelength (roughness) from longer wavelength (waviness) domains. In a typical data analysis sequence filtering is performed after reference geometry removal and before parameter analysis.

Filtering most commonly involves passing an averaging function through the profile data. This can be as simple as the five-point average illustrated in **Figure 69**.

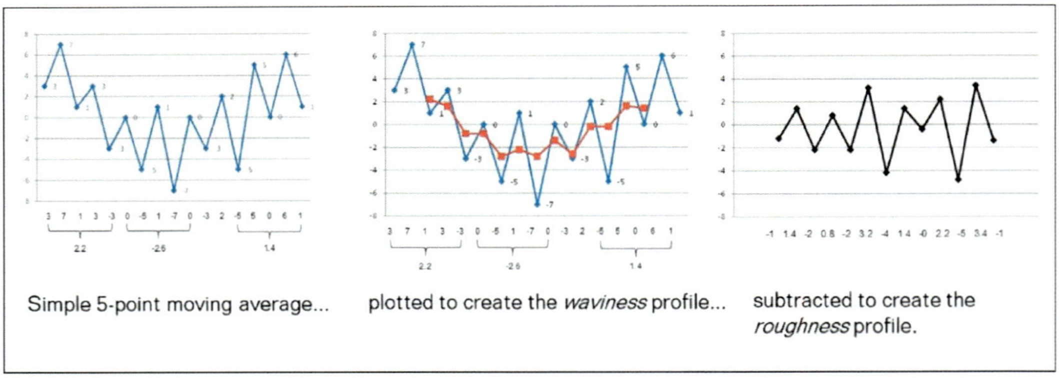

Figure 69: Illustration of a simple 5-point moving average creating *waviness* and *roughness* profiles

A better filter uses a weighted average such as the Gaussian function illustrated in **Figure 70**. The Gaussian filter is the most common because it is critically-damped, phase correct, expressible in both the spatial and frequency domains, and digitally implementable. The width of the moving average window is adjusted based on the filter cutoff wavelength, which is selectable by the user and is very important in the measurement of surface texture.

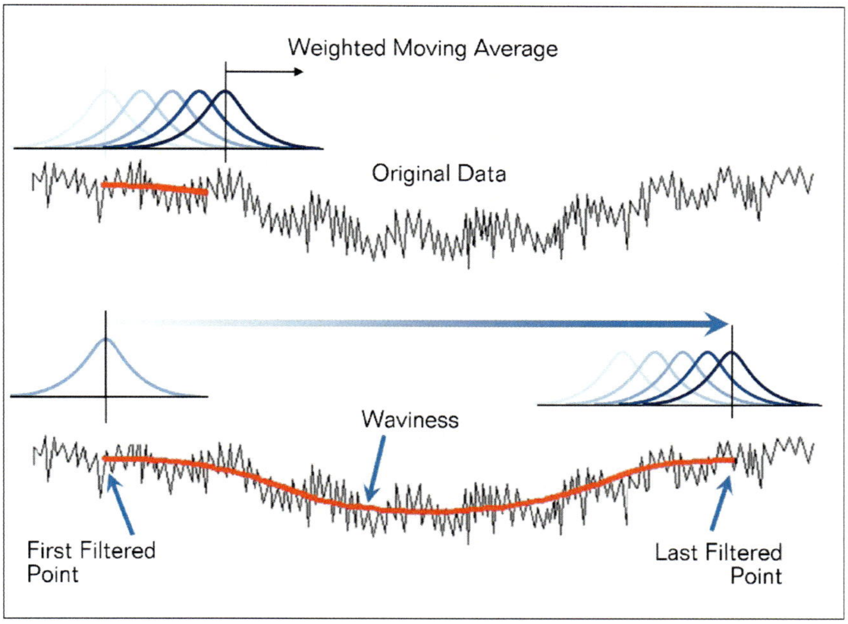

Figure 70: Illustration of a Gaussian filter generating a *waviness* profile

How is a filter used to separate roughness from waviness?

The difference between short wavelength "roughness" and longer wavelength "waviness" is a fundamental concept for understanding and controlling surface texture. The separation of long and short wavelengths is made possible by a filter (see the section, **What is "filtering?"**). A "filter cutoff value" determines the wavelength at which "roughness" is separated from "waviness." The *primary* profile is filtered using the roughness cutoff (lambda-c, or λc). The passage or "convolution" of this filter generates the *waviness* profile. The *waviness* profile is subtracted from the *primary* to generate the *roughness* profile (see **Figure 69** and **Figure 71**).

The component designer should choose the value of λc to fit their definition of roughness and waviness, and they should assign roughness and waviness parameters and limits accordingly. In cases where the roughness cutoff wavelength is not specified, tables are provided in standards such as ASME B16.1-2009 or ISO 4288-1996.

Figure 71 through **Figure 73** illustrates the importance of this choice. For the *primary* profile shown in **Figure 73**, as λc increases from 0.08 to 25 mm, the amplitudes associated with the prominent 2.5 mm wavelength feature are gradually moved from the waviness domain into the roughness. This can be seen in the location of the filter transmission curves in the wavelength content graphs in **Figure 72**. With this shift, Ra increases by a factor of six while Wa decreases by a factor of 100 (see **Figure 73**). There is no inherently "right" value for Ra—0.17 μm is as correct as 1.02 μm for this profile—but a supplier and customer would have difficulty communicating about this data if the roughness cutoff value wasn't clearly stated.

When choosing the cutoff wavelength, the designer needs a knowledge of the function of the part and of the system it operates within as well as, if possible, the manufacturing process. The filter should "fit" the surface and unambiguously place features into roughness or waviness. For example, λc = 2.5 mm is a poor choice for the profile in **Figure 73** because it places half the amplitude of the prominent feature into roughness and half into waviness (see **Figure 72**). The choice of cutoff must include consideration of the parameter selection so all relevant wavelengths are controlled. For example, if no waviness parameter is specified, special care must be taken to be sure the roughness domain includes all important features. ISO and ASME provide guidelines for cutoff selection (see page **145**), but functional and manufacturing considerations should always take precedence over these default values.

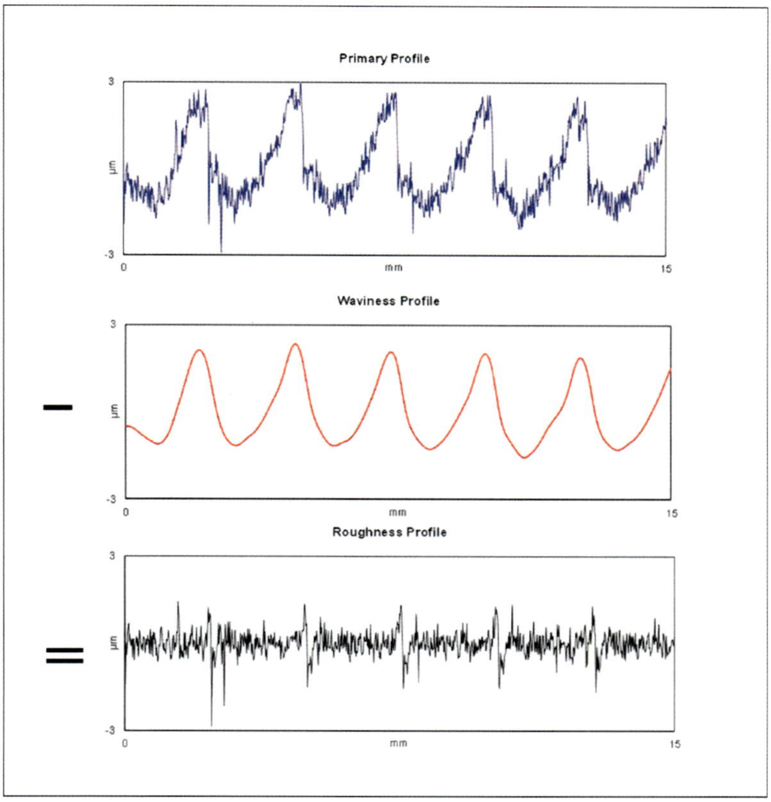

Figure 71: *Primary* profile minus *waviness* equals *roughness*

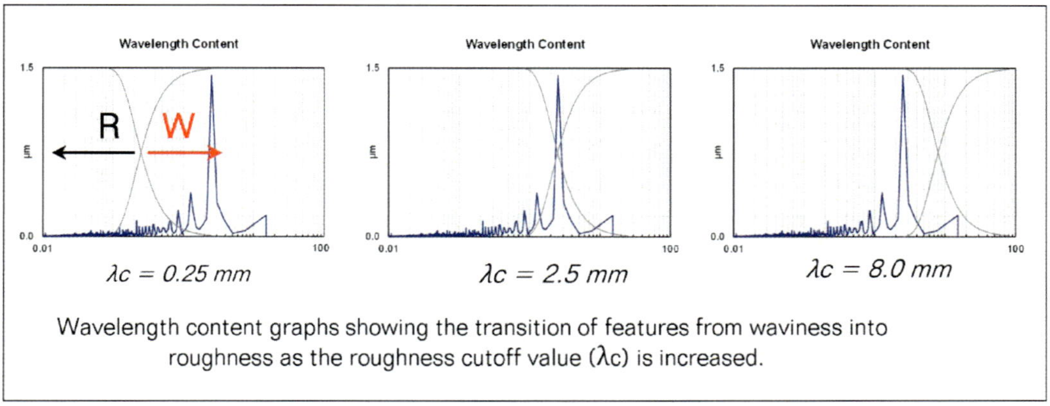

Figure 72: Increasing λc moves waviness to roughness, thereby affecting parameter values.

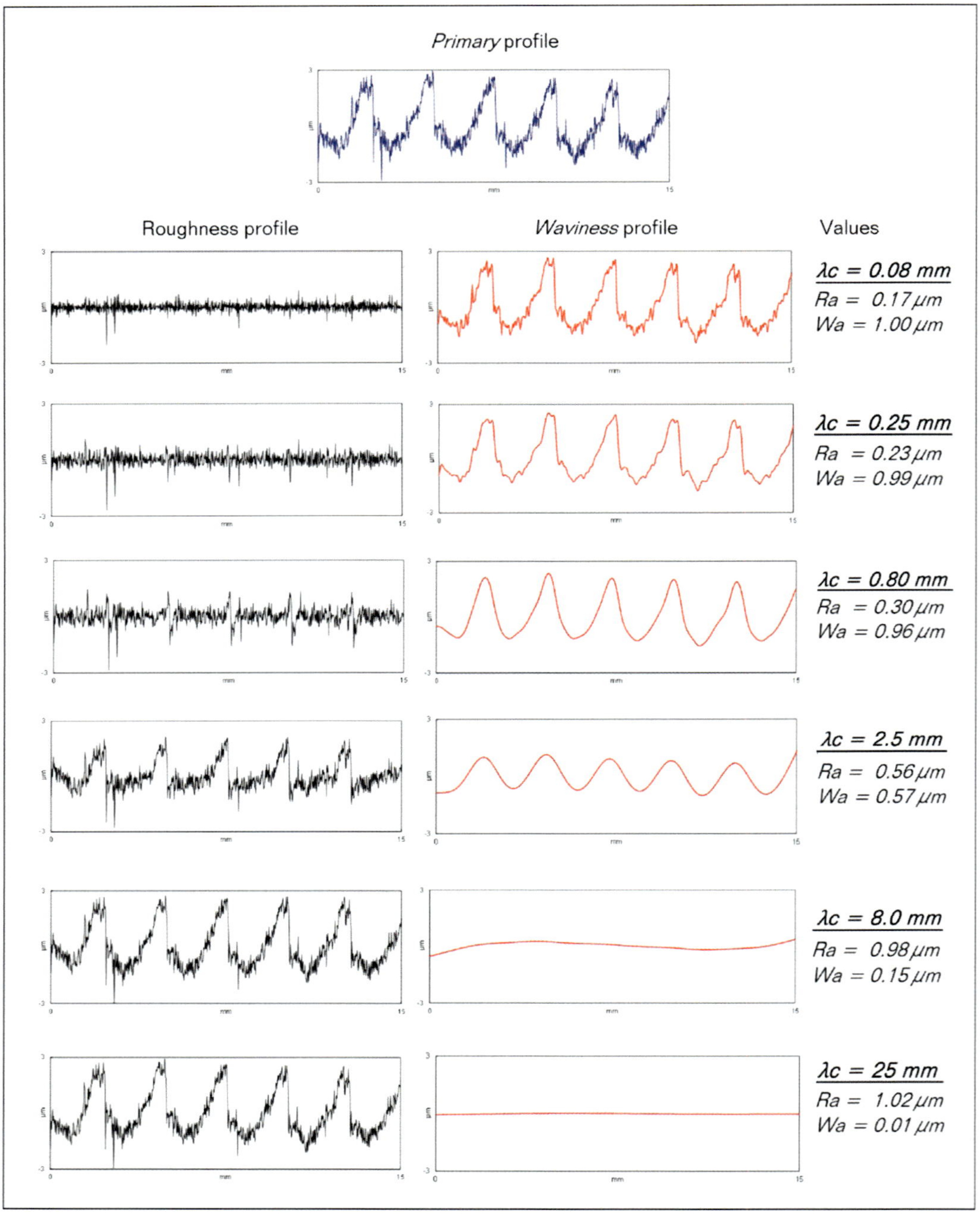

Figure 73: Influence of λc value on *roughness* and *waviness* profiles and parameters

What do all of those parameters mean?

Surface texture parameter names are combinations of the **profile** designation and the **characteristic.** The profile designations are P for the *primary* profile, W for *waviness*, and R for *roughness* (see page **320** for more on these three profiles). The characteristic is derived from that profile. For example, Ra is the average (a) height of the *roughness* (R) profile. The same characteristic can often be derived from each profile. For example, the average amplitude (a) parameter can be derived from primary, waviness, and roughness, creating the parallel set of parameters Pa, Wa, and Ra, as shown in **Figure 74**. The profile is written in uppercase; the characteristic is written in lower case (*not* as a subscript).

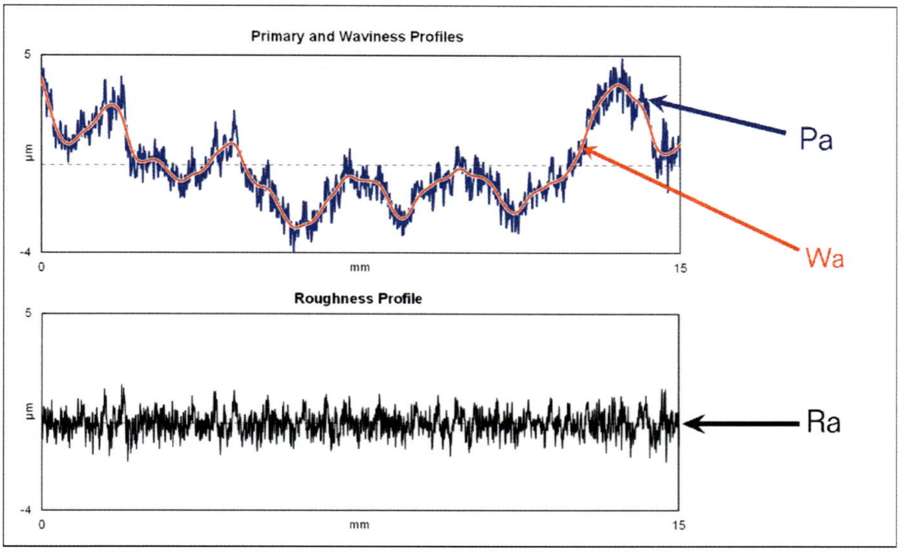

Figure 74: *Primary*, *waviness*, and *roughness* profiles

Profile characteristic dimensions (e.g., heights and depths) are measured relative to a reference or zero point. That reference is a "mean line" for each of the profiles—the mean line being the long-pass output of the relevant filter. For more information about mean lines see the section, "**What is the zero line in my profile graph based on?**" on page **122**.

Hundreds of parameters have been proposed, and most can be grouped into a small number of types or classes. Definitions for many of these can be found in the Definitions and Terminology section staring on page **291**, as well as in ASME B46.1, ISO 4287, the upcoming ISO 21920, and other national or international standards. While the following list contains mostly roughness parameters, the same mathematics can be applied to the *primary* (P) and *waviness* (W) profiles. While this is possible, it is not always common in practice.

Averaging parameters (Ra, Rq, Rc, Rsk, Rku, etc.)

The averaging parameters use all of the data points, so many of them are relatively stable and repeatable. This positive aspect can also be a negative one: these parameters don't report on the extremes that may be important for a component's function. In addition, many of these parameters use the absolute values of the surface features and therefore don't differentiate between peaks and valleys.

Peak-to-valley parameters (Rt, Rp, Rv, Rz, R3z, Rpm, Rmax, etc.)

The peak-to-valley parameters, as the name implies, focus on the more extreme points of the profile. Because of this, they are sensitive to outside influences and outlier features such as noise, debris, and non-statistical scratches, making it difficult to use them to control manufacturing processes. Parameters such as Rz or Rpm incorporate some averaging to improve the stability. Others, such as R3z, improve stability by ignoring a few of the largest peaks and valleys.

Slope, spacing, and counting parameters (Rsm, Rdq, Rcl, Rvc, Rpc, Rhsc, etc.)

While the averaging and peak-to-valley parameters mainly assess the height of the profile, the slope, spacing, and counting parameters address the lateral spacing of the features. Many of these parameters include the x-axis data spacing in their formulas so their values are sensitive to that spacing, and comparisons of surfaces can only be made when this spacing is consistent.

Material ratio curve parameters: (Rmr, Rk, Rpk, Rvk, Rpq, Rvq, Rmq, Rdc, etc.)

The material ratio curve (Abbott-Firestone curve) is a representation of the amount of material at various depths within a surface. Several parameters have been derived to quantify different characteristics of this curve. Additional information can be found on pages 340 and 346.

Motif parameters (ISO 12085) (R, Rx, AR, W, Wte, Wx, and AW)

The motif parameters were developed by the French automotive industry in an attempt to correlate surface texture to component function. Additional information on these parameters is given on page 354.

Spectral moment parameters (Rm0, Rm2, Rm4)

The spectral moment parameters are occasionally used in analyses of the tribological interactions between rough surfaces. At the time of this writing, these parameters have not been adopted by the standards organizations and are therefore not included in most instruments' software.

Morphological filter parameters (Wvoid, Wvdd, Wcvx)

The morphological filter parameters are derived by applying either an opening or a closing morphological filter to a profile. The profile used is typically the *waviness* profile since the features of interest are most often those with these intermediate wavelengths. The resultant parameters are useful for assessing functional issues such as creating a seal between two surfaces or the stress concentration from a local high spot. At the time of this writing, these parameters have not been adopted by the standards organizations and are therefore not included in most instruments' software.

Optical "slope" parameters (Wlslp, Wlslpl, Weslp, Weslpl)

The optical "slope" parameters are specific to the optics industry and are used to assess profile slopes within a defined segment length. There are no standard naming conventions for these parameters.

Parameter names—sources of confusion

Historically, parameters were calculated based on the sampling lengths. This is currently the case in some standards (for example, ISO 4287-1997). In this approach, the parameters are calculated for each sampling length and then averaged over the evaluation length[32]. Other standards (for example, ASME B46.1–2009) define parameters as being calculated over the "evaluation length." For many parameters the results are similar, but there are some (most notably Rp and Rv) where the results can be very different.

Note: At the time of this writing, discussions are taking place in ISO regarding the use of evaluation length as the default means of calculation. Under this proposed approach, only the partitioning-based parameters such as Rz would use sampling lengths.

One unfortunate result of this shift is a change in terminology and in the definitions for some parameters, a change which can be very confusing. A prime example of this is Rp. In the evaluation length based approach (for example, ASME B46.1-2009), Rp is the highest peak in the whole evaluation length, while in the sampling length system (ISO 4287-1997) Rp is the average height of the highest peak from each sampling length. Thus, if the evaluation length is made up of five sampling lengths, five peaks will be averaged (one from each sampling length). The Rp value in the sampling length system (ISO 4287-1997) is equivalent to Rpm (average maximum peak height) in the evaluation length system (ASME B46.1-2009).

Other parameters may have multiple designations, either due to this shift in meaning over time or due to different national and international conventions. Examples of this include Rtm and post-1997 ISO Rz, and Rmr and tp. **Table 21** on page **292** lists several parameters with different names but equivalent meanings.

Because some parameter names can be confusing, care should be taken to ensure that an engineering document using these parameters is unambiguous. Stay consistent with whichever specification system is in use, and clearly note any differences. In addition, not all instruments will use the same definitions, either by choice of the supplier or because the instrument's software is not up to date. Check your measurement system's software and operating manual if a parameter definition is unclear. Ultimately, the instrument and its software will dictate the calculation and numerical results that you see—regardless of what standards are saying.

---

[32] Note that ISO standards do not explicitly state that the individual sampling-length values are to be averaged. However, this seems the obvious approach, except for parameters such as Rq (or those incorporating Rq); Rq is a standard deviation, and is properly combined as the root mean square.

What is the zero line in my profile graph based on?

Characteristics such as peak height and valley depth are measured relative to a reference, the "zero" of a profile graph. That reference is a "mean line" for each of the profiles—the mean line being the long-pass output of the relevant filter.

Mean lines are illustrated in **Figure 75** and **Figure 76**. **Figure 75** is an example with one-sided waviness (no λf), while **Figure 76** is that same profile with bandpass waviness (*λf = 8 mm*). The mean line for the *primary* profile is the reference geometry (a least squares line in **Figure 75**) or the *form* profile (**Figure 76**). The mean line for one-sided waviness is the reference geometry and for bandpass waviness the *form* profile from the λf filter. For roughness, the mean line is the *waviness* profile, the long-pass output of the lambda-c filter. As illustrated in **Figure 71** on page **117**, the *waviness* profile is subtracted from the *primary* profile to generate the *roughness* profile, so the "wavy" *waviness* profile (mean line) shown in **Figure 75** (e) becomes the straight mean line (at *y = 0*) in **Figure 75** (f). Parameters are then calculated relative to the appropriate mean line: height parameters as the vertical distances from this line, and spacing parameters from the points where the profile crosses this line.

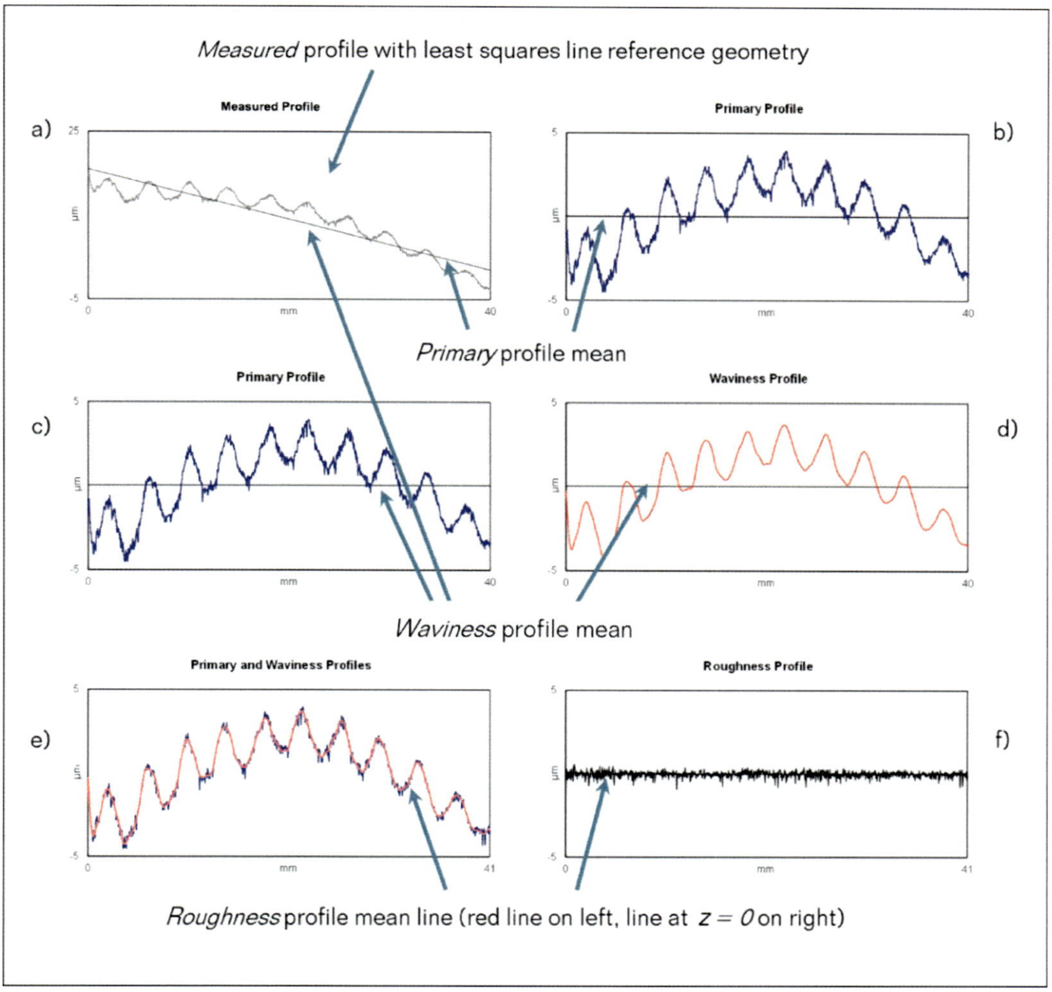

Figure 75: Mean lines for the P, W, and R profiles with one-sided waviness (no λf)

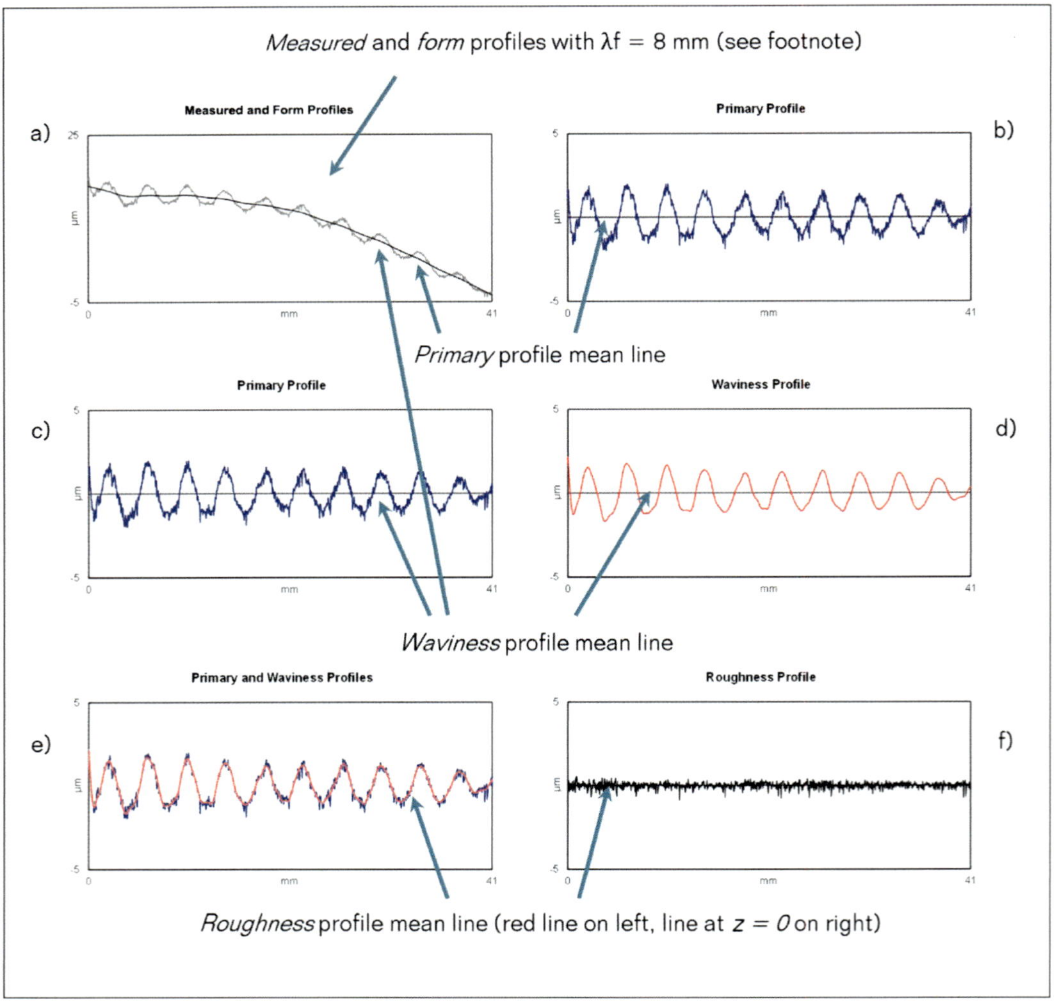

Figure 76: Mean lines for the P, W, and R profiles with bandpass waviness[33]

[33] The *form* profile in **Figure 76** (a) may not be strictly correct. Analysis software may level the profile before applying the λf filter, in which case the *form* profile would be a leveled version of what is shown here. However, **Figure 76** is adequate for the intended purpose of illustrating the *primary*, *waviness*, and *roughness* profiles' mean lines.

Note that this "mean" line will not necessarily always indicate the mean, or average height, of the profile. With some filters, especially the robust filters, the profile can be shifted relative to the mean line. In extreme cases the *waviness* profile, for example, may lie totally above or below the mean line, resulting in negative values for the valley or peak parameters (see the turned profile in **Figure 77**). In less extreme situations the mean line shift can still result in significantly different parameter values (**Figure 78** for Rpm).

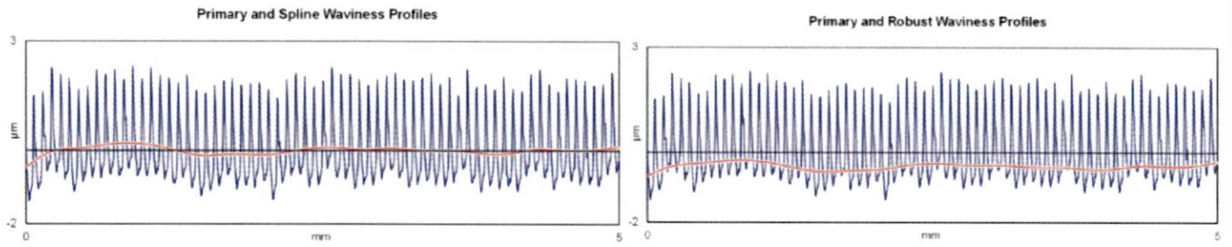

The spline *roughness* filter on the left results in a *waviness* profile lying on the mean line. The robust filter on the right yields a *waviness* profile lying below the mean line, with a negative Wp

Figure 77: An example of a *waviness* profile offset from the mean line with a robust filter

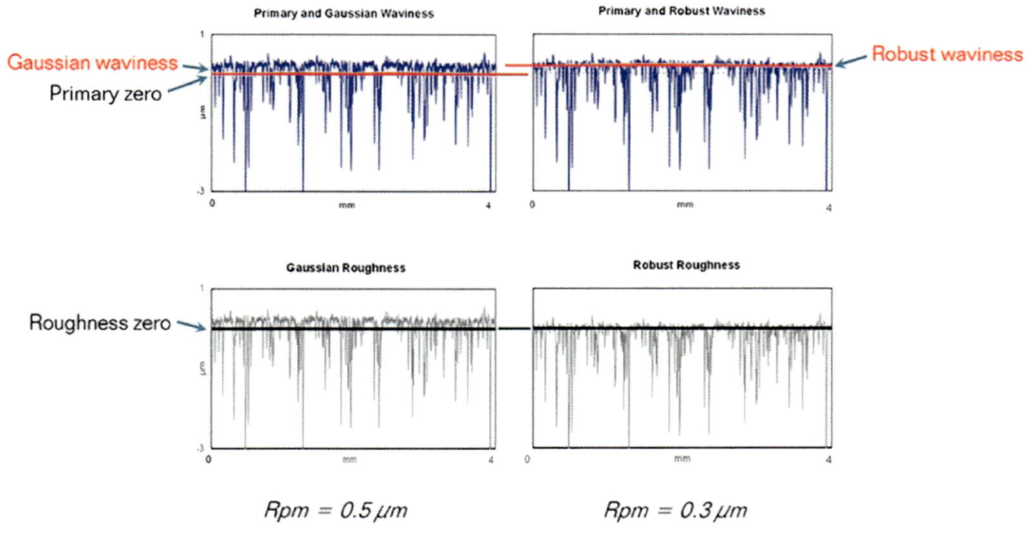

Figure 78: The *waviness* profile is offset for a plateaued surface

How do I decide which type of filter to use?

Filtering is the process of separating the measurement data into short wavelength (e.g., roughness) and long wavelength (e.g., waviness) domains. This involves moving an averaging function through the data, with the function width related to the transition wavelength between domains. Numerous functions are available, selected because of their ability to not distort the data, for their computational efficiency, for historical reasons, to avoid creating artifacts, or because they simulate two-surface interactions. The following filter types will be addressed in the next sections:

- Gaussian, both 50% and 75% transmission (page **127**)
- Spline (page **131**)
- Robust (page **134** and page **136**)
- Valley suppression, ISO 13565 (page **137**)
- ISO (page **141**)
- 2CR/2RC (page **142**)
- Phase corrected (page **144**)

My system has different Gaussian filters. Which should I use?

The filtering process separates data into roughness and waviness domains, based on the specified transition wavelength or cutoff. The filtering cannot be abrupt (i.e., a step function at the cutoff wavelength) because this would introduce an undesirable "ringing" into the data. Gaussian filters provide the required gradual transition while minimizing distortion. The Gaussian filter is described in ISO 16610-21.

As a filter is passed through a set of data the very short and very long wavelengths are separated into the roughness and waviness domains respectively. However, those wavelengths close to the cutoff value are only partially transmitted, with some of the amplitude becoming part of the *waviness* profile and the rest becoming roughness. At the cutoff wavelength a 50% transmitting filter will split the amplitude evenly between waviness and roughness, while a 75% roughness filter will put 75% of the amplitude into roughness and only 25% into waviness. An illustration of the 50% transmission of a Gaussian filter is shown in **Figure 79** for a sine wave with a wavelength equal to the cutoff value. The transmission function is shown in the wavelength content graph in **Figure 80** as the two S-shaped curves; the black, left-to-right falling curve defines the transmission for the *roughness* profile and the red, left-to-right rising curve is the transmission for waviness. The curves cross at the cutoff value (here, 0.8 mm).

Early digital filters simulated the analog 2CR filters which have 75% transmission at the cutoff wavelength. These filters were inconvenient, as a 75% transmitting roughness filter and a 75% transmitting waviness filter were used. This required two separate filters when performing roughness and waviness analysis. As Gaussian filters were implemented there was interest in maintaining some level of correlation with the 2CR filter, so a 75% Gaussian filter was used by some instrument manufacturers. However, it is more intuitive to evenly divide amplitudes into roughness and waviness at the cutoff wavelength, so the 50% Gaussian filter became standard, and is most commonly used today. The 50% transmitting approach is useful in that the *waviness* profile can be derived by the filtering process and the *roughness* profile can be derived by a simple subtraction of the *waviness* profile from the *primary* profile.

If a non-standard filter is to be used it should be clearly stated in the technical documentation or report, since it can introduce significant differences into the parameter values.

Primary profile for a 0.8 mm wavelength sine wave with a mean-to-peak amplitude of 1 μm.

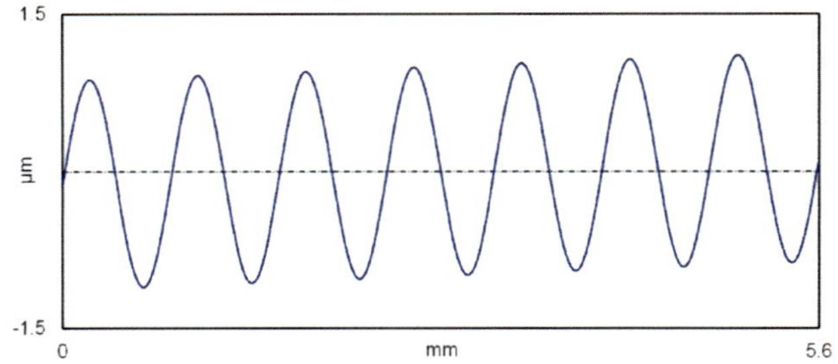

Roughness profile with a 0.8 mm roughness cutoff and a 50% Gaussian filter. The mean-to-peak amplitude is 0.5 μm.

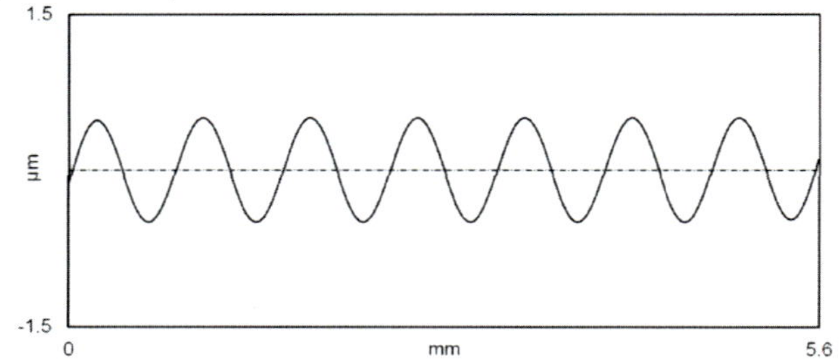

Waviness profile with the same shape as the roughness profile. The 50% filter partitioned half the sine wave amplitude into the roughness domain and half into the waviness.

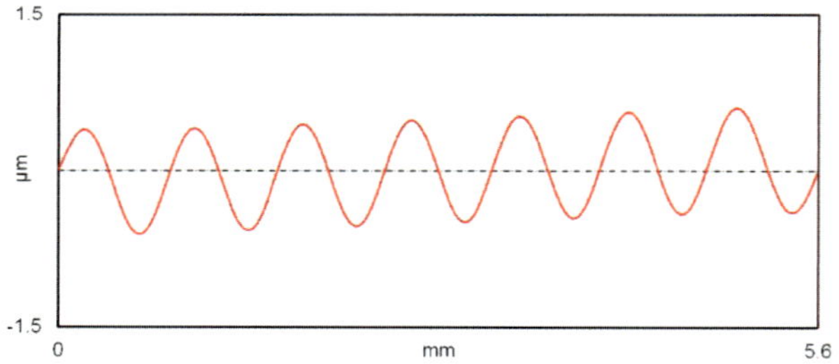

Figure 79: Illustration of waveform separation by a 50% Gaussian filter

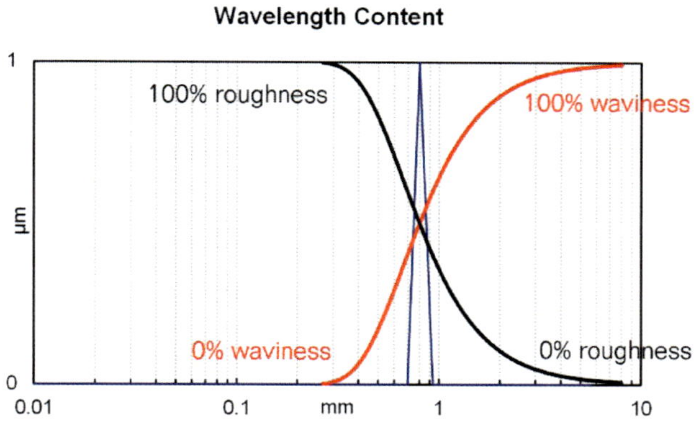

Figure 80: Filter transmission curves on a Wavelength Content graph

The Gaussian filter is the default filter type in that it has the unique mathematical property of being "critically damped." In other words, the Gaussian filter provides the sharpest separation between roughness and waviness in terms of the transmission characteristics. If a filter is created with a steeper/sharper transmission, the resulting profile will exhibit "ringing" or "overshoot." For example, a perfect "brickwall" filter is shown in **Figure 81**. This filter can create strange "ripples" in the resulting profiles (see **Figure 82**), ripples which are referred to as the *Gibbs effect* in Fourier analysis. These can result in misleading profile graphs and improper roughness parameter values.

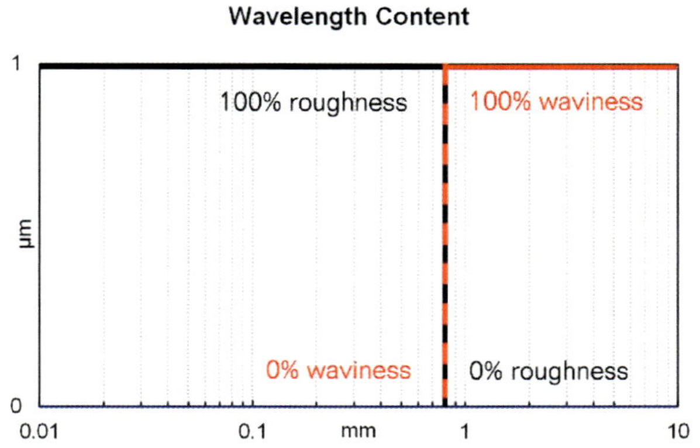

Figure 81: A perfect "brickwall" separation between roughness and waviness

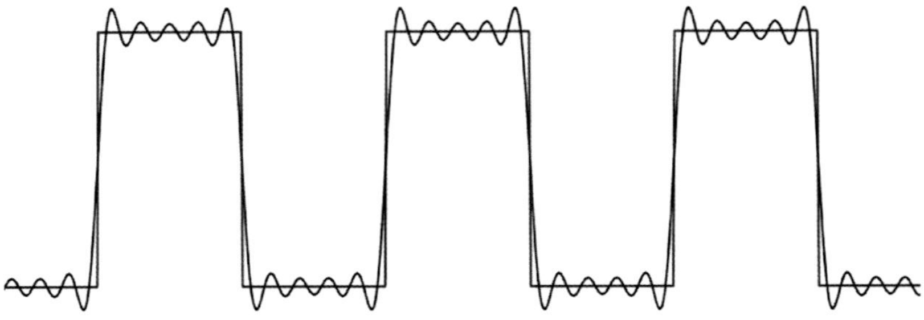

Figure 82: Ringing from a brickwall filter

How do I decide whether to use the Gaussian or the spline filter?

Gaussian[34] and spline[35] filters give very similar results. If a spline filter is available it is recommended that you use it unless you are trying to match an analyses for a calibration specimen or some other surface for which the Gaussian filter was known to be used.

Spline filters offer the following advantages relative to Gaussian:

- Very fast computation
- No discarding of end regions (see **Figure 83**). This no-discard aspect of a spline filter is especially important when measuring short features where the Gaussian end discards can be a significant fraction of the measured length.
- Avoidance of certain distortions at the ends of a profile. Different mathematical approaches for creating the Gaussian filter can cause different kinds of end effects. **Figure 84** shows one example of such a Gaussian filter implementation. These effects must be discarded as they are not representative of the surface; this is accomplished with the spline filter.
- More representative *waviness* profiles for profiles with steep slopes at the ends (see **Figure 85**)

Spline filters incorporate a tension parameter (β) to control how tightly the curve fits through the data points. A tension of $\beta = 0.625242$ provides a Gaussian-like behavior, which may be desirable when trying to match a Gaussian filter output. However, this can produce distortions in the *waviness* profile, especially when there is significant curvature or slope at the profile ends. For profiles such as these a form-following tension of $\beta = 0.0$ is often preferable for waviness and form analyses (see **Figure 85**).

It should be noted that some spline filtering methodologies (such as 2nd order spline filters) incorporate local polynomial or "curve" removal. This can be useful for some geometries; however, the curve removal may have an influence on the wavelengths of interest.

[34] Gaussian filters are defined in ISO 16610-21

[35] Spline filters are defined in ISO 16610-22

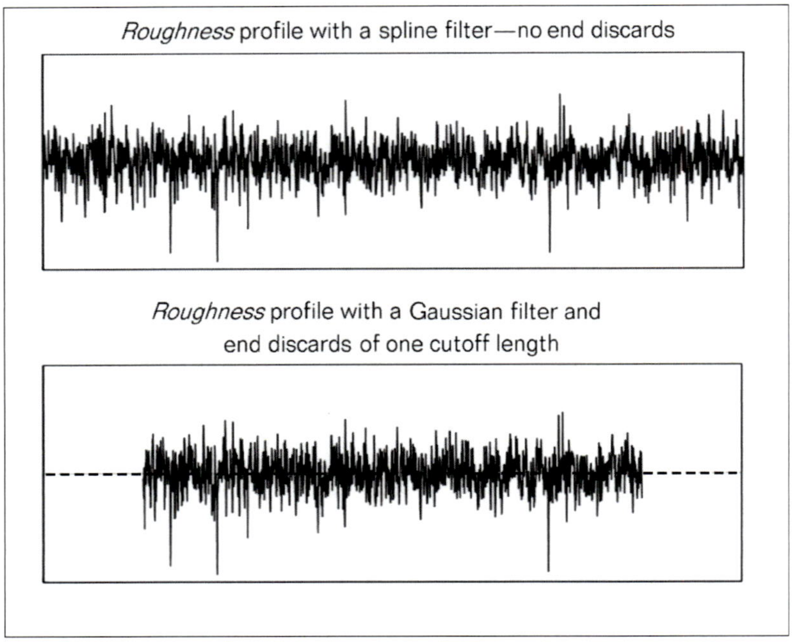

Figure 83: Spline and Gaussian filter effect on end discards

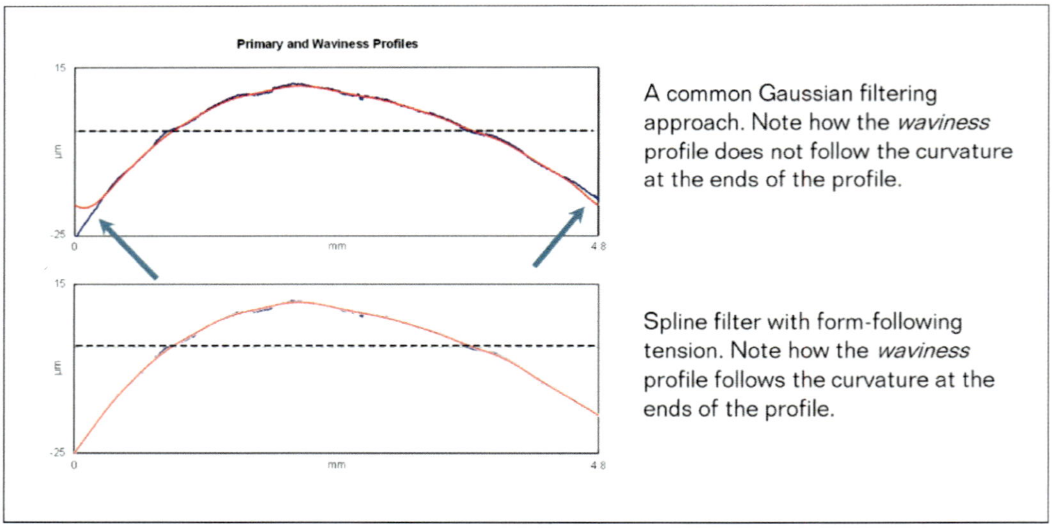

Figure 84: Comparing Gaussian and spline *waviness* profiles with steep end slopes

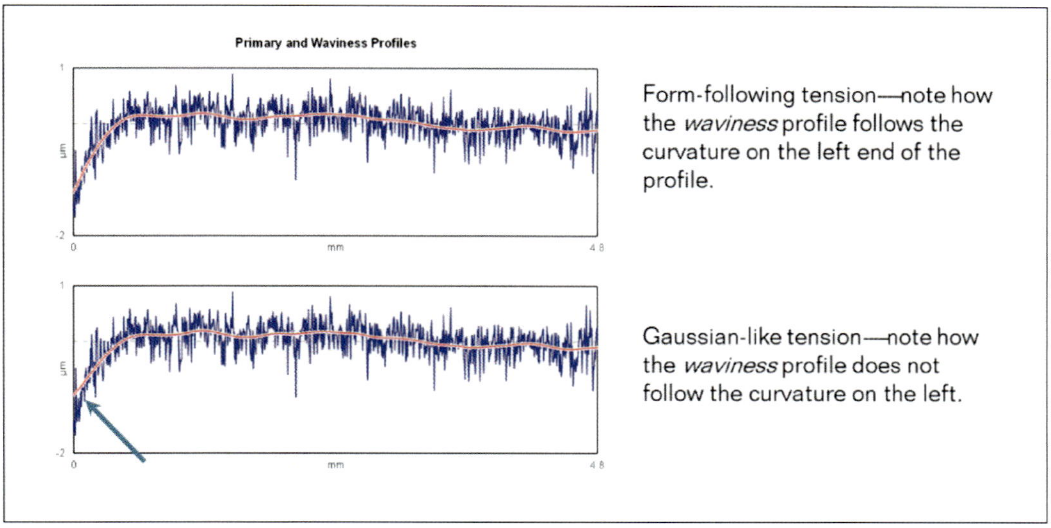

Figure 85: Examples of the effect of spline filter tension on end regions with high curvature

When should I use the robust filter?

The robust filter is designed to minimize the creation of artifacts in the *waviness* and *roughness* profiles due to extreme peaks and valleys. For profiles without these outliers the robust filter will often give results very similar to the spline (but see the section, "**Should I use the robust filter for everything?**" on page **136**).

When a Gaussian or spline filter encounters an extreme feature, such as a deep valley, the *waviness* profile gets "pulled" into the feature, as in the red profile in **Figure 86** (a). This adds a "dip" to the *waviness* profile that is not related to the underlying shape of the actual waviness. Furthermore, the waviness "dip" is typically much wider than the outlying profile points. This introduces areas on either side of the valley that become "pushed up" when roughness is calculated, as shown in **Figure 86** (b). The pushed-up areas on the *roughness* profile are often misinterpreted as the raised material that occurs when a surface is gouged or plowed. However, we can see from the primary profile in **Figure 86** (a) that there were no such peaks in the surface. (Note: This emphasizes the point that it is often best to view *primary* rather than *roughness* profiles.)

The robust filter down-weights the outliers and provides a *roughness* profile and roughness parameters much more representative of the actual surface (profiles (c) and (d) in **Figure 86**).

While **Figure 86** shows the problems related to profile valleys, it should be noted that peaks have the same effect on the *roughness* and *waviness* profiles, but in the opposite direction.

The robust filter is defined in ISO 16610-31, 32.

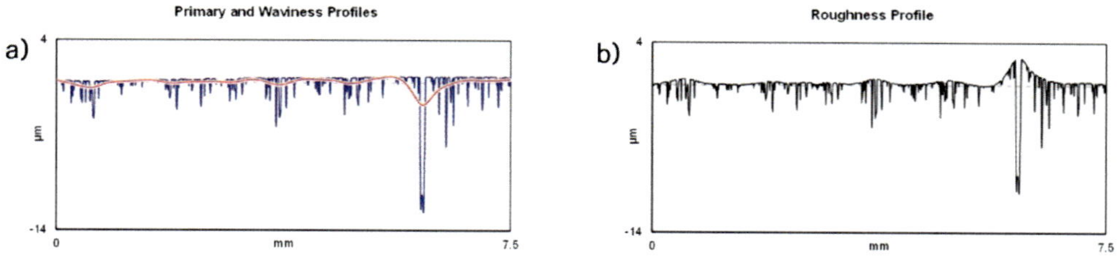

With a Gaussian or spline filter, the *waviness* profile (red line) on the left is pulled down into the deeper valleys. When this is subtracted from the *primary* profile (blue line) to create the *roughness* profile on the right, the data points above the *waviness* profile appear as raised material. *Ra = 0.46 µm, Rpm = 0.66 µm*

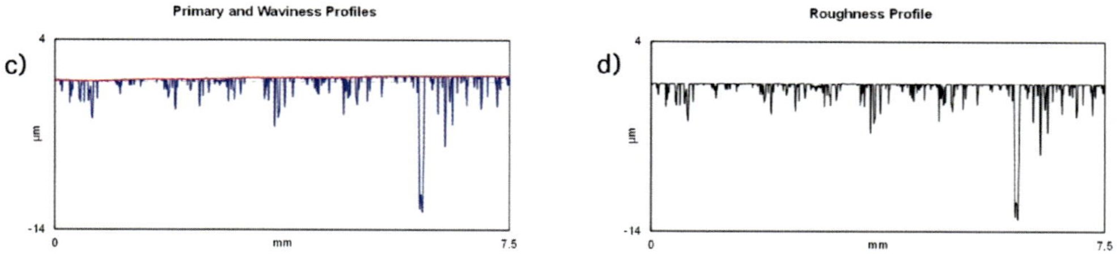

The robust filter is less sensitive to the outliers (left-hand graph) and therefore produces a *roughness* profile more representative of the actual surface. *Ra = 0.34 µm, Rpm = 0.05*

Figure 86: Influence of the robust filter on data with valley outliers

Should I use the robust filter for everything?

Robust filters are very powerful in cases where surfaces have outlying features such as porosity, nodules, extreme valleys, dirt/debris, etc. (see the previous section, **When should I use the robust filter?**). However, they should not be generally applied to all surfaces. For example, a turned profile consisting of relatively sharp peaks will cause the robust filter to drive itself away from the peaks, treating those peaks as outliers, and driving the *waviness* profile into the valleys (see **Figure 87**). This will result in inappropriate values for many of the roughness and waviness parameters. Further discussion of this phenomenon can be found in the section, **What is the zero line in my profile graph based on?**, specifically **Figure 77** and **Figure 78**.

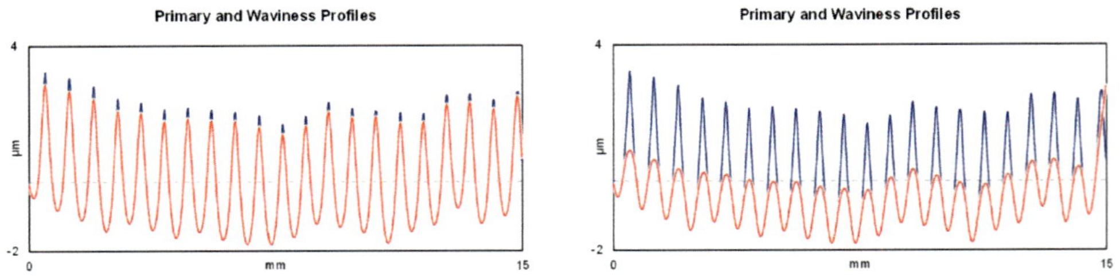

Form-following spline filter. The waviness profile (red line) follows the primary profile This is an appropriate filter for this type of profile. $\lambda c = 0.25\ mm$

Robust filter. The robust filter treats the peaks as outliers, "locking" the *waviness* profile into the valleys. This filter is NOT appropriate for this type of profile. $\lambda c = 0.25\ mm$

Figure 87: Example of inappropriate use of the robust filter

When should I use the valley suppression filter (ISO 13565-1)?

The valley suppression filter documented in ISO 13565-1 is similar to the robust filter (see page **134**) in that it desensitizes the *roughness* profile to the presence of deep valleys. However, it does not remove the influence of the deep valleys as well as the robust filter. Furthermore, it has several other disadvantages discussed below. For these reasons it is recommended that, whenever possible, the robust filter be used instead of the valley suppression filter when dealing with surfaces containing outliers. At the time of this writing, discussions are taking place regarding the removal of this filter from ISO standards since the robust filter is strongly preferred.

Disadvantages encountered with the valley suppression filter include:

- Significant issues with debris on the surface (see **Figure 89**)
- Insufficient outlier removal (see **Figure 90**)
- Removal of valley features that are not truly outliers (e.g., valleys that are part of the plateau region on a plateaued surface)
- The discard of data equal to one full cutoff length from each end of the trace, reducing the amount of data included in the analysis or forcing longer traverse lengths

Much of this behavior can be understood from the way the ISO 13565-1 filter is applied. The concept is illustrated in **Figure 88**.

1. Determine the mean line (*waviness* profile) on the original profile using a standard Gaussian filter and the specified cutoff wavelength.
2. Remove all profile features that lie below this mean line. When applied to something like a plateaued surface (the type of surface for which this filter is commonly specified), this removes not only the deep valleys but the plateaus' valleys as well.
3. Recalculate the mean line (*waviness* profile) for this *valley suppressed* profile.
4. Apply this second mean line to the original *primary* profile and derive the *roughness* profile and calculate all parameters using this combination.

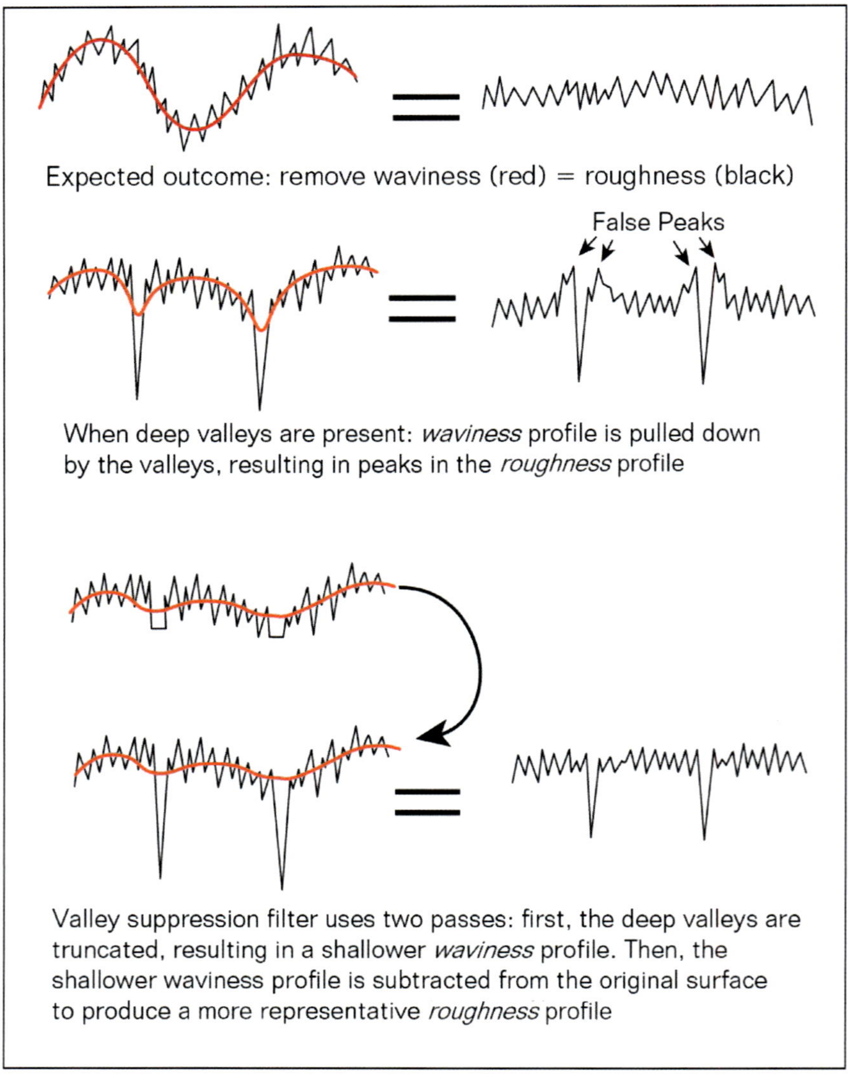

Figure 88: The concept behind the ISO 13565-1 valley suppression filter

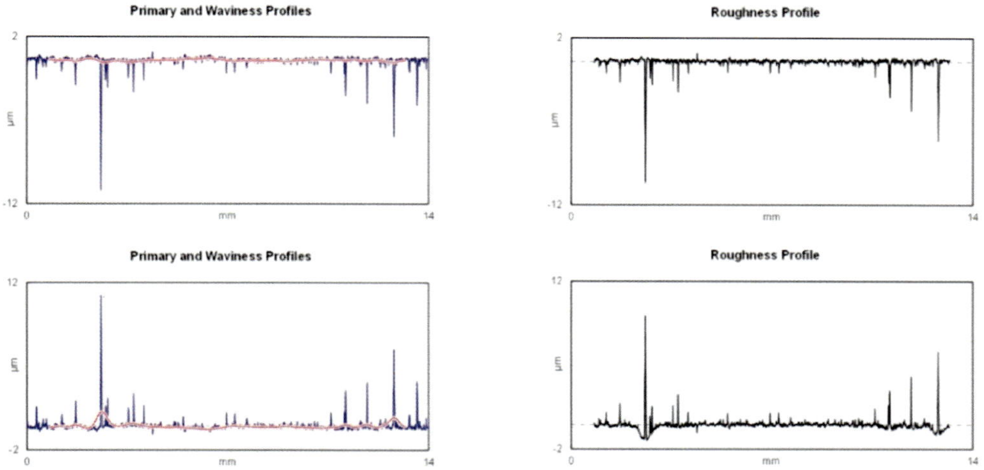

Unlike the robust filter, the valley suppression filter responds differently to peaks (e.g., debris) than it does to valleys. This is shown here, with the same profile inverted to represent both types of features. The distortion of the *waviness* profile (red lines in the left hand graphs) and the negative impact on the *roughness* profile (right hand profiles) is much greater if the feature is a peak rather than a valley.

Figure 89: Difference in response of the valley suppression filter to peaks and valleys

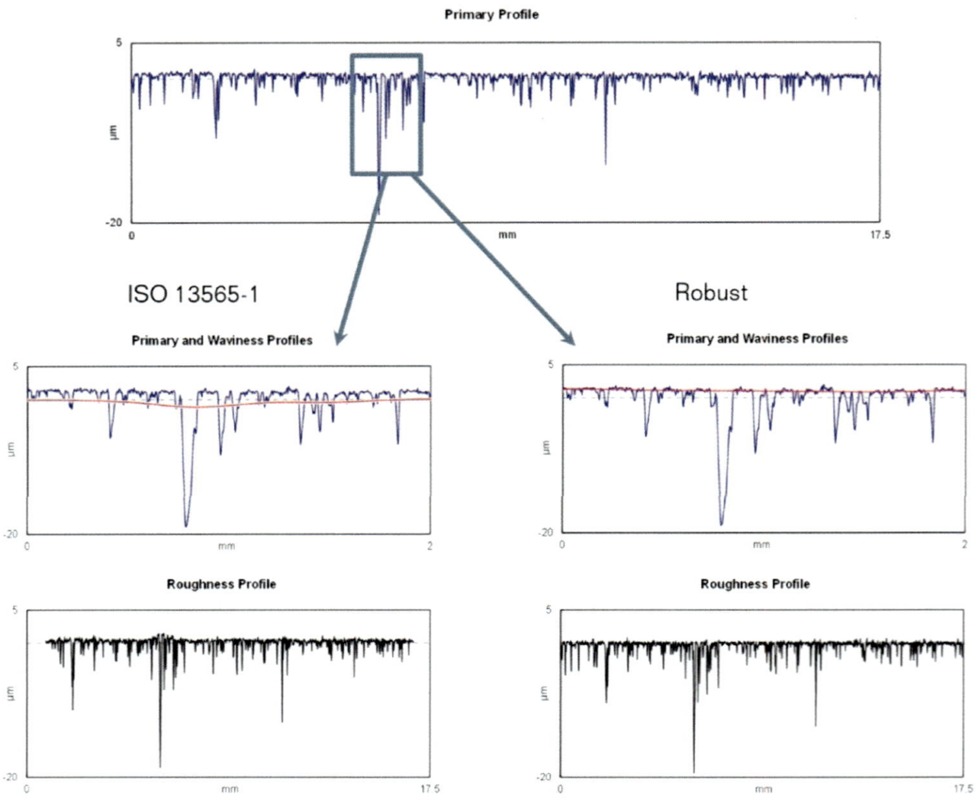

Poorer removal of outliers by the valley suppression filter still leaves "raised metal" artifacts on the *roughness* profile (plots on the left). These are not created by the robust filter. Notice also the difference in end discard shown in the *roughness* profiles.

Note: The plots of *primary* plus *waviness* profiles in the center are of a short section of the trace to show the difference in *waviness* profiles. Full-length *roughness* profiles are shown at the bottom.

Figure 90: Comparison of the ISO 13565-1 and robust filters for a trace with deep valleys

It should be noted that in many commercial measuring systems the valley suppression filter is automatically applied to the parameters in the Rk family (Rk, Rpk, Rvk, Rmr1, Rmr2, etc). Depending on the instrument's software, this filter setting may not be controllable by the user.

The notion that certain parameters (i.e., the Rk parameter family) require a certain filter does not follow the logic that the filtering should be chosen to extract the features of interest. If a surface profile is of a shape that requires a particular filter type and cutoff wavelength, it stands to reason that the same filtering conditions should be used for all parameters (not just those of the Rk family). In any case, it is strongly recommended that the robust filter be used in place of the valley suppressing (ISO 13565-1) filter.

Should I use the "ISO" filter on my system?

ISO has historically taken an approach where the best practice was standardized. In recent years, ISO has moved to more of a "toolbox" approach from which many analysis options are available for the user. Today we have the ISO 16610 series of standards which provides many different filters which could all be considered "ISO" filters.

Some older instruments use the term "ISO filter" for various filter types ranging from the electronic (phase lagging) 2CR filter all the way to the Gaussian filter. Your best bet is to look at your instrument's manual for a description.

What is a 2CR filter? Should I use it?

No, you should not use the 2CR filter, unless you are comparing to historical data which used the filter.

Early surface texture instruments used electrical filters to separate the roughness data from the longer wavelength features. These systems (prior to the introduction of digital computers) would plot the *roughness* profile directly to a strip chart while measuring, and display an average roughness value on a dial. The first internationally accepted filter was the set of two capacitors and two resisters shown in the electrical schematic in **Figure 91**; it was called the 2CR (or in some cases the 2RC) filter. User-selectable values for the resistors and capacitors created the different roughness cutoffs. This type of filter was used for almost 30 years and was implemented as a mathematical approach in early digital instruments.

One problem with 2CR filters is that they create a phase shift in the output data[36], with each wavelength being shifted by a different amount. The shift is illustrated in **Figure 92**, showing a *waviness* profile overlaid upon the *primary* profile with an obvious offset between the two. The distortion is especially noticeable around abrupt features such as scratches or peaks. This disqualifies the 2CR filter for some parameters, especially the extreme value parameters such as Rp, Rpm, and Rv.

A second issue with the 2CR filter is that, unlike with a Gaussian filter, the *roughness* and *waviness* profiles cannot be recombined to produce the *primary* profile. This is due to a lack of sharpness, with shorter wavelength waviness features contributing to the *roughness* profile and longer wavelength roughness features contributing to the *waviness* profile. This filter should therefore not be used to evaluate waviness.

2CR filters have historically produced a 75% transmission at the cutoff wavelength, in contrast to today's filters which have 50% transmission at the cutoff wavelength. This 75% transmission (which was set at 70.7% in some standards) was originally introduced to imitate manual (pencil and paper) methods for leveling profile sections.

Digital phase-correct versions of the 2CR filter have been developed, but the adaptation of the inherently phase-correct Gaussian filter has replaced this filter in modern instruments. Except when checking correlation with older analog instruments, there is no advantage to using the 2CR filter with modern instruments, so its specification and use should be avoided.

---

[36] The electronic filters are applied in real time to the data (the electronic signal) being accumulated. Since this data accumulation can only look backwards in time, it can only react based on the past information. This results in a phase shift or lag. Modern digital instruments filter the data after the measurement is complete and can therefore look "forward" and "backward" in time. This allows a "phase correct" output.

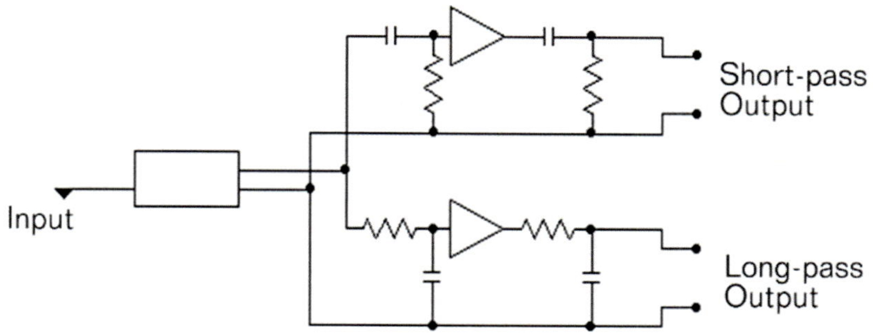

Figure 91: Schematic of 2CR analog filter

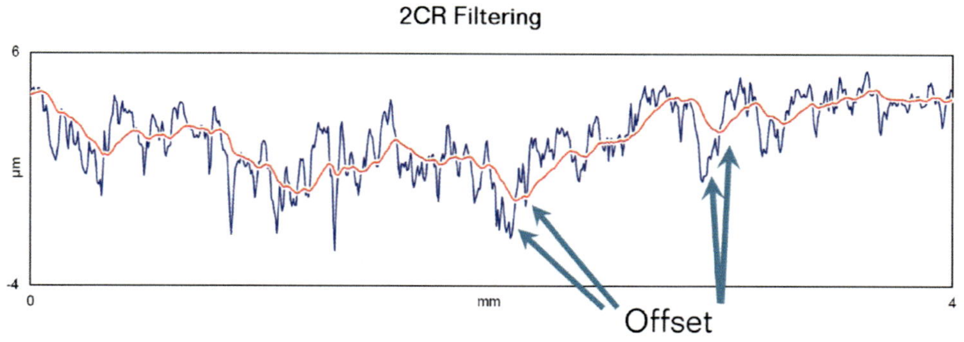

Figure 92: *Waviness* profile is offset from *primary* profile due to 2CR filter phase error

What is the difference between Phase-Corrected and Gaussian filters?

The 2CR filter only looks backward as it is convolved (see the previous section, and particularly **Figure 92**). This produces a phase shift in the output data, with the amount of shift proportional to the wavelengths of interest.

The Phase-corrected (PC) filter was developed by some of the instrument manufacturers to address this phase shift. The PC filter has a similar (75%) transmission as the 2CR, but it looks forward and backwards as it smooths its way through the data. This produces an output that is in phase with the input data.

The PC filter is seldom used now, having been supplanted by inherently phase correct filters such as the Gaussian. The Gaussian filter also typically provides a 50% transmission at the cutoff wavelength, instead of the 75% for the PC.

If a cutoff wavelength is not specified on my drawing, what should I use?

The correct answer to this question depends on whether it is being asked by an inspector or a designer. These situations will be addressed separately.

Inspector—scenario #1

In the first scenario, an inspector is given a manufactured component and a surface texture specification that consists only of parameters and their limiting values. In this situation the inspector must assume, rightly or wrongly, that the specification was written with the default analysis settings in mind. The inspector must therefore apply those default settings. If these appear to be inappropriate, the inspector can point that out to their customer, but the inspector is <u>not</u> at liberty to change from those default values, which are selected using the process discussed below.

A complete analysis of surface texture data can involve three different cutoff filters:

- The short cutoff the
- The roughness cutoff, and
- The (optional) form cutoff.

The first choice is the roughness cutoff. ASME B46.1 and the ISO standards 3274 and 4288 provide default roughness cutoff wavelength values based on the specified properties of the sample (see **Table 13** through **Table 15**). For non-periodic[37] profiles (see **Figure 93**) a cutoff is chosen based on the specified Ra or Rz (see **Table 13** and **Table 14**). The cutoff for a periodic[38] profile (see **Figure 94**) is based on the spacing parameter Rsm (see **Table 15**).

After selecting the roughness cutoff, if the short filter wavelength (lambda-s, or λs) is not specified it is determined as follows:

- 300:1 bandwidth for values of λc of 0.8 mm or larger
- 100:1 for λc = 0.25 mm, and
- 30:1 for λc = 0.08 mm.

Note that most measurement systems automatically apply the appropriate λs, but this should be confirmed with each measurement system.

If bandpass waviness is specified, there is no nationally or internationally mandated bandwidth, but a value of 10:1 is common (e.g., if λc = *0.8 mm*, then λf = *8.0 mm*).

[37] Processes such as grinding, honing, and lapping typically create non-periodic profiles.
[38] Processes such as turning, boring, and milling create periodic profiles.

Inspector—scenario #2

A second "inspector" scenario involves being given a component but no drawing. This situation invokes the process described in section 3-3.20 of ASME B46.1-2009 and section 7.2 of ISO 4288:1996, summarized as follows:

1. Make an initial measurement of the surface, normal to the lay, and visually inspect the unfiltered profile to determine if the surface is non-periodic or periodic (see **Figure 93** and **Figure 94**).
2. Estimate the appropriate roughness parameter from this initial profile: Ra or Rz if the profile is non-periodic, Rsm if it is periodic.
3. Estimate the appropriate roughness cutoff from **Table 13–Table 15**.
4. Measure the Ra, Rz, or Rsm value using the estimated cutoff.
5. Iterate on steps **3** and **4** until the combination of cutoff wavelength and parameter value is consistent with the table.
6. If you have not yet tested the next cutoff shorter than that identified in step **5**, measure Ra, Rz, or Rsm using this shorter cutoff. If this combination of cutoff wavelength and parameter value is consistent with the table, then this becomes the correct cutoff. If not, then the correct cutoff is the one identified in step **5**.

In this scenario it is essential that all filter settings be documented in the measurement report.

Designer

In this scenario a designer is developing a specification. They may be modifying an existing drawing, developing a specification based on functional testing, or reverse engineering a component. Even if a technical drawing exists, many older drawings do not explicitly state the filter settings or cutoffs. **Figure 73** on page **118** illustrates the potential impact of the roughness cutoff on the reported parameter values, so care must be taken in selecting cutoffs, and in documenting the selection in the final product description. The remainder of this section applies to this scenario.

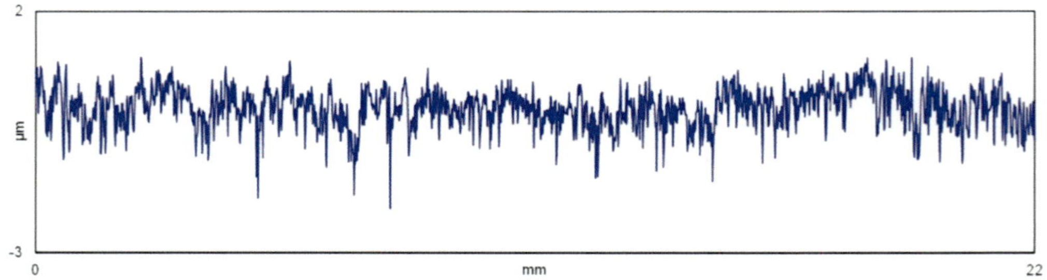

Figure 93: Example of a non-periodic profile

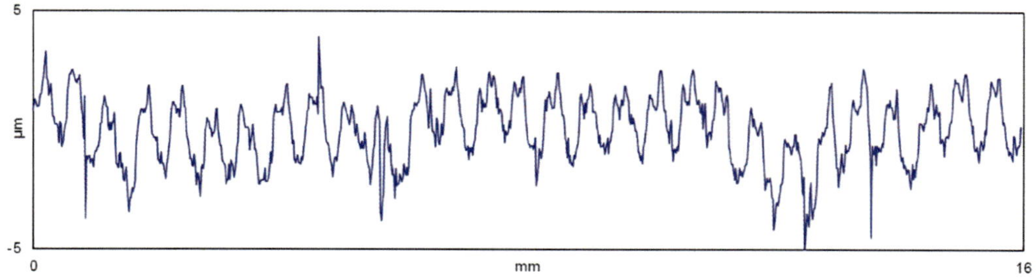

Figure 94: Example of a periodic profile

Table 13: Default roughness cutoff (λc) values for non-periodic profiles based on Ra

Average Roughness (Ra)				Cutoff Value	
Over		Up to (including)		λc	λc
μm	μin	μm	μin	mm	inch
-	-	0.02	0.8	0.08	0.003
0.02	0.8	0.10	4.0	0.25	0.010
0.10	4.0	2.0	80	0.80	0.030
2.0	80	10	400	2.50	0.100
10	400	-	-	8.00	0.300

Table 14: Default roughness cutoff (λc) values for non-periodic profiles based on Rz

Average Peak to Valley Roughness (Rz)				Cutoff Value	
Over		Up to (including)		λc	λc
μm	μin	μm	μin	mm	inch
-	-	0.1	4	0.08	0.003
0.1	4	0.5	20	0.25	0.010
0.5	20	10	400	0.80	0.030
10	400	50	2000	2.50	0.100
50	2000	-	-	8.00	0.300

Table 15: Default roughness cutoff (λc) values for periodic profiles based on Rsm

Mean Spacing (Rsm)				Cutoff Value	
Over		Up to (including)		λc	λc
mm	inch	mm	inch	mm	inch
0.013	0.0005	0.04	0.0016	0.08	0.003
0.040	0.0016	0.13	0.0050	0.25	0.010
0.130	0.0050	0.40	0.0160	0.80	0.030
0.400	0.0160	1.30	0.0500	2.50	0.100
1.300	0.0500	4.00	0.1600	8.00	0.300

The designer's choice of cutoffs should be determined by product function. If component test data is available, cutoff wavelengths should be chosen that provide a correlation between parameter values and part performance. It is even better if that correlation extends to the manufacturing process(es), with the parameter values and bandwidths relating to specific process variables; this provides a clear line of communication between manufacturing and function.

The next step is to examine the *primary* profile or the wavelength content graph to be certain the selected filters are capturing all wavelengths of interest. As shown in **Figure 95**, inappropriate cutoff selection can miss significant features, or it can split the wavelength content between roughness and waviness and thus dilute the impact on the parameter values. Parameter selection must coincide with this choice of filter. For example, a specification without a waviness parameter would provide very poor control over the situation shown in **Figure 95** (c).

A visual examination of the profile also allows an assessment of how well the cutoff wavelength "fits" the shapes of interest. In **Figure 96** the *waviness* profile best fits the shape with a λc of 0.8 mm (center image). The shorter and longer cutoffs aren't consistent with an intuitive sense of how the roughness and waviness (or form) should be separated.

Finally, using the default cutoffs shown in the tables may be a perfectly acceptable design choice. However, the tables' recommendations should always be considered secondary to the functional and visual guidance discussed above.

The standard cutoff values (0.25, 0.8, 8 mm, etc.) are adequate for the majority of analyses. There may, however, be situations in which non-standard cutoffs do a better job of differentiating between acceptable and unacceptable textures. **Figure 97** shows an example, where surface A (top profile) performed acceptably while surface B did not. The *roughness* profile peak count (Rpc) was determined to provide the best correlation with performance, with higher values performing better. However, the predominant features on each surface were at different wavelengths: ≈ 0.3 mm for surface A and 1 mm for surface B, as shown by the wavelength content graphs. A 0.8 mm cutoff was therefore not appropriate, and an 8 mm cutoff would include some long wavelengths on surface A that weren't part of the issue. It was therefore decided to use a non-standard 3 mm roughness cutoff, generating distinctly different peak count values to correlate with the different behaviors. The Rpc threshold was ±1 µm. A 25 µm λs was used to accommodate the 4.6 µm data spacing.

Note that your instrument's software may not allow non-standard cutoffs.

Whatever filter settings are chosen, they need to be clearly and unambiguously stated in the final design document (typically an engineering drawing). This must include the filter cutoff value(s), and it may be necessary to include the type of filter (Gaussian, robust, etc.) and various other supporting parameters. Default values can be assumed as long as there is no potential for misinterpretation. Further information on documentation can be found in the section, "**How do I specify surface texture requirements on a drawing?**" on page **194**.

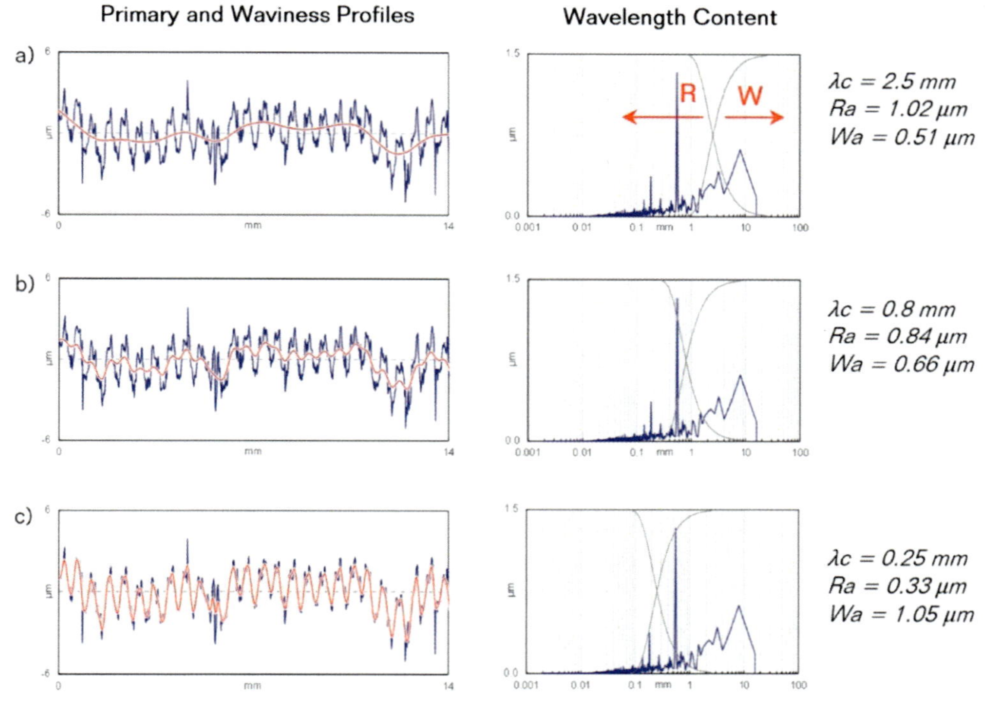

The 0.55 mm wavelength feed rate on this turned surface is contained within the *roughness* profile on the top plot with a λc *of 2.5 mm*. With λc = *0.8 mm* (center plot) some of the wavelength content is in roughness and some in waviness, diluting its impact on both. The bottom plot shows the feature almost totally in waviness when λc = *0.25 mm*.

Figure 95: Cutoff selection influenced by wavelength content

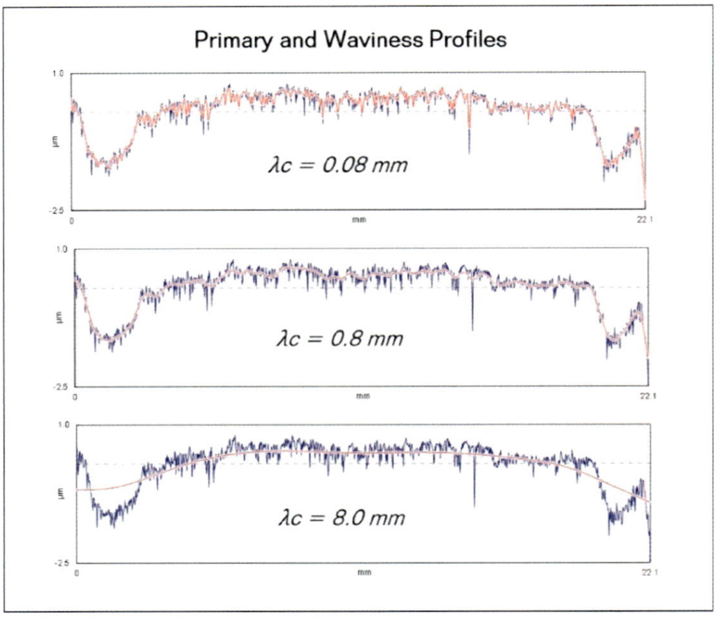

Figure 96: An intuitive feel for cutoff selection

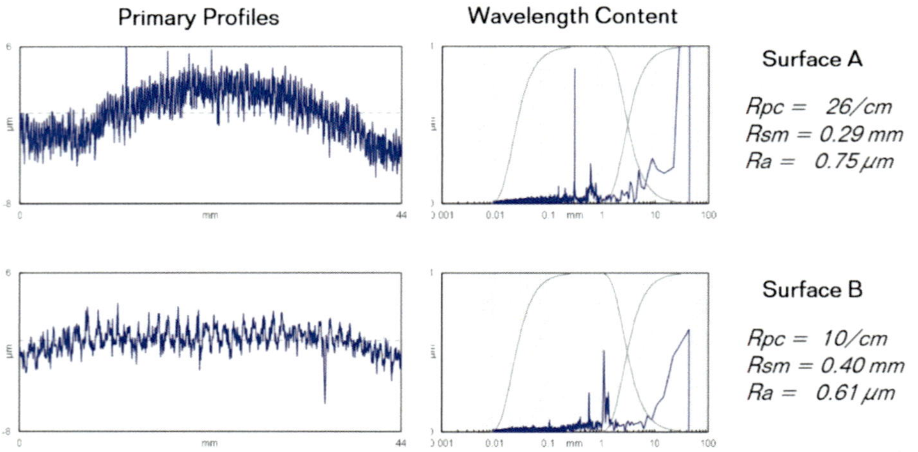

The roughness profile peak count (Rpc) for surface A (top profile) is over twice that of surface B when using the filter settings of λc of 3 mm and a peak count threshold of ±1 µm. A standard λc of 8 mm is not preferred because it would include the ≈ 7 mm wavelength features on surface A that are not relevant to its performance.

Figure 97: Using non-standard cutoffs to differentiate surfaces

Why does my instrument only have certain cutoff values?

One of the earliest surface measurement procedures was based on an optical system that reflected a small slit of light off the surface. It is believed that the length of the slit along the surface was 0.8 mm, and this became the "default" length in which roughness was measured. Over time, additional 0.8 mm lengths were added to obtain a longer profile graph. Since each length was independently measured, a procedure was developed whereby each length was leveled prior to connecting the next length (see **Figure 98**). This had the effect of a filter which removed waviness. Electronic methods (e.g., 2CR filters) and eventually digital methods (e.g., Gaussian filters) were later introduced, and the 0.8 mm sampling length was translated to a 0.8 mm wavelength.

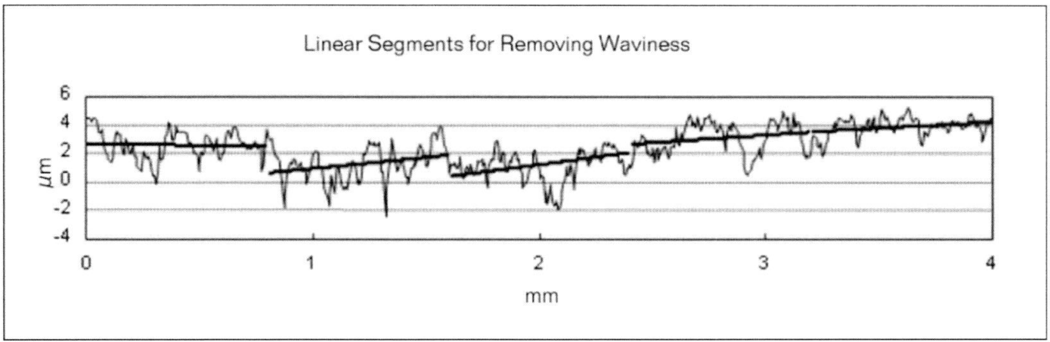

Figure 98: Illustration of leveling by applying linear segments to each sampling length

As the need arose to consider additional wavelengths, a systematic method was required to standardize the selection of cutoffs with adequate resolution but without too many choices. A logarithmic system of decades and half-decades serves that purpose. The actual cutoff values approximate the half-decade intervals to keep things simple. Therefore, while 2.5 is not exactly a half decade away from 0.8, it is much more memorable. The final result is the set of roughness cutoff values of 0.08, 0.25, 0.8, 2.5, 8, and 25 mm. The sequence can continue in either direction as the need arises. These standardized values and their short-pass transmission curves are shown in **Figure 99**.

In the spirit of keeping the cutoff values simple and memorable, the Imperial (inch) equivalents (0.003", 0.010", 0.030", etc.) of the above sequence are not exact conversions. 0.8 mm is 0.031496 inches, not 0.030". They are, however, close enough for any practical application. In fact, it should be noted that most measurement systems perform their mathematics in metric and then scale the computed results to Imperial. Thus, the 0.8 mm filter is actually being used when the 0.030" filter is selected.

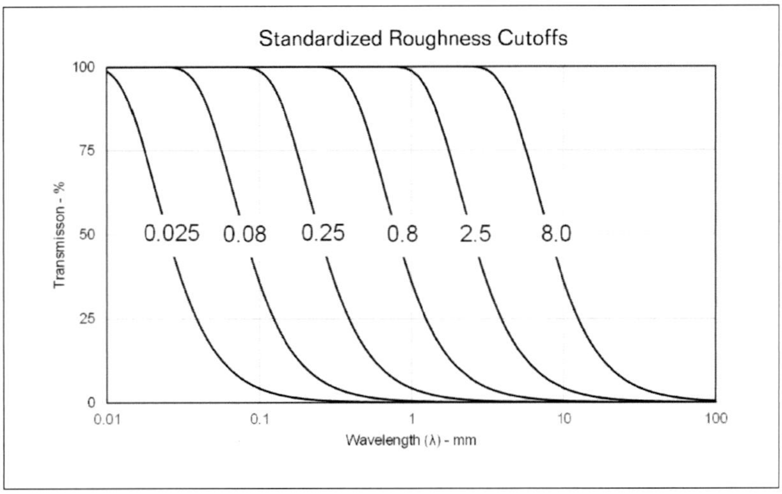

Figure 99: Transmission curves for the standard roughness cutoffs (in mm)

Why is five the default number of roughness sampling lengths?

If an evaluation length is not specified, the default number of sampling lengths in a roughness measurement is five. The origin of this number is unknown. Perhaps it is because we have five fingers on each hand.

Note: Care should be taken when measuring short surfaces that do not accommodate 5 cutoffs/sampling lengths. See the section, "**I have a short feature. Do I still need five sampling lengths?**" on page **62**.

How many data points should I have in my measurement?

The total number of data points in your measurement is determined by the number per unit length and the overall length. The overall length is, of course, determined by the size of the component or the capacity of the measuring instrument. The required number of points per unit length, or data density, is determined by the filters being applied. Five data points per filter "length" are usually considered to be the minimum number required to define a feature (although some software will accept three points per feature). Therefore, if a 2.5 μm short filter wavelength (lambda-s, or λs) is specified, the maximum allowable data spacing is 0.5 μm (2.5 divided by 5). This requirement is documented in standards such as ASME B46.1 and ISO 3274, as shown in **Table 4** on page **21**. If the data spacing is too large, the filter will not be applied. The analysis software may indicate when this occurs, but most software/instruments do not. Some software may change the λs to conform to the data spacing.

Some parameters, such as the average and RMS slopes, include the data spacing in their calculation and are therefore directly dependent on that spacing (see **Table 8** on page **47** and the slope equations on page **376**). If slope parameters are being compared for different surfaces it is imperative that a consistent data spacing be used.

Other parameters can be sensitive to spacing as well. These include, for example, parameters based on extreme peaks and valleys (Rp, Rv, Rz). With a wide spacing, a measurement point is less likely to fall on the highest point of the peak or the deepest point of the valley, so the height will be underestimated (see **Figure 28** on page **48**)[39]. Furthermore, a wide data point spacing often means a more aggressive antialiasing filter, which will further reduce the extreme surface heights. (Note: For these, and other, reasons, extreme value parameters are typically not recommended, and should be used with caution.)

The section, "**Can I evaluate roughness, waviness, and shape with one long trace?**" on page **54** provides further information about the relationship between data spacing, the λs filter, and parameter values.

---

[39] ASME B46.1 notes that, assuming a cosine waveform, the recommended 5 points per wavelength data density (N) will detect at least 90% of the total amplitude, per the equation $[1 + \cos(\pi/N)]/2$. (ASME B46.1-2009, section 7-4.2.1).

DATA INTERPRETATION

DATA INTERPRETATION

You've carefully made your measurements, selected all the right filters, and calculated the specified parameters. Now what does it all mean?

The easy part is determining whether the parameter values fall within the specified limits. If quality control is the sole purpose of the measurement, and if the numbers meet the specification, then this go/no-go decision may be enough (assuming consideration has been given to the measurement uncertainty). But if you are trouble-shooting a component issue, or fine-tuning a manufacturing process, or if you've been handed some data and you don't know exactly where it came from, there is much more that can be learned by using all the information from the measurements—the numerical and the visual. This is the purpose of this section on data interpretation.

Here are some keys to getting as much information as possible from surface texture measurements.

- Look at the actual traces, not just the numbers. The parameters and their numerical values are essential for communicating your requirements. However, the actual graphs can tell you much more. Is the surface homogeneous, or are some regions much rougher or smoother than others? Are there high peaks and/or deep valleys that may be masked by an averaging parameter? Is there a periodicity that reveals process details or that might explain a noise complaint?

- Look at the *primary* profile. It is common to only look at the *roughness* profile, since your specification may only be for roughness parameters. However, roughness is, at best, only one-third of the story. Waviness and shape may be as or more significant to the component function or for process trouble-shooting. The *primary* profile contains all of this information—roughness, waviness, and shape—showing you what the surface actually looks like. (This of course assumes the shape hasn't been removed from the *primary* profile with the reference geometry or by applying a form filter.)

- If you have been given the results, check for some indication that the data was analyzed properly. Are the filter settings correct? Is the trace long enough?

- If you were only sent numerical values by some far-off supplier, ask for the raw data. Most modern instruments can save the raw unfiltered data in a format readable by third party software.

- Consider calculating parameters other than those on the drawing. If you can't make sense out of the numbers you have, create some more and look for values that may explain the phenomenon you're investigating. The parameters on the drawing may not have been chosen as carefully as they should have been. Often, surface texture parameters are carried over from previous drawings and they may not be functional for a current design.

- Is there something about the data that just doesn't make sense? Instrument problems or measurement errors can create artifacts that are not "real"—features that don't actually exist on your surface.
- Exploring and understanding the shapes of your surface (rather than reporting a few parameter values) is the beginning of truly understanding your surface.

How do I know if my data is accurate?

There are several ways to gain confidence in your data—some quick and simple, some more sophisticated. The easiest way is to ask, "Does it make sense?" "Does it look right?" A quick look at the *primary* profile can reveal features that shouldn't be there, or that can't be created by the manufacturing process being used. The section, "**I see unexpected things in my data. Are they real?**" starting on page **69** provides some examples.

A careful review of the choice of data analysis filters and form removal is also important. Often this information is provided along with the parameters and graphs that are reported by your measuring system. Are you considering all relevant wavelengths? Are you creating artifacts by an inappropriate choice of filter type? Are filter cutoffs at the correct wavelengths relative to the dominant wavelengths?

If you have concerns about the data accuracy, and if you have access to multiple surface texture instruments, you can compare data from each to look for inconsistencies. This might reveal, for example, a damaged stylus, a noisy system or environment, or poor fixturing. Sometimes, simply flipping the surface from right-to-left can quickly indicate if the features rotated with the part or if they stayed in the same position and are related to the instrument.

To dig deeper into your data integrity, plot a wavelength content graph or power spectrum. Look for peaks in unexpected places on this graph. These peaks indicate the presence of strong periodic features. If your surface is random, for example from grinding, these spectral peaks may reveal subtle instrument problems since random machining processes should not leave periodic/repeating features on the surface.

While the actions listed above provide a qualitative look at data quality, the best (quantitative) assessment of the goodness of your data is to determine its uncertainty. Uncertainty answers the question, "How wrong can I be?" This information can then be used to evaluate or improve a measurement system, develop inspection procedures and control plans, and create more practical product specifications.

Uncertainty comes from two sources:

- The measurement system
- The variability of the surface

Measurement system uncertainty is discussed on page **85**, and includes things like system calibration, noise, measurement procedure, and frequency response. This uncertainty is inherent to your instrument. Applying your measurement system to a specific component will add factors like alignment, the perpendicularity of the probe to the surface, surface cleanliness, and the proper selection of instrument parameters (range, traverse speed, etc.). These latter factors are under the control of the operator and should ideally have only a small contribution to the uncertainty.

While instrument uncertainty is important, surface variability may actually be the larger factor in determining how representative a measurement is. The type of manufacturing process, the interaction of that process with varying feature geometries, planned and unplanned changes in the process over time (e.g., tool changes, process drift), and the measurement location, direction, and length can all add variability. Some surface texture parameters will show much more variability than others. For example, averaging parameters like Ra will be relatively stable, while extreme value parameters such as Rp, Rv, or Rmr without a reference material level can be very noisy.

Consider the following example, using assumed but realistic uncertainty values for a measurement of Rz:

- Inherent instrument uncertainty = 7%
- Operator-controlled setup uncertainty = 2%
- Surface variability = 15%
- Overall uncertainty = $\sqrt{0.07^2 + 0.02^2 + 0.15^2}$ = 16.7%

A single Rz measurement on this surface has a 95% likelihood of being within 17% of the true value—with nearly all of that uncertainty coming from the surface variability. A 17% "guard band" would therefore be required when comparing this result to an Rz specification.

The influence of the surface on the overall uncertainty can be reduced by:

- Averaging more measurements. This will improve the uncertainty by $\sqrt{(1/n)}$. Note: This assumes that the mean value is the measurand.
- Targeting the measurements for the worst-case region of the surface. The operator's skill level in stylus placement will be a factor here. Note: This applies when the "worst value" is the measurand.
- Modifying the manufacturing process to reduce the variability induced by part geometry, process changes, etc.

What can I learn about the manufacturing process from a surface texture trace?

Many manufacturing operations leave a signature on the surface that can provide information about that process, including the type of machining method, details about that method, how many processes were involved, or even hints about the base material. The signature can also guide troubleshooting when there are unwanted surface characteristics. The following are some examples, showing the profile, the wavelength content graph, and the amplitude distribution function (ADF):

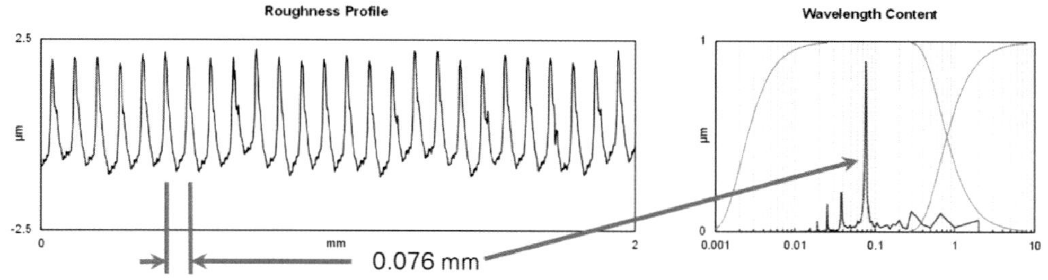

Turned or bored surfaces have a distinct periodicity determined by the tool feed rate and the tool or workpiece rotation rate. This can be identified by measuring the distance between the features (here, 0.076 mm). This same information is shown on a wavelength content graph, here showing a distinct peak at that same 0.076 mm. The amplitude distribution plot for the *roughness* profile is non-Gaussian and positively skewed, indicating larger peaks than valleys.

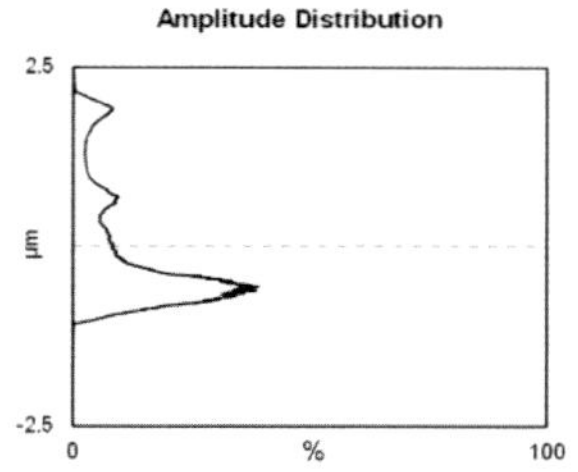

Figure 100: Typical turned surface

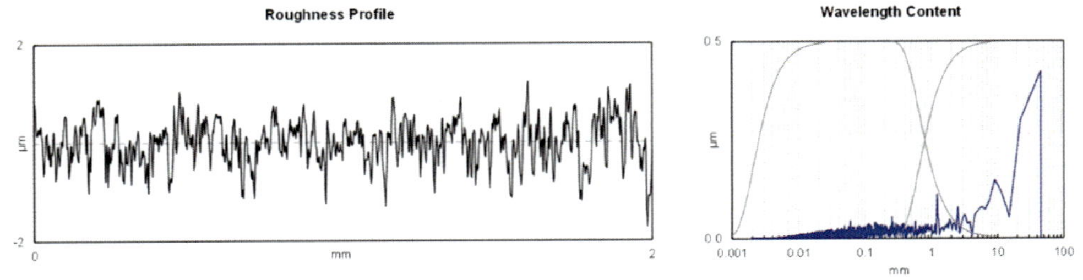

Ground surfaces do not show a repeating pattern and therefore do not have distinct peaks on the wavelength content graph. The amplitude distribution of the *roughness* profile shows a Gaussian distribution of profile heights, with peak heights and valley depths being approximately equal.

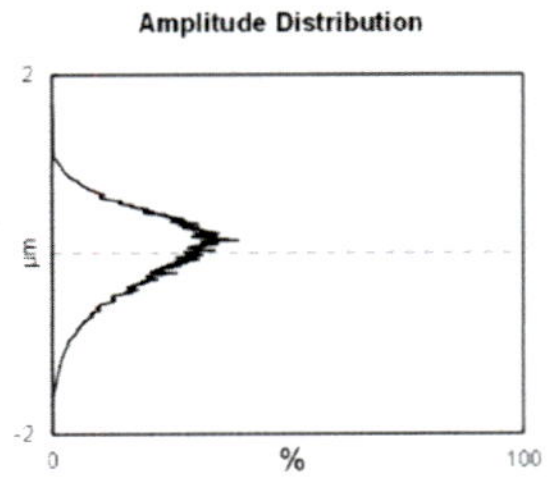

Figure 101: Typical ground surface

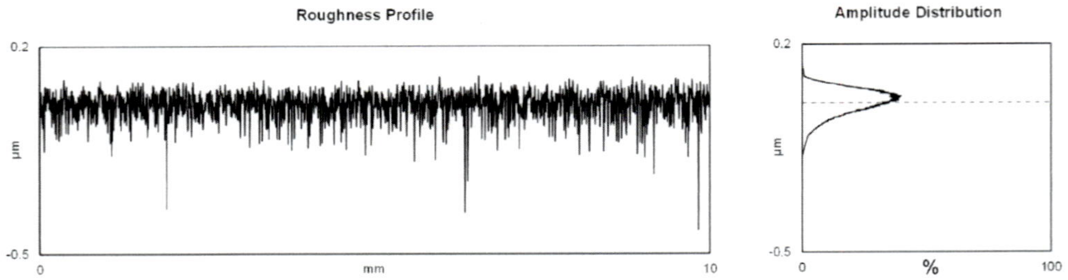

Lapping creates a random surface similar to grinding, but with lower overall roughness. Lapping may leave remnants of the prior ground surface when the amount of material removed is small or when the grinding process is relatively rough. This creates a negatively skewed surface such as shown here (*Rsk* = *-1.3)*, with the valley depths greater than the peak heights (note the longer tail at the bottom of the amplitude distribution than at the top).

Figure 102: Typical lapped surface

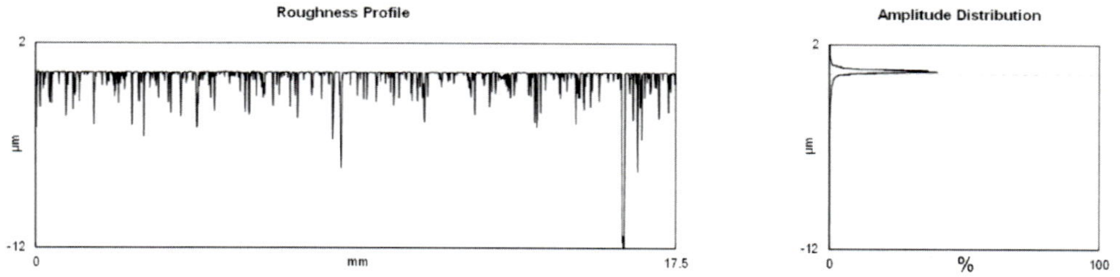

Plateau honing is an extreme example of a two-step process, creating what is sometimes called a two-process surface. The surface is negatively-skewed (this example has *Rsk = -7)*, with almost all of the surface lying at or slightly above the mean line. The amplitude distribution curve for this type of surface has a large peak near the mean line and a long tail below the line (difficult to see at this scale).

Figure 103: Typical plateau-honed surface

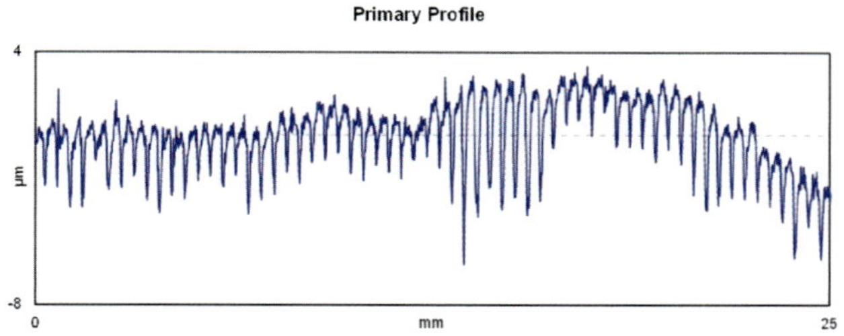

This surface is from a through-hole originally created by a boring operation, followed by a mechanical smoothing process to reduce the peak heights. In this example the mechanical smoothing was achieved by pressing a slightly-oversized ball through the hole.

Figure 104: A bored surface that has been mechanically smoothed

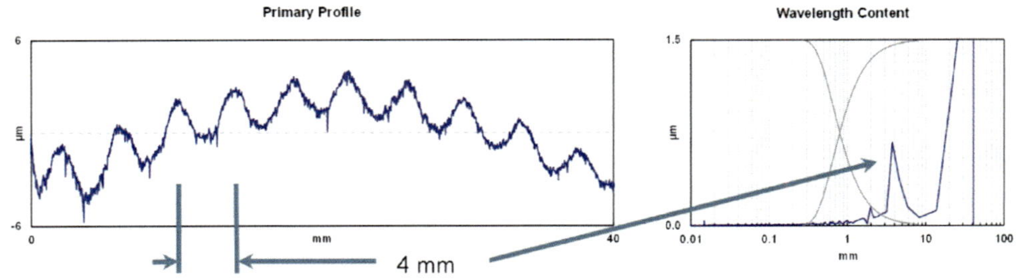

Machining processes may superimpose a waviness component over the roughness. The cause of the waviness can be the wheel dressing operation (for grinding), tool or environmental vibration, tool guide errors, etc. Determining the frequency or peak spacing from the wavelength content graph can help trouble-shoot the process and identify the source.

Figure 105: A ground surface with waviness

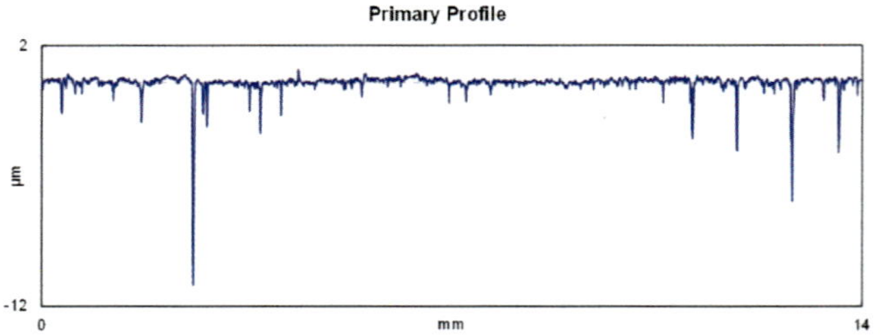

Measurements of porous surfaces will occasionally encounter deep pores that represent the base material structure and not necessarily the surface finishing process. Powder metal components are one example of this. Machining of graphitic cast iron may pull out some of the graphite particles, also creating a surface with random deep features. A decision must be made on whether or not these features need to be controlled by the surface texture parameters.

Figure 106: Porous surface

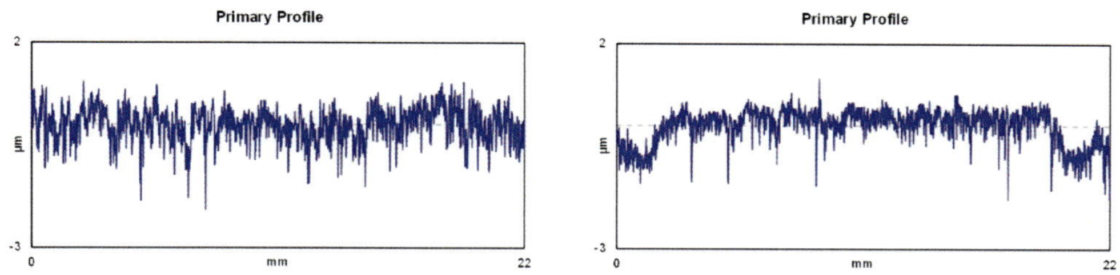

Polishing a surface that has a varying hardness can result in surface relief, with the softer material being preferentially removed. The surface on the left was ground with a very stiff CBN wheel and shows minimal profile error. The trace on the right is from the same surface after polishing with a Mylar abrasive tape backed by nylon shoes. The ~ 1 µm drops at the edges are due to that material being much softer than the ~ 18 mm long hardened zone in the center.

Figure 107: Profile changes from polishing a surface with variable hardness

Why do my roughness traces show metal that I can't polish away?

Roughness profiles from surfaces with porosity or deep valleys can show what appears to be raised metal beside those valleys. In these situations the *primary* profile should always be examined to determine whether this raised metal is real. As discussed on page **134**, a normal Gaussian or spline filter is pulled down into deep valleys, creating a roughness artifact above the mean line. This can be addressed by using a robust filter. An example of this behavior on samples with and without actual raised metal is shown in **Figure 108**. Although it doesn't work as well, a valley suppression filter can be used if a robust filter isn't available (see page **137**).

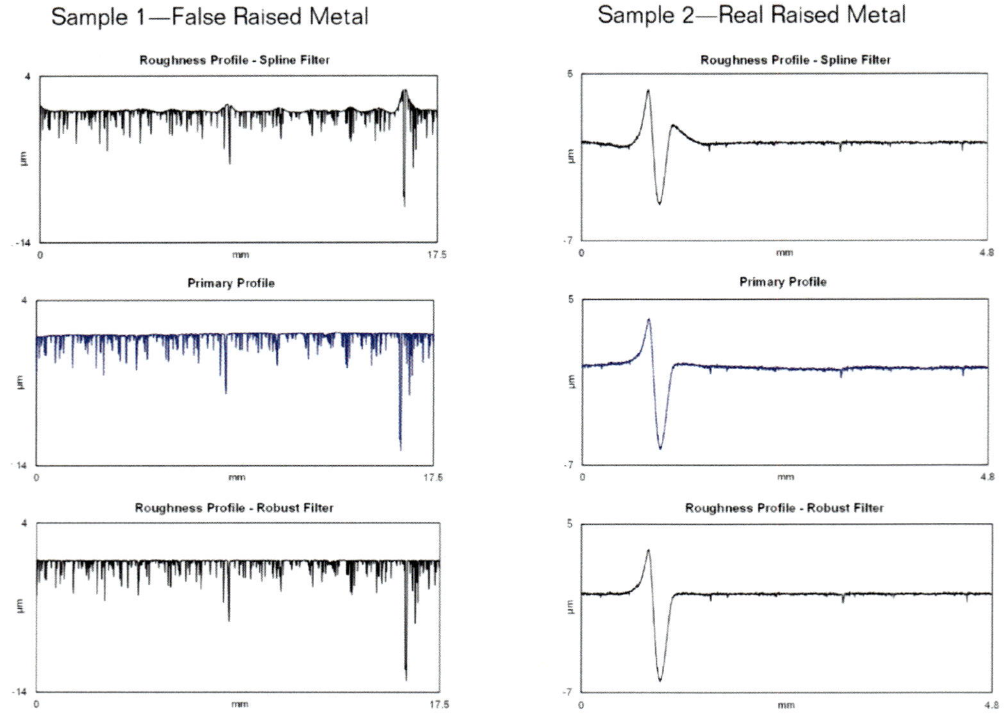

The top *roughness* profiles show what appears to be raised metal beside the deep valleys. The middle *primary* profiles show the actual situation: Sample 1 does not actually have raised metal, while Sample 2 does. The bottom *roughness* profiles were generated with a robust filter and reflect the actual surface conditions.

Figure 108: Examples of surfaces with and without raised metal

Why don't my traces look like the photos of my surface?

Surface texture profiles do not look like the actual surface because of the aspect ratio used when plotting the data. The actual vertical heights of surface features are usually much smaller than their widths. In order to see these features, measurement data is typically plotted with an x-axis scale 500 to 1000 times larger than the z-axis. Surface features with slopes of one or two degrees now appear to have slopes of 80 degrees or more. Thus, these graphs give very distorted, and incorrect, impressions of what the surface is really like.

When examining surfaces with an SEM or an optical microscope, surface features appear with their actual 1:1 aspect ratio. Many features do not create enough contrast to be visible in the photomicrographs, especially in SEM secondary electron images. More extreme features, such as the grooves in the plateau-honed surface shown in **Figure 109**, will be visible, but you will see the actual contours rather than the distorted aspect ratio seen in the surface texture trace.

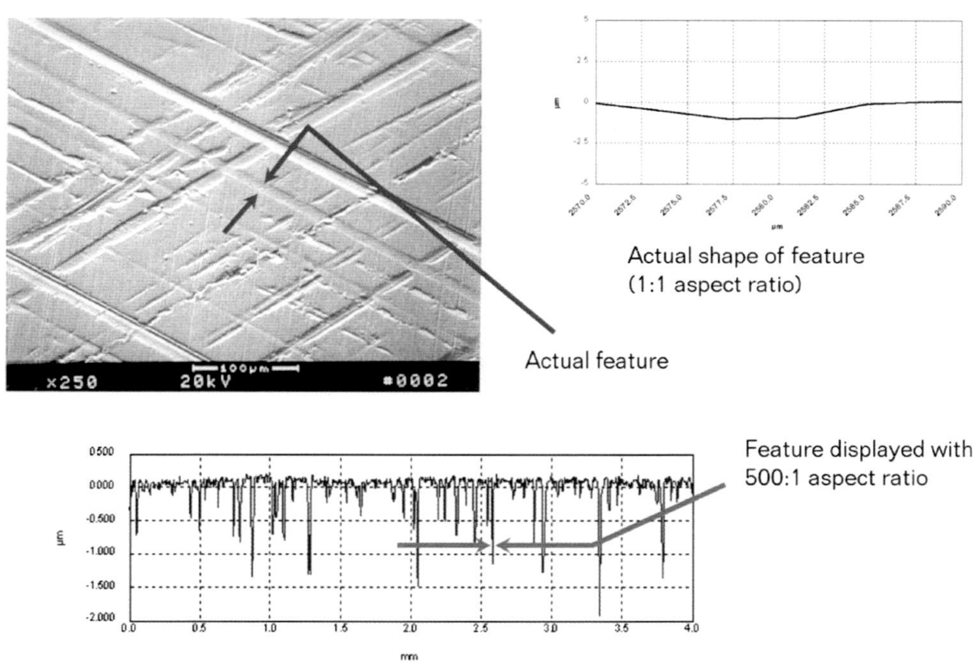

Figure 109: Photomicrograph of a plateau-honed surface with typical surface texture data

I have an Rz requirement. Can I just scale my Ra result to get an Rz?

That is not a good idea.

You may see pocket guides or published tables of Ra and Rz numbers implying a quantifiable relationship. However, such relationships are approximate at best, and they are only true for specific manufacturing processes and types of surfaces. **Figure 110** shows two surfaces whose Rz to Ra ratios differ by a factor of six. As a general rule, there is no numerical relationship between Ra and Rz, other than the not-very-useful one that Rz is always larger than Ra.

The reason for the large difference in the Rz:Ra ratios can be seen in the definitions of the two parameters and the shapes of the *roughness* profiles. Ra is the average of the absolute value of *all* the data points while Rz is the average of only a few points—those with the largest peak-to-valley distances. A profile with gradual transitions from peaks to valleys will provide more similar values of the two parameters than one with distinct extreme peaks or valleys.

The top example in **Figure 110** is a sinusoid waveform from a calibration sample with a nearly constant peak-to-valley distance of approximately 10 µm. The lack of extreme peaks or valleys results in a stable relationship between Rz and Ra[40].

The bottom example in **Figure 110** has an Ra value dominated by the smooth plateau part of the profile since most of the data points lie there. The Rz value is dominated by the small number of deep valleys. This results in the large difference between the two parameters. One can envision even larger ratios by careful manipulation of the depths and locations of the valleys.

---

[40] The Rz:Ra relationship for this particular waveform is actually predictable, based on its sinusoidal nature. The positive half of a sine wave (or the full length of a rectified wave) with a mean-to-peak amplitude of A has an average height of $(2/\pi)A = 0.637A$ and a ratio between the peak-to-valley height and the average height of π. With $A \approx 5$ µm for the top example in **Figure 110**, this gives an $Ra \approx 3.2$ µm and a Rz:Ra of $\pi = 3.14$, values very consistent with those shown in **Figure 110**.

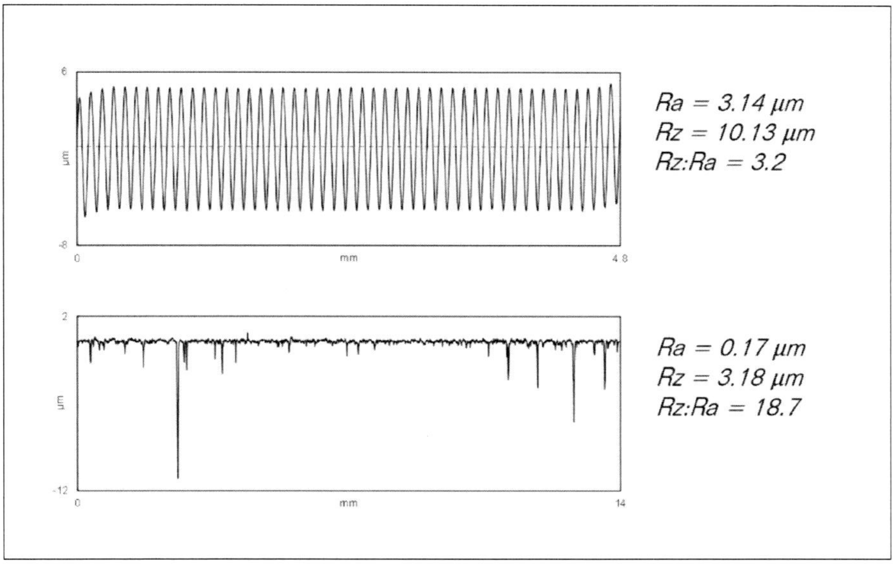

Figure 110: Rz:Ra ratios for two different surfaces

Figure 111 provides additional examples of the varying relationship between Rz and Ra for different types of surfaces. Each of these surfaces has the same Ra but a different Rz.

Quantitative relationships between Ra and most of the other roughness parameters are as tenuous, or even more so, than that between Ra and Rz.

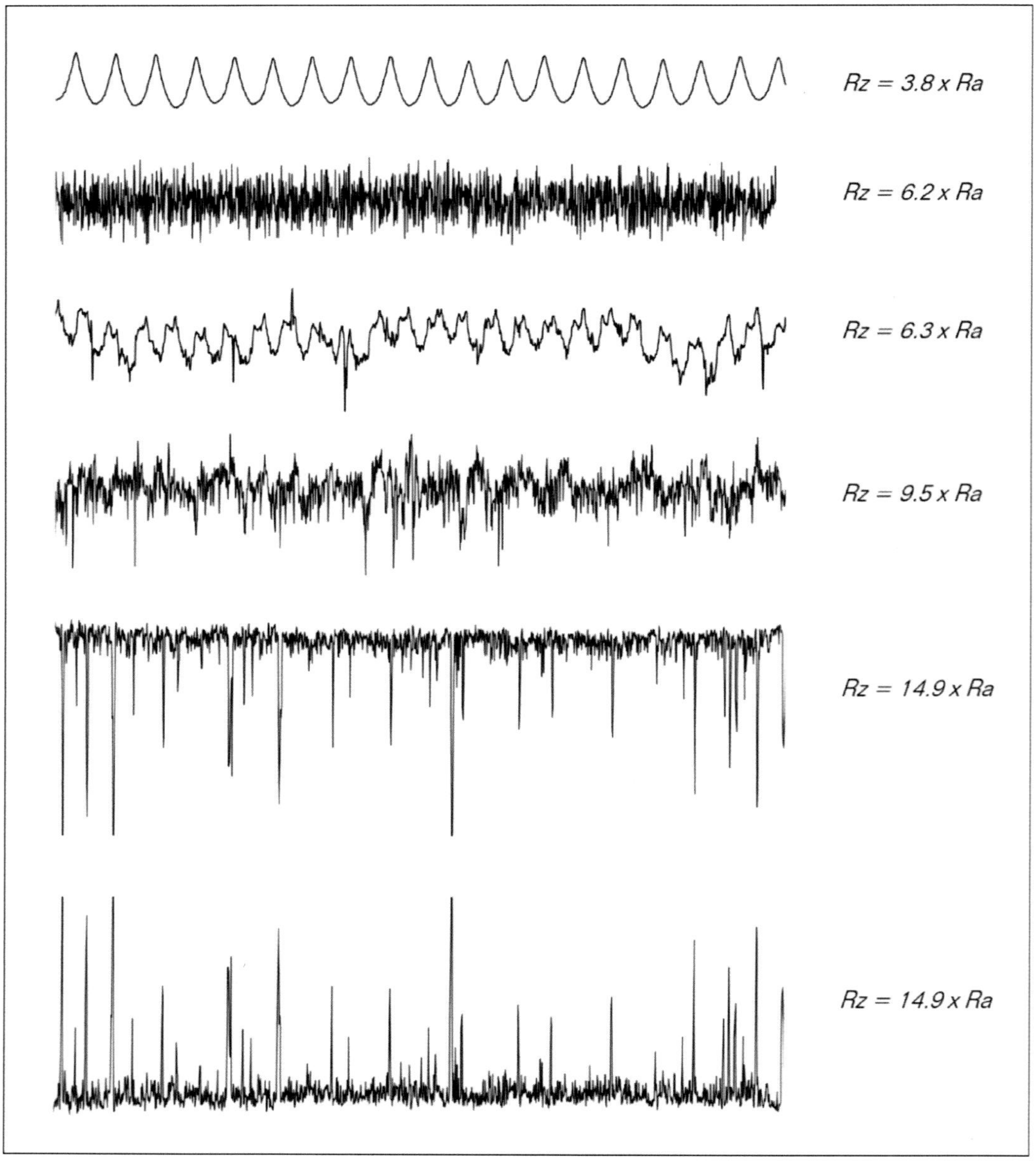

Figure 111: Rz–Ra relationship for different types of surfaces, all with the same Ra

My supplier consistently reports values of Rz smaller than mine. Why?

Rz, the average peak-to-valley roughness, can be confusing because there are actually two different Rz parameters, and the one being referred to by the "Rz" designation has varied over time[41].

RzDIN[42] was originally developed in Germany and is defined as the average over the evaluation length of the distance between the highest peak and the deepest valley in each sampling length (see **Figure 112**). When the evaluation length consists of the default five sampling lengths, the total number of points represented by Rz is ten, with the "z" coming from *zehn*, the word for ten in German.

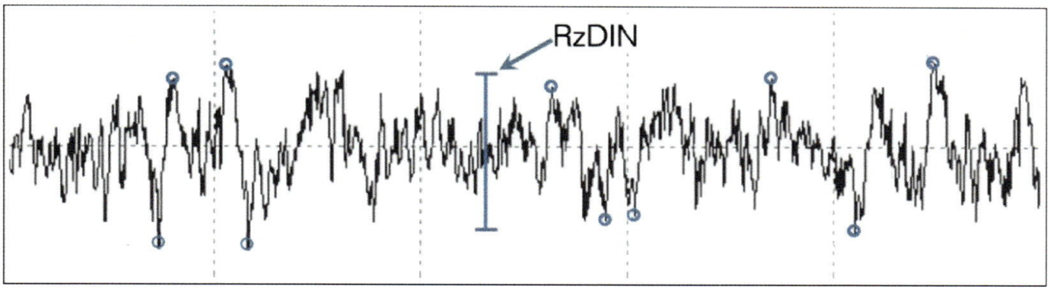

Figure 112: Points used for RzDIN calculation—one peak and one valley per sampling length

RzJIS was developed in Japan as a refinement to RzDIN. RzJIS refers to the five highest peaks and the five deepest valleys in *each* sampling length (see **Figure 113**). It involves five times as many points as RzDIN and will always have a smaller value than RzDIN. Therefore, if inconsistent Rz values are being reported by two different laboratories, it may be that they are referring to two different parameters.

[41] For more information about the history of Rz see, for example, Alex Tabenkin, "The ABCs of Rz," *Quality 46*, no. 11 (2007): 40-43.

[42] For this discussion, and occasionally throughout this book, the unofficial names "RzDIN" and "RzJIS" will be used to distinguish between these two different definitions of Rz. This is consistent with the convention followed by some software packages.

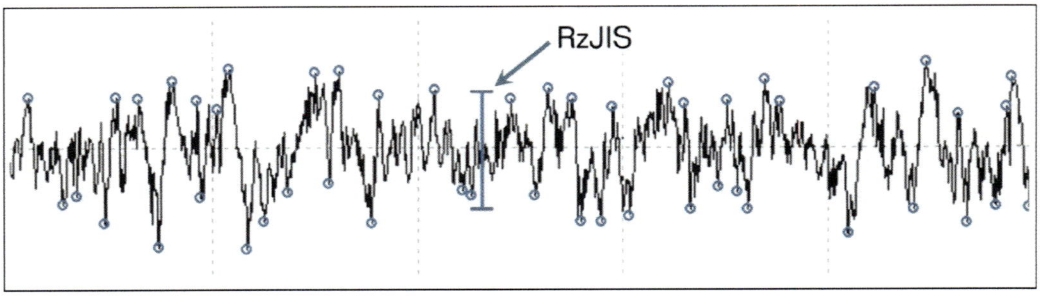

Figure 113: Points used for RzJIS calculation—five peaks and five valleys per sampling length

Note that prior to 1997 ISO and ASME standardized the Rz parameter as RzJIS. After that date the Rz parameter refers to RzDIN. If an old drawing, an old instrument, or old software is being used, the parameter being requested or being reported may be different from that of newer drawings or instruments.

If there is the possibility of confusion in a specification about which Rz is intended, it is recommended that some clearer form of identification be used, such as the unofficial names RzDIN and RzJIS.

Should I be concerned with waviness?

Why can't I just look at roughness?

Surface texture specifications almost always include roughness parameters, but many do not include waviness. This may result in features being uncontrolled that are detrimental to a component's performance. It is therefore advisable to consider all wavelengths when developing a specification and only ignoring waviness if it isn't functionally important or if it is precluded by the manufacturing process.

There are several reasons why waviness does not receive as much attention as roughness.

- Measurement reports often include a *roughness* profile, a natural companion to the roughness parameters that are most commonly requested. Since the *roughness* profile does not show the longer wavelengths, this reporting tradition leaves people unaware that there is more to a surface than roughness.
- Waviness analyses require long trace lengths that include multiple samples of the wavelengths of interest. Instruments may only have short drives, adequate for roughness but not for waviness. Even if the instrument is capable of making longer measurements, productivity and tradition often dictate making short traces (e.g., five roughness sampling lengths).
- Skidded instruments are more affordable and easier to use. These instruments inherently suppress waviness from the measured profiles.
- Waviness may be considered to be part of a GD&T form control (e.g., as part of straightness) and therefore controlled by that gaging. However, surface texture instruments may be better able to detect and quantify waviness than other types of measurement systems.
- Waviness is often "buried" in the general shape of the part. In fact, many commercial instruments still only evaluate waviness with a long-pass filter, making it difficult to discern or quantify shorter wavelength features that might be detrimental. Bandpass waviness analysis is a better approach for analyzing surfaces with both a nominal form and waviness.
- There appears to be a general lack of appreciation of the significance of waviness. There are documented situations where inspectors saw waviness in their measurements but ignored it because they had never been told it might be important.

The following are examples of where surface waviness rather than roughness correlates with the behavior of the component or system.

Leakage

A gasketed interface may leak due to waviness features, even though the roughness is as good as another interface that does not leak. An example of this is shown in **Figure 114**, where the waviness parameter Wvoid (waviness void area) differentiates between a surface that leaks and one that doesn't. See page **243** for more information about this type of analysis.

Journal bearing performance

Circumferential waviness (lobing or chatter) has long been known to have a negative effect on journal bearing performance[43]. Axial waviness can also be detrimental, as shown in **Figure 115**, where the tops of the waves on the steel shaft have been preferentially polished by the mating bronze bearing.

Seizure

Journal bearing distress such as that shown in **Figure 115** may eventually lead to seizure. An even more detrimental situation is where there is a step in the *waviness* profile (see **Figure 116**). The total amount of waviness (Wt) may be acceptable, or even desirable, if it occurs gradually over the length of shaft (e.g., a crowned axial profile), but it quickly leads to failure if it is concentrated over a short distance. This can be detected by analyzing the total waviness (Wt) within a short window that is moved along the length of the feature.

Bearing or gear noise

Waviness features can often set up vibrations and noise when a mating surface rolls or slides over them. For example, the noise level from a system of gears correlated with the magnitude of the gear tooth waviness shown in **Figure 117**.

Stress concentration in a hard contact

Rolling element bearings, gears, camshafts, and other Hertzian contact systems often operate under very high nominal stresses, and are sensitive to any features that further intensify that stress. The same is true for other systems that do not have a conformable material (e.g., a journal bearing) between two hard surfaces. These detrimental features may appear as a convex "bump" that is best characterized by a waviness analysis. Wcvx (waviness convexity) is a proposed waviness parameter that quantifies those bumps using a morphological opening filter (see **Figure 118**; see also page **239**). Careful selection of the ball radius for the morphological filter allows the establishment of Wcvx limits that can minimize the stress increase without rejecting innocuous profile deviations.

---

[43] For example, A. DeHart and J. Smiley, "Imperfect Journal Geometry – Its Effect on Sleeve Bearing Performance," SAE Technical Paper 640837, 1964

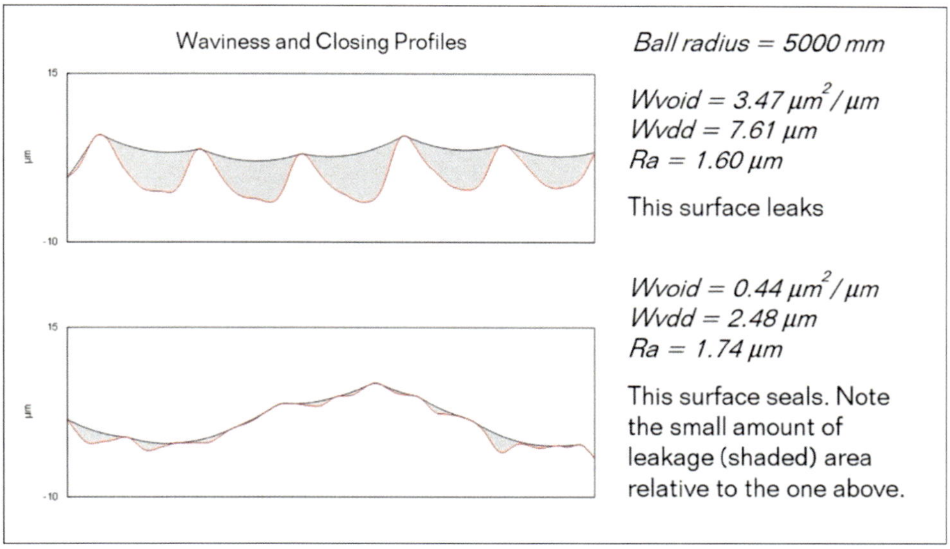

Figure 114: Correlation between waviness and gasket leakage

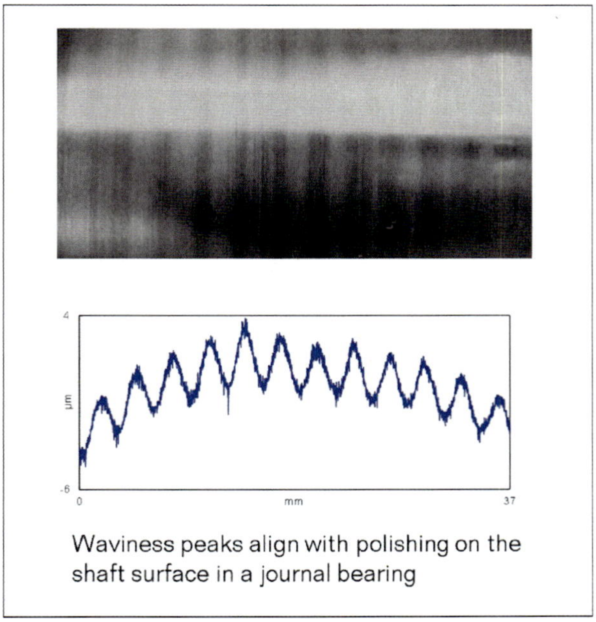

Figure 115: Correlation between axial waviness and shaft polishing

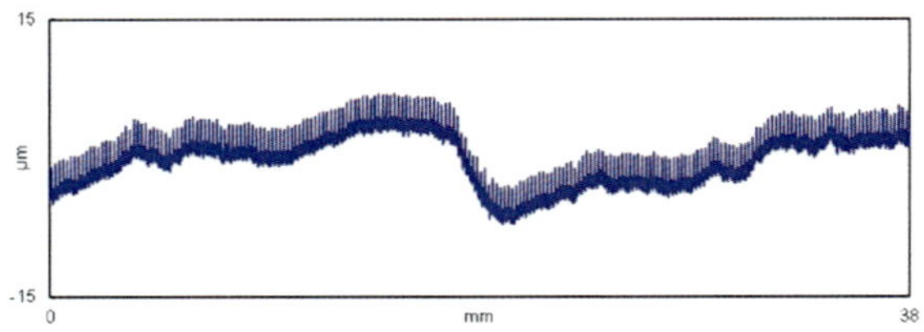

This localized waviness is more detrimental than if the same amplitude is spread over a longer length. It is therefore better characterized by a waviness parameter than by an overall form/straightness requirement.

Figure 116: Step in shaft surface that can cause bearing seizure

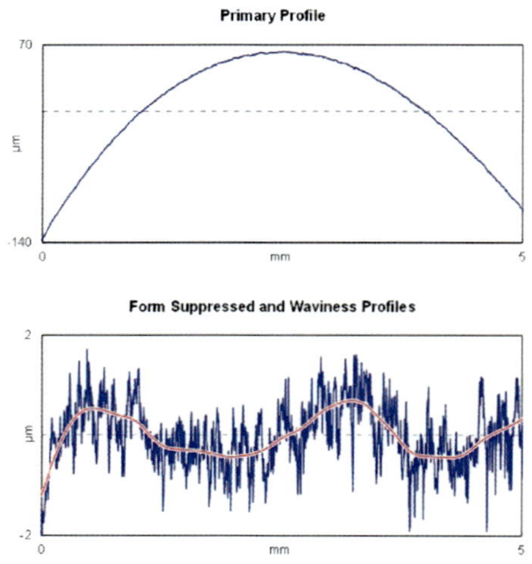

Gear tooth profile

Gear tooth waviness revealed by suppressing the nominal shape with a 4th order polynomial. The audible noise level correlated with the Wt parameter when Wt was calculated ignoring the profile extremes at the ends of the trace.

Figure 117: Gear tooth profile with waviness that correlated with noise level

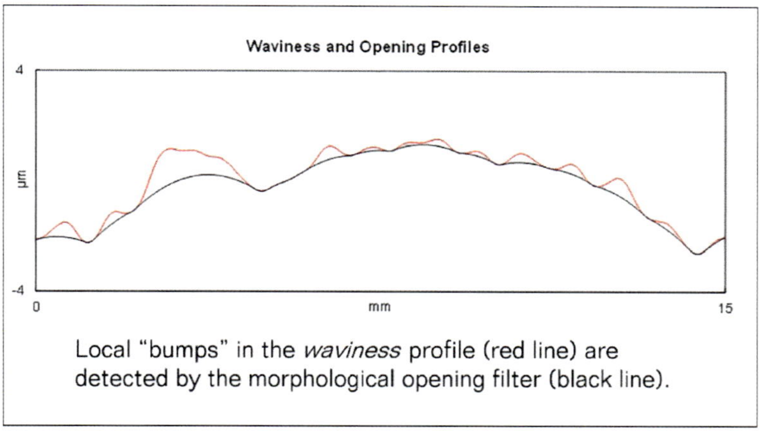

Figure 118: Bumps on surface characterized with a morphological opening filter

Why do parts measure the same but function differently?

In many cases, particularly for surface with only an "Ra" callout, we find ourselves with parts that measure the same, but function differently. This is primarily due to the fact that Ra is simply an average deviation of all profile points, which may not relate to the actual function of the surface. When these cases arise, it is best to explore the actual profile shapes, as discussed in the introduction to the section, "**Should I be concerned with waviness?**" on page **173**.

In some situations, people will go to the other extreme and report all of the parameters available from their instrument's software to see if there is a correlation between any of them and the component's function. While this can reveal more than looking only at Ra, the final result can be misleading if a parameter is selected which is not related to the physics of the surface's function. It is important to consider the surface attributes that are related to function and then select a parameter that best characterizes these attributes. Whenever possible, it is also important to choose a parameter with a known relationship to the manufacturing process. It is not helpful if the manufacturer doesn't understand how to adjust their process variable(s) to meet the specified parameter.

My instrument reports roughness and waviness. Where is form?

Many books and specifications refer to surface texture as being made up of three wavelength domains: roughness, waviness, and form. In these contexts, the term "form" is meant to describe the longest wavelengths on the surface—those wavelengths that are longer than waviness (see **Figure 119**).

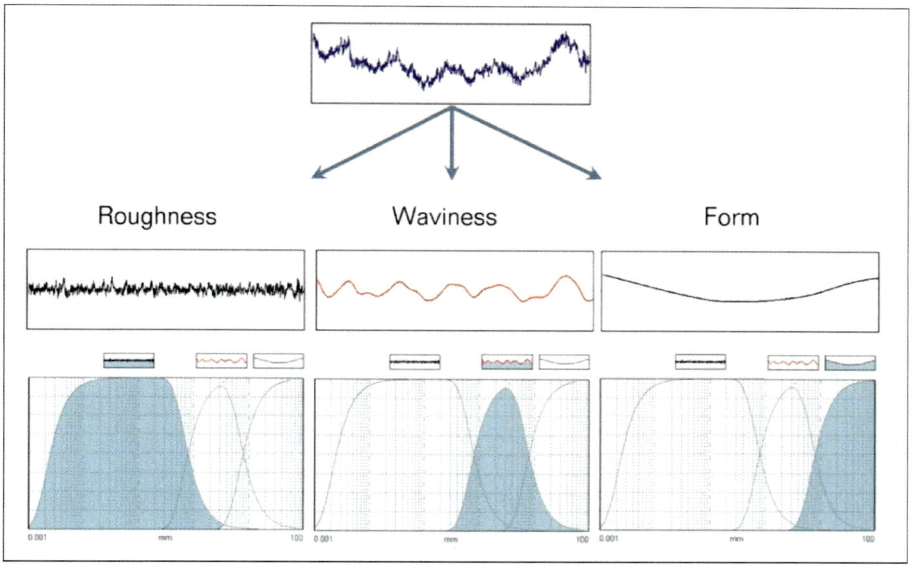

Figure 119: Illustration of the three wavelength domains for a sample profile

The term "form" can be confusing as it is also used in some Geometric Dimensioning and Tolerancing (GD&T) standards, such as ASME Y14.5, to refer to features like straightness, roundness, flatness, etc. While there is ongoing debate on the subject, in the context of Y14.5 "form" is generally intended to describe all of the surface deviations regardless of wavelength. This is quite different from the generally accepted use of the term "form" in the context of surface texture.

This section will discuss "form" in the surface texture context. Page **181** provides additional comments on the relationship to "form" in GD&T.

As illustrated in **Figure 119**, true form, under the technical surface texture definition, only exists when the analysis includes bandpass waviness, and consists of those wavelengths longer than the form cutoff (lambda-f, or λf). However, many surfaces have a long wavelength shape that is captured in a surface texture measurement, a shape we are tempted to call form (for example, the convex shape in **Figure 120**). These long wavelengths are typically removed from measured data prior to analyzing the roughness and waviness because, in general, longer wavelengths have larger amplitudes, and these would mask the surface characteristics of interest. The removal process involves subtracting a "reference geometry" from the *measured* profile, creating the *primary* profile. This reference geometry is a user-defined or best-fit shape; it can be a straight line, an arc, a polynomial, a spline, or some other geometry. This process is illustrated in the pairs of images in **Figure 120** and is discussed in greater depth on page **109**.

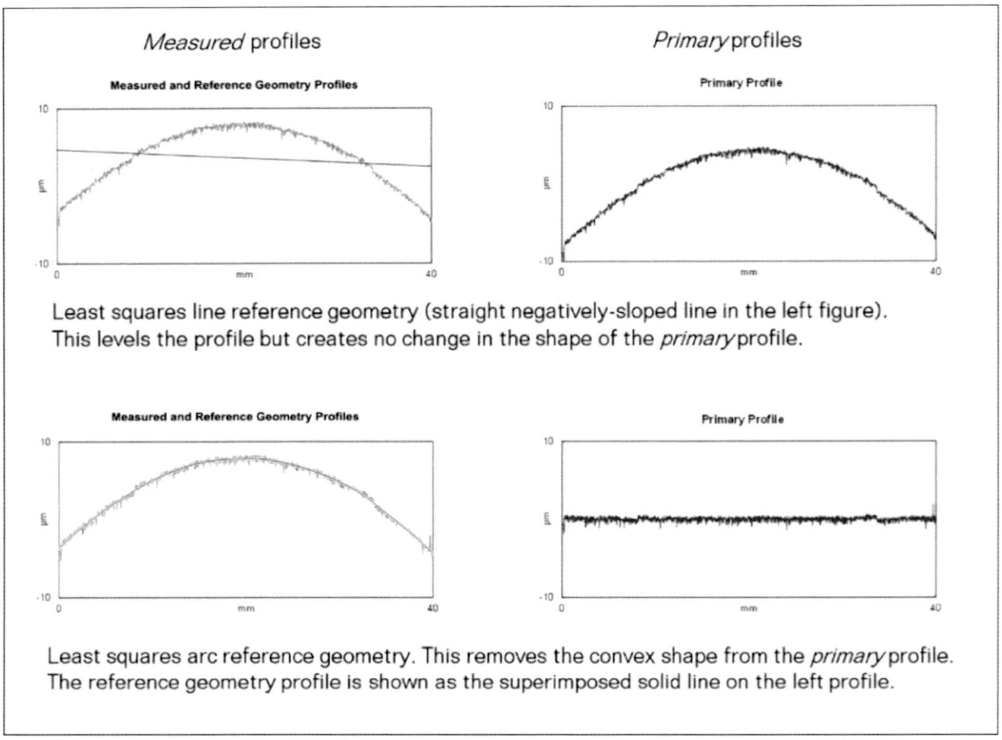

Figure 120: Reference geometry profile representation of shape

The reference geometry is often incorrectly called "form" but is more appropriately called the *reference geometry* profile. The characteristics of this reference geometry, such as the radius of a least squares arc, can describe the shape of the surface. However, to avoid confusion, these characteristics should not be called "form," and the numerical values should not be called "form" parameters. There is only one true form parameter in common use: that is Ft, the total height of the profile (the distance from the highest peak to the deepest valley). Ft is only calculated in conjunction with bandpass waviness.

Note that while the *measured* profile contains information about a surface's shape, it unfortunately also contains errors from measurement misalignment. The error may take the form of slope when the part is not level, or curvature if the traverse is off-axis to a cylindrical shape. The reference geometry can remove the effect of this misalignment from roughness and waviness assessments by, for example, suppressing a least squares line to level the profile or a least squares arc to remove curvature. However, it is impossible to separate the misalignment from the actual shape using this procedure. The resultant reference geometry characteristics are therefore only as useful as the care taken in sample fixturing and alignment. More information on the influence of alignment is given on page 79.

How do GD&T and surface texture specifications relate to each other?

Geometric dimensioning and tolerancing (GD&T) specifications that address form (e.g., straightness, circularity, profile of a line, flatness) are influenced by similar considerations as those involved in surface texture. Specifically, it must be decided what size (i.e., wavelength) features are to be included. Unfortunately, at the time this book is being written, there is no universally-accepted convention as to what those wavelengths are. Depending on the standards organization you're complying with, or your geographical location, there are several options:

- Under ASME Y14.5, all wavelengths are included, including roughness and "defects" such as scratches or pores. Y14.5-2018, Section 4.2, Fundamental Rule "S" states: "Unless otherwise specified (UOS), elements of a surface include surface texture and flaws (e.g., burrs and scratches). All elements of a surface shall be within the applicable specified tolerance zone boundaries." To add some confusion to this situation, ASME Y14.5.1M indicates the need for smoothing, and references ASME B46.1 for determining the amount of smoothing to be applied to the measured data. However, ASME B46.1 does not define the amount of smoothing for GD&T features.

- In ISO, it was once proposed in the standards for straightness, roundness, flatness, and cylindricity that these features be based on removing wavelengths below the roughness cutoff (lambda-c, or λc). This proposed approach was voted down as some of the committee members argued for even more filtering. The end result was that ISO states that when GD&T features of form are specified, the wavelength domain must be specified as well.

- Practically speaking, for some applications straightness is considered to be those wavelengths longer than waviness[44].

- From a different practical perspective, it is physically impossible to gather and analyze data with sufficient density from a large feature to meet, say, the ASME Y14.5 requirement[45]. See also: **https://digitalmetrology.com/flatness-in-the-usa/.**

---

[44] This may happen, for example, if a feature has roughness and waviness controlled by a surface texture callout and form controlled by a GD&T specification. Waviness may then be excluded from the form assessment to avoid counting it twice.

[45] For example, if roughness is to be included in a flatness measurement, the data spacing would need to be 0.5 μm in both directions to support a 2.5 μm λs. This is 4 *million* data points per square millimeter—obviously an unmanageable amount of data for most engineering features.

GD&T tolerances address the extremes of the line or surface; i.e., the single highest peak to the single deepest valley. Therefore, depending on the applicable standard or convention, straightness would be similar to Pt, Wt, or Ft (see **Figure 121**).

Given this confusion, no clear direction can be given to an inspector when measuring a component relative to one of these GD&T callouts. Each measurement laboratory should develop a standard methodology so its measurement results are consistent over time and between inspectors. The capabilities of the inspection method and equipment and the size of the component will influence what features are included. Historical data, component function, and customer expectations should also guide the inspection. The component designer should include as much information as possible on the engineering drawing or in supporting documents so the inspector isn't faced with making these decisions.

On a separate note, if roughness, or roughness and waviness, are excluded when evaluating a GD&T callout, an interesting situation arises relating to GD&T rule 1, often stated as "perfect form at MMC." Rule 1 requires a feature of size to have perfect form at its maximum material condition (MMC), but it allows the form to vary to the same extent that the size is less than MMC. If roughness, or roughness and waviness, features are excluded from an assessment of form, then, theoretically, you can have surface texture peaks that extend outside the form tolerance. Most components have size tolerances that far exceed the surface texture features, so this discrepancy is not important. However, for some high precision components, size and form tolerances can be on a similar order as the texture. In these cases, the interpretation of form may accept components that might not be functional. This is illustrated in the profiles on the right side of **Figure 121**.

One implication of GD&T rule 1 that should be considered is that GD&T is based on the concepts of "fits" and "tolerance stack-ups." For example, the entire surface of a pin must "fit" in a bore. The measured size of the pin should therefore include all of the unfiltered surface features (peaks, valleys, etc.). However, when we move beyond the topic of size and start to consider shapes, it becomes very useful to separate shapes into separate domains such as roughness, waviness, and form.

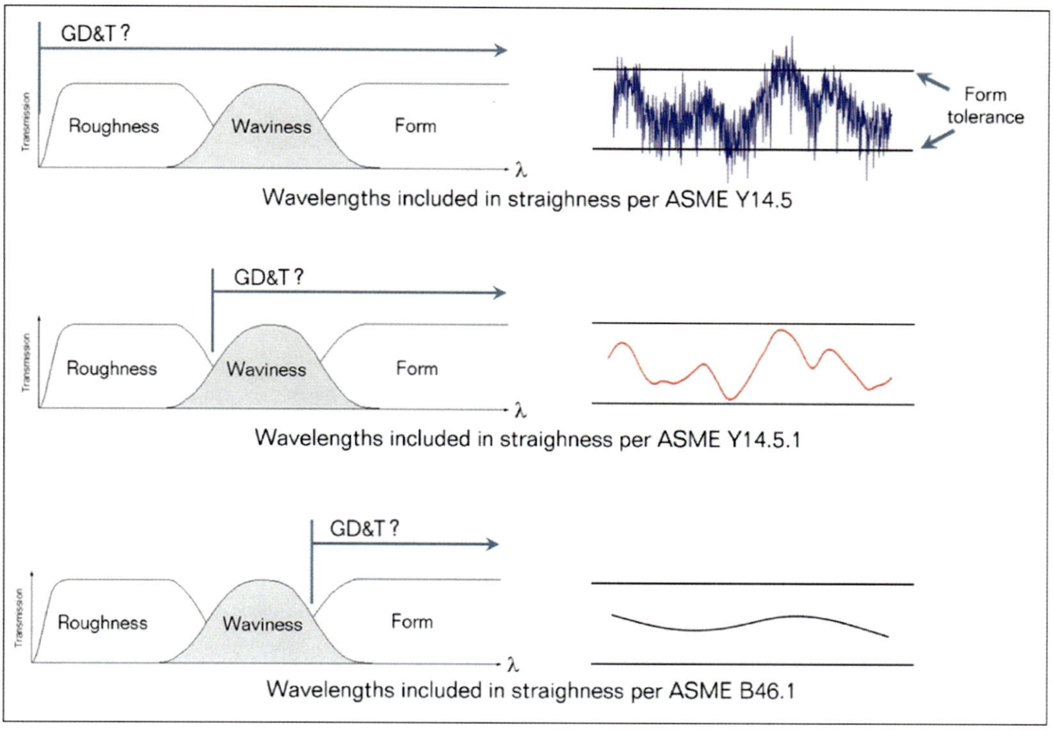

Figure 121: Schematic showing possible interpretations of GD&T

Which parameters should I use to detect scratches and dents?

None! Normal surface texture analyses are not designed to detect individual flaws. Optical techniques are the preferred method for detecting flaws, either manually or with an automated vision system. Standards have been developed by specific industries and by the standards organizations to support visual analyses; for example, see ISO 8785.

There are numerous reasons for not using surface texture methods for flaw detection:

- Surface measurements are meant to be taken at random locations (within the constraints of the technical drawing), so the chance of finding random, localized damage is small.
- In many work environments it is human nature for an inspector to look for a good area to make his or her measurements, so there will be a bias against tracing over defects.
- Even if the inspector targets an area of handling damage, it is extremely difficult to position the measurement stylus so that it travels across the worst part of the flaw. You may detect the flaw, but you cannot be certain you have its worst-case measurement, so any comparison to a defect specification or go/no-go limit is of dubious validity.
- Many surface texture parameters are an average over the full evaluation length, so localized defects may be diluted by a long trace and have minimal impact on the final parameter value, even if they will have a significant impact on performance.

Surface texture equipment and analysis *can* be used to characterize defects to aid in determining if they are detrimental or not (e.g., is there raised metal?). **Figure 122** is an example, showing the results of tracing through handling damage to determine its height and depth. In this case the measurement was made as an aid in creating visual standards for sorting parts. However, this type of measurement must be done under very controlled conditions with very specific data analysis, and it does not translate into a quality control technique for most production components.

Since a stylus can only trace a small line on a surface, it is much more common to use optical or imaging systems for the detection of defects over large areas.

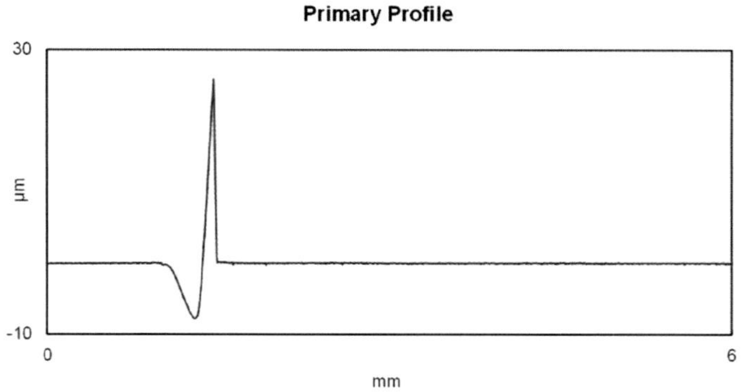

A nick in a surface with associated raised metal. This measurement showed that the raised metal was about 25 µm high. It was subsequently used to establish a tactile and visual guide for sorting defective parts.

Figure 122: Surface texture characterization of a handling defect

What does 300:1 or 100:1 mean on my printouts?

The 300:1 or 100:1 is the ratio of the roughness filter cutoff wavelength (lambda-c, or λc) to the short filter cutoff wavelength (lambda-s, or λs). It is sometimes included in reports of the surface texture parameters (see, for example, **Figure 123**).

The standard value for the short filter cutoff is commonly determined by applying this ratio to the roughness filter. Therefore, if a 0.8 mm roughness filter cutoff is used, the appropriate λs is 0.8/300, or 0.0025 mm. If λc is 2.5 mm, λs is 0.008 mm. If λc is 0.25 mm, a 100:1 ratio is used because the shortest generally available λs is 2.5 µm. A more complete description of the filter selection process is given on page **145**.

It should be noted that these bandwidth ratios are not actual computed values; rather, they should be considered as "names." For example, a 0.0025–0.8 bandwith is actually a ratio of 320:1. Similarly, a 0.008–2.5 bandwidth is actually a ratio of 312.5:1. We use the naming convetion of 300:1 to simplify the description.

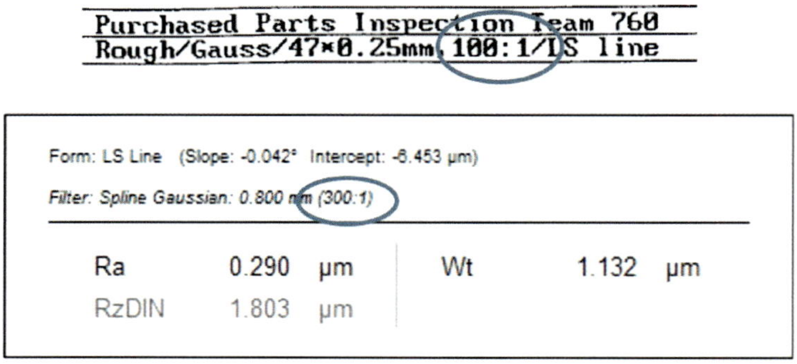

Figure 123: Analysis reports showing a λc : λs ratio

What does "max" mean in a surface texture callout?

A surface texture specification conforming to ISO 1302 may include "max" after the parameter name, as shown in **Figure 124**. "Max" is a quality control instruction; it does not influence the analysis of the data but rather guides the response to any out-of-specification results. It does *not* indicate that the adjacent numerical value is a maximum limit.

ISO 4288 describes two responses to any out-of-specification surface texture measurement. The default response is termed the 16%-rule, where the workpiece is considered acceptable if no more than 16% of the measured values fall outside the specification limits. If the surface texture is normally distributed for the parameter of interest, the 16%-rule corresponds to a mean plus one-sigma control on component quality. Guidelines for implementing the 16%-rule in a quality control plan are summarized on page **188**.

Critical surfaces that require tighter control can apply the "max" suffix to the relevant parameters. This requires all measurements to meet the specification limits.

An ISO 1302 callout assumes the 16%-rule unless the max-rule is explicitly invoked.

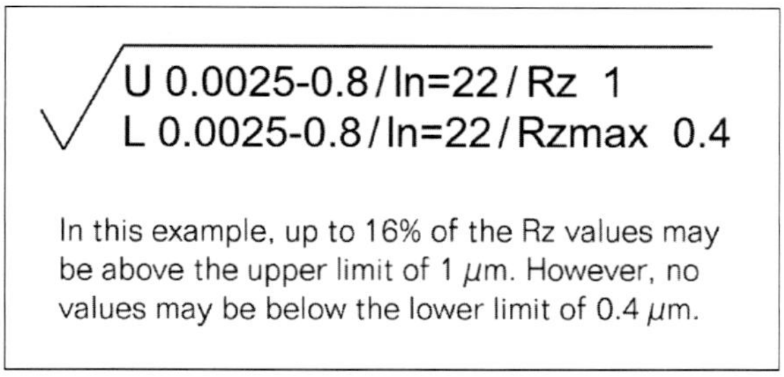

Figure 124: Example of the 16% and max rules

Note that some instruments can now report parameters such as Ramax and Rzmax. These are defined as the maximum parameter value of any sampling length within the evaluation length—a very different meaning than the parameter with the max-rule, as described above. This is unfortunate, since there is no obvious way to differentiate the Ramax parameter from the max-rule Ramax when it appears on a drawing. Notes on the engineering drawing or explanatory supporting documents will likely be required to avoid confusion.

For more on this topic see: https://digitalmetrology.com/whats-with-the-max/.

How do I implement the 16%-rule in practice?

ISO 4288 provides two approaches for dealing with surface texture measurements that are outside specification: the 16%-rule and the max-rule. These rules are discussed on page **187**.

The 16%-rule states that a component is acceptable if no more than 16% of the measurements on a surface are outside the specified limits. In a production quality control plan this might be implemented as follows (derived from Annex A of ISO 4288):

Based on measurements over the specified evaluation length, a component is acceptable if:

- the first measurement does not exceed 70% of the specification limit
- the first three measurements do not exceed the specification limit
- no more than one of the first six measurements exceeds the specification limit
- no more than two of the first twelve measurements exceed the specification limit
- the mean plus one standard deviation of the measured data does not exceed the specification limit.

If these conditions are not met, the workpiece is to be rejected.

Consider the component specification shown below in **Figure 125**. The actions in **Table 16** show how the third option listed above would be applied to the four sample parts.

These guidelines can be modified by mutual agreement of the supplier and the customer. For example, expensive parts could be measured twenty-four times with up to four values being allowed to exceed the specification limit.

Note that there are no standardized procedures for calculating manufacturing capability (Cpk or Ppk) values for a characteristic with the 16%-rule. From a statistical point of view, the 16%-rule implies that acceptance is based on the 84th percentile value. If six measurements were taken, this would mean the second-highest value is used to represent the workpiece. Alternatively, one could say that the surface is best represented by the average of all the measured values. Given the lack of a standardized approach, the supplier and customer should develop a mutually agreeable method for assessing process capability.

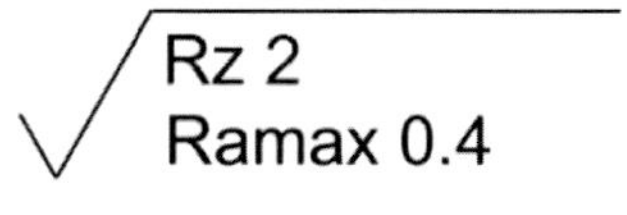

Figure 125: Example specification combining the 16%- and max-rules

Table 16: Sample application of 16%- and max-rules based on the specification in **Figure 125**

Part	Measure	Rz	Ra	Action	Disposition	Comments
1	1	0.44	0.12	None	Accept	Parameters meet the specification
2	1	2.05	0.31	Measure more	Accept	No more than one of the six Rz values exceed the limit so Rz meets the 16% -rule. All values of Ra meet the limit and comply with the max-rule. The part is accepted.
	2	1.95	0.23			
	3	1.62	0.19			
	4	1.49	0.25			
	5	1.83	0.30			
	6	1.76	0.22			
3	1	2.57	0.37	Measure more	Reject	More than one of the six Rz values exceed the limit so the part fails to meet the 16%-rule. The part is rejected.
	2	1.96	0.31			
	3	2.11	0.30			
	4	1.89	0.25			
	5	2.02	0.33			
	6	1.71	0.23			
4	1	1.22	0.43	None	Reject	Ra exceeds the limit. The max-rule applies and the part is rejected

DOCUMENTATION

DOCUMENTATION

The technical drawing and associated product documents are the primary communication channels between a component's design, engineering, manufacturing, and quality control personnel. As with any of the dimensional, material, or mechanical property requirements, the surface texture specifications must be communicated clearly and unambiguously so that:

- Measurements are made in the desired locations and in the correct direction
- The applicable analysis options are recognized and selected during processing of the raw data
- The desired surface texture parameters are reported and compared to their respective limits
- The expected response to out-of-specification results is clear.

Ideally, all of the relevant information will be included on the face of the drawing or in readily available supporting documents. This section addresses questions about how to include this information when creating a technical drawing, and how to extract that information from an existing drawing.

How do I specify surface texture requirements on a drawing?

The drawing is the primary means of communicating design intent to the manufacturing facility. This communication must be clear and unambiguous so, given its complexity, the surface texture callout has its own language and syntax, described in various national and international standards such as ASME Y14.36 and ISO 1302.

ISO 1302 provides the most complete drafting methodology, giving an unambiguous description of the analysis characteristics that affect the parameter values[46]. Examples are shown in **Figure 126** and **Figure 127**; these describe the same specification but with alternate means of stating the evaluation length. **Figure 126** states that the evaluation length is to be 20 sampling lengths, or 16 mm (20 lengths x 0.8 mm). **Figure 127** specifies the 16 mm directly, using the abbreviation for evaluation length of “ln”.

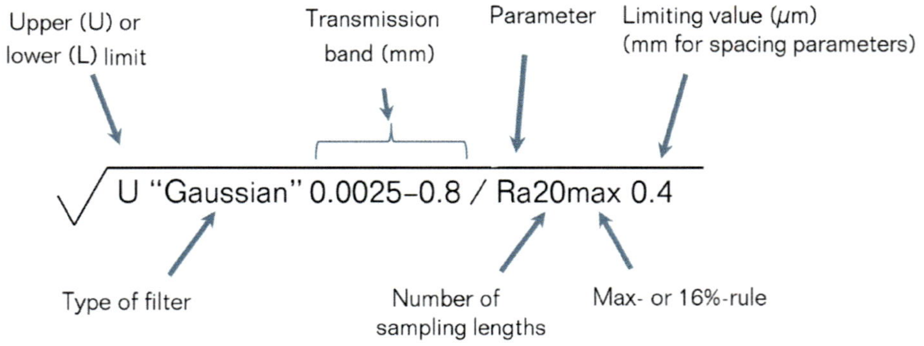

Figure 126: A surface texture callout according to ISO 1302

[46] Note that as this is being written ISO is considering changing some of the ISO 1302 syntax. ISO 21920-1 is planned to supercede 1302 in the future.

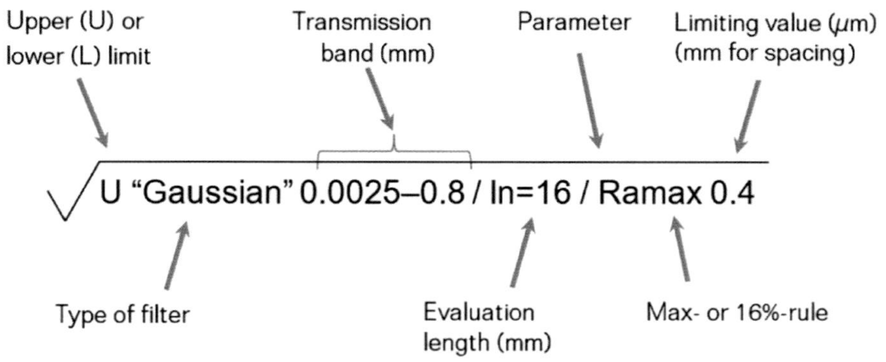

Figure 127: A surface texture callout according to ISO 1302

A complete callout for a roughness material ratio specification is shown in **Figure 128**.

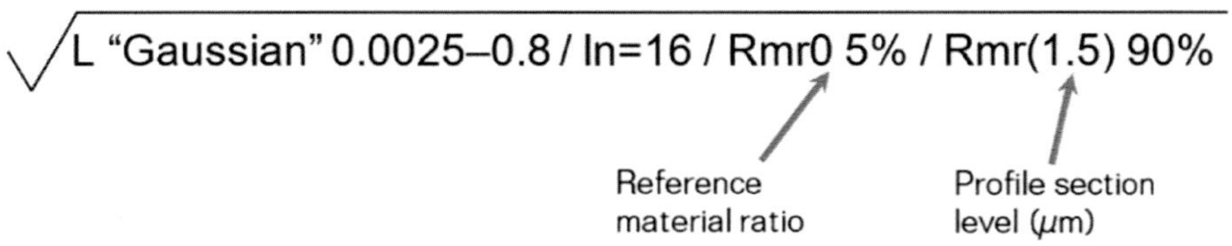

Figure 128: A callout for a roughness material ratio according to ISO 1302

The callouts are placed under the surface texture radical symbol, with all of the callouts stacked under one radical. The default values do not need to be included. An example of a complete surface texture description, with both roughness and waviness specifications, is shown in **Figure 129**. In addition to the parameter and analysis definitions discussed above, this shows the inclusion of optional additional process information:

- the type of process (here, superfinished)
- the direction of the lay (the "X" indicates cross-hatched—reference **Table 17**)
- the fact that material removal is required (the horizontal bar on top of the "V" of the radical symbol)
- the minimum material removal (the 0.005 mm to the left of the radical)
- the all-around symbol (the small circle at the junction of the two legs of the radical) indicating that the specification applies to all surfaces around a workpiece outline.

The complete list of lay symbols is given in **Table 17**, derived from Table 2 in ISO 1302-2002 and Figure 4 in ASME Y14.36M-1996. These symbols define the lay relative to the drawing plane which contains the surface texture callout.

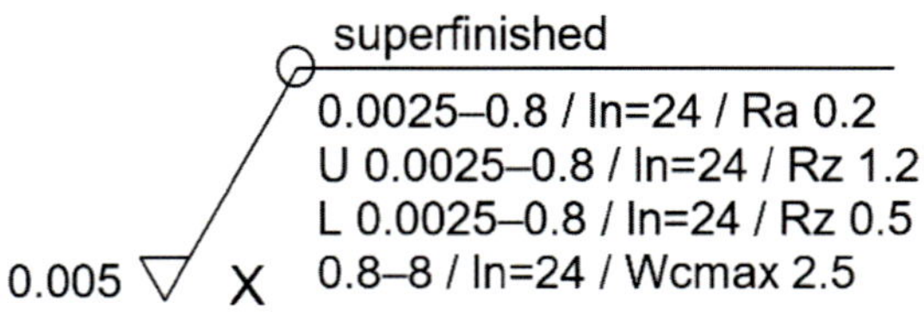

Figure 129: An example of a complete surface texture callout according to ISO 1302

Table 17: Symbols for indicating surface lay

Symbol	Interpretation
=	Parallel
⊥	Perpendicular
X	Crossed in two oblique directions
M	Multi-directional
C	Approximately circular relative to the center of the surface
R	Approximately radial relative to the center of the surface
P	Particulate, non-directional, or protuberant
All of these directions are relative to the line representing the surface to which the symbol is applied	

Some of the specification parameters have default values and need not be included on the drawing: these are listed below. If all of these default values are appropriate, a callout can be as simple as that shown in **Figure 130**. However, it is recommended that a callout be as unambiguous as possible, so it is often appropriate to include some of this information to avoid confusion.

Default values

- Type of limit: Upper (U) is default. However, if there are specified both an upper and a lower limit for a parameter, inclusion of the U and the L can be helpful.
- Type of filter: Gaussian is default, and, given the similarity between Gaussian and spline, either can be left off. Options like robust need to be stated.
- Transmission band: Default values of lambda-c and the λc:λs ratio apply to various ranges of surface roughness, based on the expected Ra or Rz value for non-periodic surfaces and

Rsm for periodic surfaces (see page **145**). However, given the importance of these cutoff values, it is recommended they be explicitly stated unless there is a very good reason not to. Transmission bands may be specified using one number, with the other limit being the default value. The hyphen is required so it is clear which value is being assumed. For example, the roughness transmission band 0.0025-0.8 could be written as -0.8, with the 0.0025 being implied from the default 300:1 roughness bandwidth. In this case the hyphen indicates that the short wavelength limit is assumed.
Note that there is no default upper limit for bandpass waviness, so this will need to be included with any bandpass waviness parameters. A callout of, for example, "0.8-" indicates that waviness is to be evaluated over all wavelengths longer than 0.8 mm. The default short wavelength limit is lambda-c.

- Evaluation length: For roughness parameters, five sampling lengths is default. There is no default length for waviness parameters so it needs to be stated. This can be as an integral number of sampling lengths if the form filter cutoff wavelength (lambda-f, or λf) is defined, or it can be specified in the "ln=xx.x" format. The evaluation length for primary parameters will be the feature length unless otherwise specified in the "ln=xx.x" format.
- Max- and 16%-rule: 16% rule is default, so this will be applied to any parameter not modified with the "max" suffix.

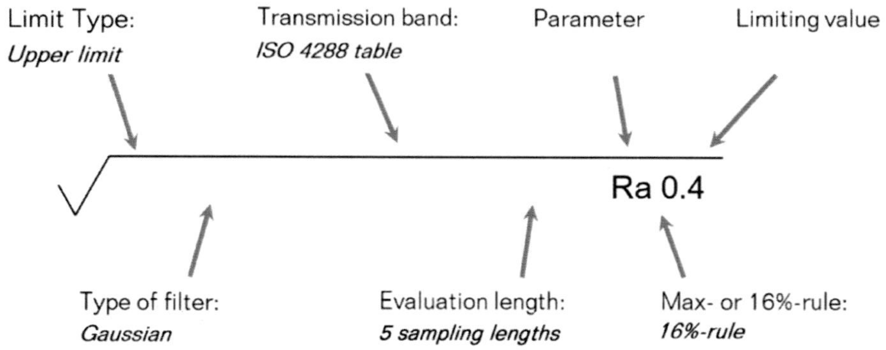

Figure 130: Simplest version of ISO 1302 callout using all the available default options

Additional information is required for the analysis of some parameters. There is no agreed-upon syntax for placing some of this information on technical drawings, but suggested methods are described on page **201**.

If a drawing must be ASME Y14.36 compliant the surface texture callout would be laid out as shown in **Figure 131**. Depending on the complexity of the specification, this approach can be ambiguous and more confusing than ISO 1302—see page **199** for a comparison of the two techniques.

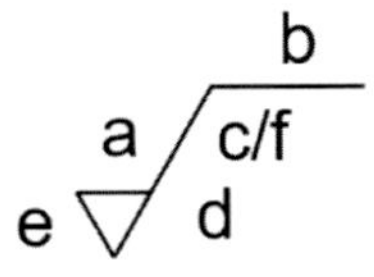

The letters correspond to the following parameters:

a = the value of Ra, in micrometers. If there is an upper and lower limit the numbers are stacked

b = production method, other text, a note reference, etc.

c = the roughness cutoff, in millimeters

d = the direction of the lay

e = minimum material removal requirement, in millimeters

f = a roughness parameter other than Ra - the symbol and the limiting value, in micrometers

Figure 131: A generic callout that conforms to ASME Y14.36-1996

How do ISO and ASME surface texture callouts differ?

The following comments are based on the 1996 revision of ASME Y14.36 and the 2002 revision of ISO 1302. At the time of this writing, a new ASME Y14.36 is being developed.

Table 18: Comparison of ISO 1302 and ASME Y14.36-1996 drafting standards[47]

ISO 1302	ASME Y14.36-1996
All parameters are placed under the radical	The Ra parameter is placed outside the radical while all other parameters are placed under the radical
Each parameter is fully described on one line, with provision for an unambiguous description of the type of filter, the transmission band, the evaluation length, and the numerical limit.	There is no provision for specifying the type of filter. An unwieldy method exists for specifying the cutoff, but with no clear method for including the two ends of a bandpass filter. There is no provision for specifying the evaluation length if it is different than the default value.
There is a well-defined set of default values that allows callouts to be shortened when appropriate.	There are no clear default settings
Additional information required for parameters such as material ratios can be included in the same single-line callout	No guidance is given for parameters such as material ratios

An example of a callout for the same specification in the two systems is shown in **Figure 132**. While the ASME version is shorter, there is less ambiguity with the ISO callout concerning:

- the evaluation length required for the parameters
- the short cutoff filter cutoff wavelength to be used
- the form filter cutoff to be used for Wa.

[47] ASME Y14.36-1996 includes the statement: "Certain measurement conditions are to be specified by notes." This would include several of the "undefined" items listed in this table. This method is more complicated, and potentially much more wordy, than the ISO 1302 practice.

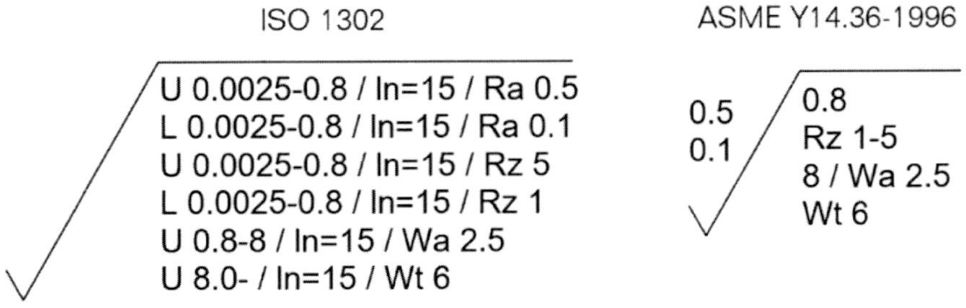

Figure 132: ISO 1302 and ASME Y14.36-1996 presentations of the same specification

Given these considerations, later versions of the ASME B46.1 surface texture standard refer to ISO1302 as the preferred method for specifying surface texture on drawings.

Note: At the time of this writing, efforts are underway to develop a revision to ASME Y14.36-1996 that is more in line with ISO 1302.

How do I specify information that is not covered in the standards?

For some of the newer parameters or analysis conditions discussed in this book, there can be a lag between when that parameter begins to be used and when it is included in the national or international standards. Some of these parameters require additional information for how they are to be evaluated. Following are some suggestions for including this information on the technical drawing. The preferred format follows the practice used for the section level for the "standardized" material ratio parameters Rmr and Pmr, with the supporting information being placed in parentheses immediately after the parameter name.

Ball radius for morphological filters (see pages 173, 239, and 243 for examples)

Include the radius in parentheses after the parameter name (preferred method[48]) or in a note. This applies to parameters such as Wcvx (waviness convexity), Wvoid (waviness void area), and Wvdd (waviness void depth). For example, within the surface texture callout it would appear as follows:

0.25– / Wvdd(750 mm) 50

or, as a note on the drawing it would appear this way:

Wvdd ball radius = 750 mm

Discrimination level for the "sm" and "c" parameters (see page 352 for more information)

Include the levels within parentheses immediately after the parameter name, with the percentage of the "z" parameter first and the percentage of the sampling length second, as z=x%, L=y%. An example for Wc is shown below.

0.8-8.0 / ln=24 / Wc(z=0%, L=0%)max 1.6

Rpm on an ISO 1302 compliant drawing (see page 336 for more information on Rpm)

Include a note on the drawing stating that this parameter is defined by ASME B46.1 The following is an example, but other formats can be used.

Rpm conforms to ASME B46.1

---

[48] The example shown here assumes it is understood that the dimension is the ball *radius*. If there is a chance this will be misinterpreted, the parenthetical material can be expanded to "750 mm ball radius."

Segment length for Wseg (segment waviness)

The Wseg segment length can be specified within parentheses immediately following the Wseg.

0.8–8 / ln=40 / Wseg(10 mm) 3

If you don't wish to reference the non-standard Wseg parameter on the drawing you can use Wt with the segment length as a qualification, as shown below. Here, the Wt specification over the 40 mm evaluation length is 5 μm but the Wseg limit with a 10 mm window is 3 μm.

0.8–8 / ln=40 / Wt 5 (Not to exceed 3 in any 10 mm segment)

Other non-standard parameters

These should be described in a drawing note or in a supporting document referenced in a note or above the surface texture radical.

How do I specify trace location and direction on a drawing?

Surface texture measurements are to be made on the surface indicated by the surface texture symbol (√). The surface texture symbol may be placed in several locations:

- directly on the surface
- on an extension line
- on a leader line pointing to the surface
- on a dimension line in connection with a feature-of-size dimension
- on a geometric tolerance frame

The surface texture specification applies to the entire surface defined by the associated dimension. Areas of transition (e.g., fillets or chamfers) are to be toleranced individually. If no specification is provided for these transitions, they are expected to conform to the roughest adjacent surface.

Unless instructed otherwise, the trace can be made at any location on the indicated surface, consistent with the specified evaluation length. The trace direction must be parallel to the plane of the drawing and can be made in either direction (left to right or right to left). Most instruments travel left to right.

If a specification applies to all surfaces around a workpiece outline, the all-around symbol may be used (see **Figure 133**).

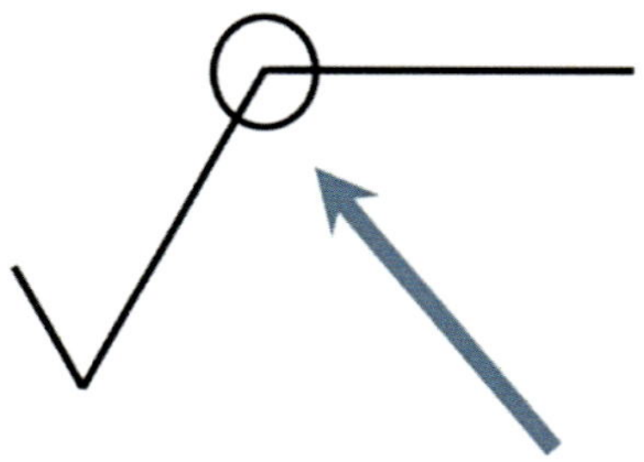

Figure 133: All-around symbol on a surface texture callout

On older technical drawings, what are the numbers above the radical?

Figure 134 is an example of a callout from an inch-based drawing that specifies waviness parameters by placing the limiting values above the radical. The first number, 0.002, is the maximum waviness height in inches, and is equivalent to Wt. The second number, 2, is the maximum waviness spacing, also in inches, and is equivalent to Wsm. Note that these limits have units of whole inches (or millimeters), unlike the limit on, say, Ra, which has units of microinches or micrometers.

The remainder of this callout follows the standard ASME Y14.36-1996 syntax: an upper Ra limit of 16 μin with a 0.010 inch roughness filter cutoff and a perpendicular lay.

Note that older drawings may have additional differences in the surface texture callouts relative to current drafting standards. These will need to be researched on a case-by-case basis, preferably by examining older versions of the relevant national or international standards.

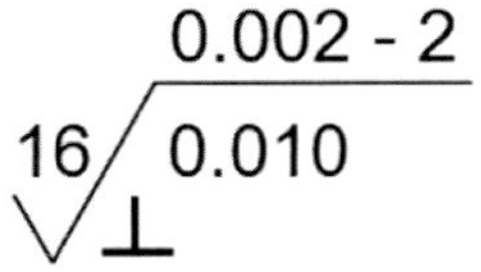

Figure 134: Sample callout according to ASME standards prior to 1995

What do triangles mean on a drawing?

Some technical drawings use a system of inverted triangles to designate the manufacturing process and the desired surface texture; an example is shown in **Figure 135**. More triangles indicate a smoother surface. This method is now obsolete, having been cancelled in ISO in 1978 and in JIS in 1994; it should therefore not be used on new drawings.

The triangles may be accompanied by a numerical value followed by the letter "S." The S parameter is equivalent to the ISO or ASME Rmax, the maximum peak-to-valley distance in a single sampling length. The meaning of the triangles and an approximate equivalent value of the Ra parameter are given in **Table 19**. Note that any correlation between Rmax and Ra is very tenuous and should be used with caution. Examples of this are shown in **Figure 136**, where the Rmax:Ra ratios range from 3 to 33 for two different surfaces.

In addition to the roughness range, the number of triangles also dictates the roughness filter cutoff. This is similar to ISO's cutoff recommendations based on Ra, Rz, or Rsm, as shown in **Table 13**–**Table 15** on page 147.

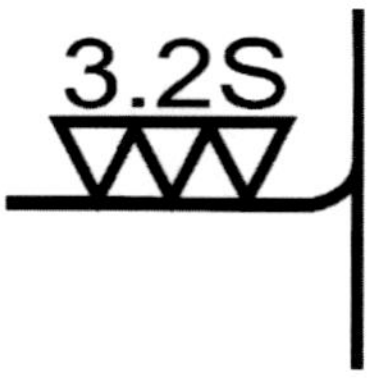

Figure 135: Example of a surface texture specification using triangles

Table 19: Explanation of the triangle-based surface texture specification system

Rmax (μm)	0.2	0.4	0.5	0.63	0.8	1.0	1.25	1.6	2.0	2.5	3.2	4.0	5.0	6.3	8	10	12.5	16	20	25	32	40	50	63	80	100
Ra (μm)	0.04	0.05	0.063	0.08	0.10	0.125	0.16	0.2	0.25	0.32	0.4	0.5	0.63	0.8	1	1.25	1.6	2	2.5	3.2	4	5	6.3	8	10	12.5
Ra (μin)	1.6	2.0	2.5	3.1	3.9	4.9	6.3	7.9	9.8	12.6	15.7	20	24.8	31	39	49	63	79	98	126	157	197	248	315	394	492
λc (mm)	0.25					0.8										2.5					8					
Finish symbols	▽▽▽▽					▽▽▽										▽▽					▽					
Working Methods	Superfinishing, lapping					Grinding, fine boring, fine turning, paper finishing										Turning, milling, broaching, boring, reaming, planing					Turning, shaping, drilling, sawing, punching, planing					

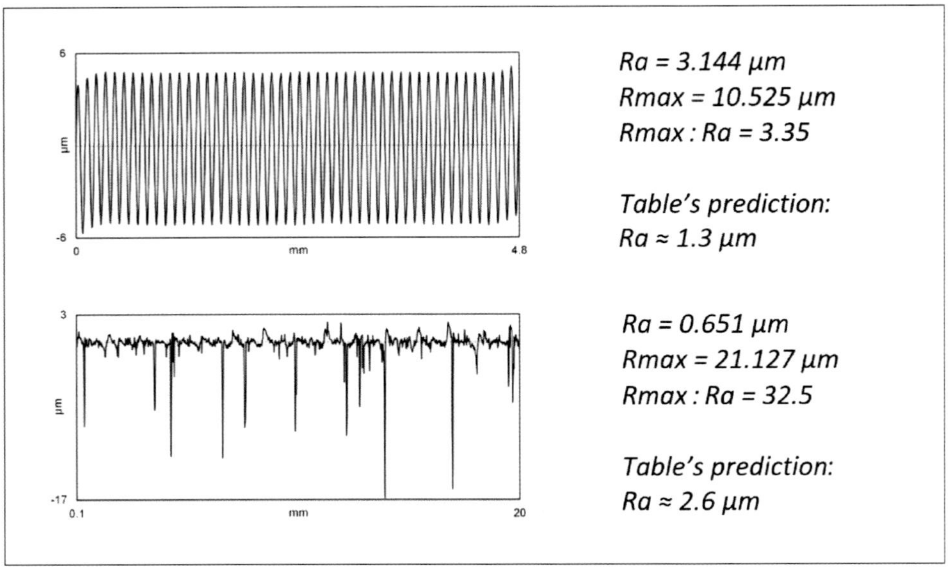

Figure 136: Comparison of Rmax and Ra for different types of surfaces

How do I specify the measurement of smaller features?

The convention of including five sampling lengths in a roughness measurement provides a reasonable sample size for calculating roughness parameters without excessively diluting short segments of higher (or lower) roughness. It is the default evaluation length for roughness measurements. However, it is valid to include fewer or more sampling lengths; this may be required when the surface of interest is short or when it is important to monitor the condition at multiple locations along the full length of a long feature.

When using the ISO 1302 drafting convention, non-standard evaluation lengths can be specified in two ways (see **Figure 137**, with both callouts specifying the same measurement):

- as an integral number of sampling lengths, as shown in the upper callout
- as a specific evaluation length, as shown in the lower callout

Technical drawings conforming to ASME standards should include the evaluation length as a note, or should incorporate the ISO 1302 techniques with a note that this particular part of the drawing does not conform to ASME conventions. See page **194** for more information about drafting conventions.

√ 0.0025-0.8 / Ra2 0.2

√ 0.0025-0.8 / ln=1.6 / Ra 0.2

Figure 137: Two equivalent methods for specifying a non-default evaluation length

Note that, for short features, the evaluation length should be shorter than the feature length to provide room to place the stylus and stop the travel. Alternatively, a probe limiter can allow the trace to start and stop with the stylus off the surface, and the data analysis software can be used to select the evaluation region.

It may be tempting to change the roughness filter cutoff to accommodate the five sampling lengths on the shorter feature (see the section, "**Are "cutoff" and "sampling length" the same thing?**" on page **327**). In fact, some instruments automatically adjust the roughness cutoff according to the evaluation length. This is inadvisable because changing the sampling length changes the roughness cutoff and therefore changes the features/wavelengths being included in the roughness and waviness analyses. **Figure 138** is an example of a feature that is too short for five 0.8 mm sampling lengths, and where it could be tempting to reduce the roughness filter cutoff wavelength (lambda-c, or λc) to 0.25 mm. **Figure 138** also lists the difference in roughness and waviness values that result from such a change. **Figure 73** on page **118** is another example, illustrating how filter changes affect the value of the Ra parameter and how they can alter a decision on the acceptability or functionality of a component.

Short surfaces can also be evaluated using the *primary* profile parameters instead of roughness. This is especially appropriate for evaluation lengths on the order of one sampling length or less. This approach avoids the sampling length issue, since the *primary* profile does not have the roughness filter applied. The purpose of the filter is to separate the profile by wavelength, and since the short profile cannot contain the long wavelengths, a *primary* profile parameter is essentially equivalent to the *roughness* profile parameter for a short surface.

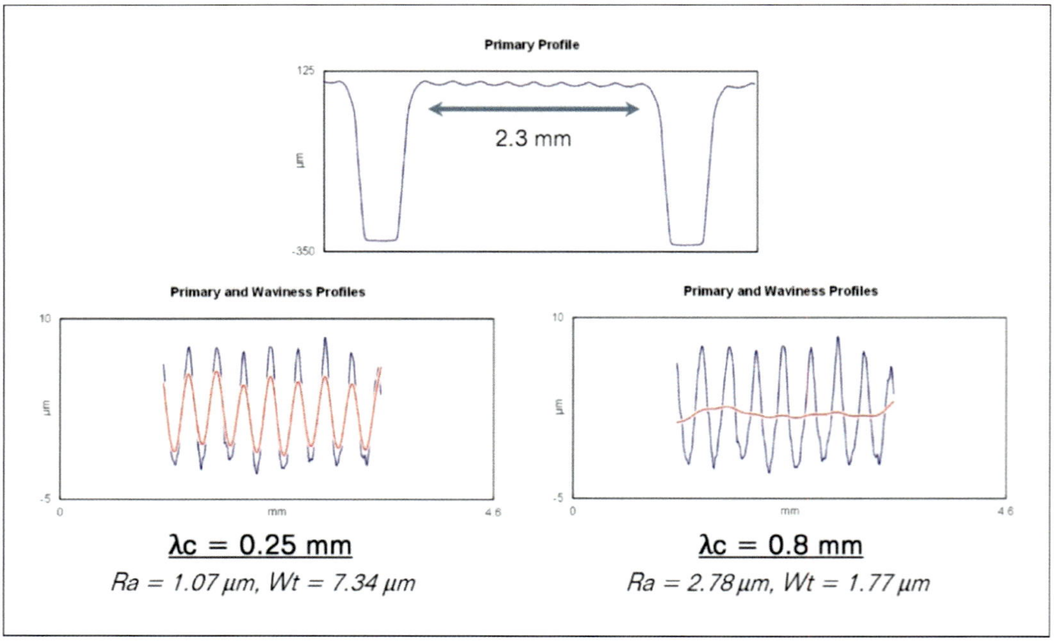

Figure 138: Effect of λc selection on roughness and waviness parameters

Most drawings only specify Ra. Is that enough?

What other parameters should I consider?

The "average roughness" Ra is the most commonly used roughness parameter, occurring in approximately 80% of the world's roughness specifications (see page **301** for a definition of Ra; see page **375** for its mathematical equation). There are several reasons for this:

- Ra is statistically a very stable, repeatable parameter
- Ra is often useful for controlling random type surfaces, such as grinding
- Ra provides a good parameter for monitoring a manufacturing process that is under control and where the conditions are always the same, e.g. tool geometry, speeds, feeds

However, Ra also comes with several disadvantages, including:

- Ra is not a good discriminator between different types of surfaces (see **Figure 139)**
- Ra makes no distinction between peaks and valleys (see **Figure 140**)
- Ra is not very informative for surfaces with heavily skewed finishes (Rsk outside ±2)
- Ra is not a good measure of sealing surfaces or other situations where waviness may be as or more important than roughness

Therefore, Ra may be useful for controlling a manufacturing process after it has been developed, but it does not do well at fully defining a functional surface. A functional description includes roughness parameters representing features that affect that function; for example, peak heights, material ratio, or the number of features. Waviness may also need to be controlled with some of the *waviness* profile parameters. Sample lists of parameters applicable to various types of surfaces and functions can be found in the series of sections starting with, "**Which parameters specify surfaces with functional requirements?**" on page **251**.

As a cautionary note, the designer should avoid "parameter overload." There are hundreds of parameters, and it may be tempting to assume that a longer list will result in a better surface. However, the component manufacturer may not know how to control multiple characteristics, and some specifications of parameters may, in fact, conflict with others. There should always be a logical basis for the parameter choices—does it make sense that the feature being controlled really has an effect on the performance of the surface?

Rz is the second most common roughness parameter, but it has some of the same disadvantages as Ra. While it can often discriminate between different types of surfaces (all the surfaces in **Figure 139** have different values of Rz), it does not differentiate between peaks and valleys (**Figure 140**). It therefore may not perform better than Ra in determining whether a part will or will not function.

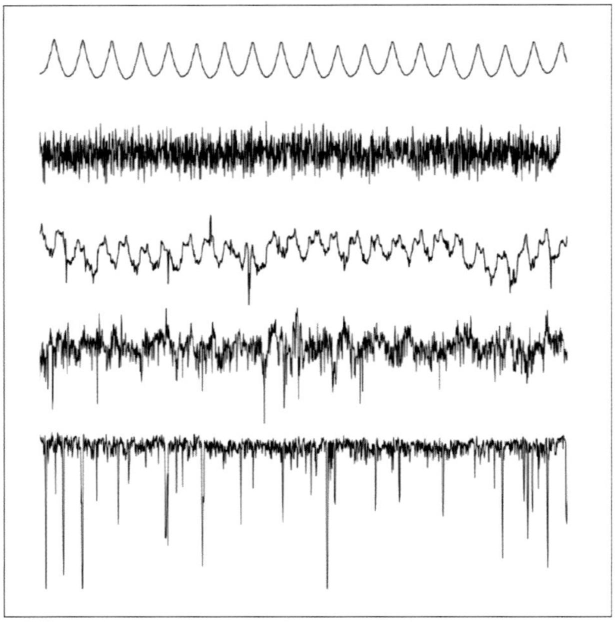

Figure 139: Multiple profile shapes with the same Ra

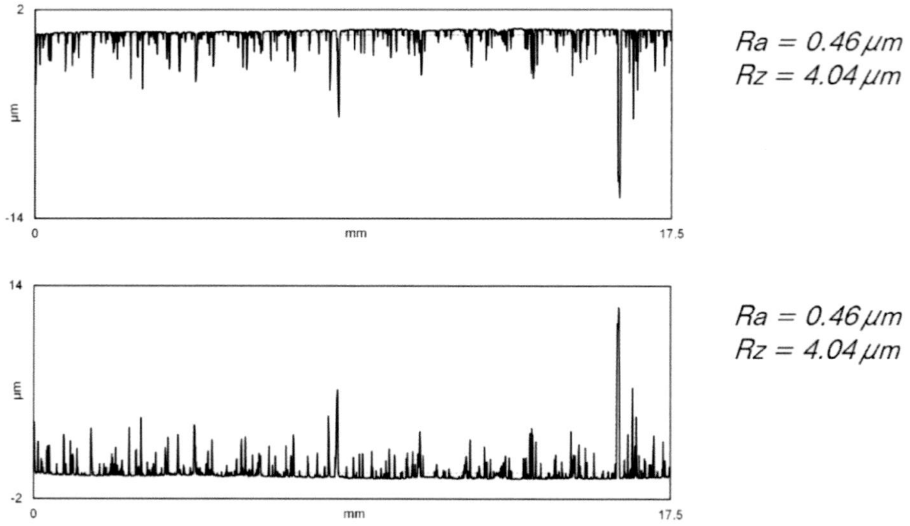

The bottom surface is an inverted copy of the top surface. The Ra and Rz values are the same, but, in most applications, the functionality would be very different. As someone has asked, "Which one would you prefer sitting on?"

Figure 140: Example of lack of discrimination between peaks and valleys with Ra and Rz

APPLICATIONS

APPLICATIONS

Much of the material in the other sections of this book deals with the fundamentals of surface texture, the measurement process, filter selection, and the definitions of curves and parameters. This section applies that information to answer specific questions and solve specific problems.

How can I measure roughness in grooves?

The answer to this question depends on whether the cross section of the groove is rectangular or semi-circular.

For the case of grooves of a rectangular cross-section, the measurement of the bottom surface's texture can be a challenge. **Figure 141** shows how the width of the stylus shaft can consume a significant amount of the groove width.

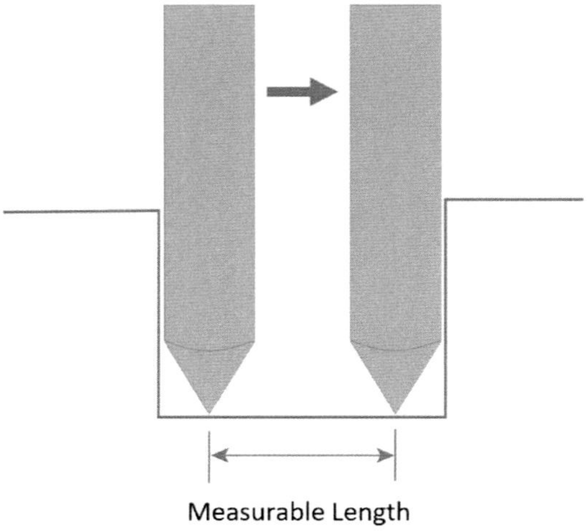

Figure 141: Groove with rectangular cross section

In this scenario, we should consider how much area can be measured relative to the functional area of the surface. For example, the full width of the groove may not be important in terms of how it contacts the mating component, such as an o-ring.

In some cases, the stylus geometry is simply too large to make an acceptable trace. In these cases, alternative approaches such as optical measurement methods or replication should be considered. It is strongly suggested that these alternate methods be validated against the reference (typically stylus) measurement. This may require the cutting/destroying of example workpieces.

Figure 142 Shows a groove with a semi-circular cross section. In the following discussion it is assumed that the whole groove cross section is being traced.

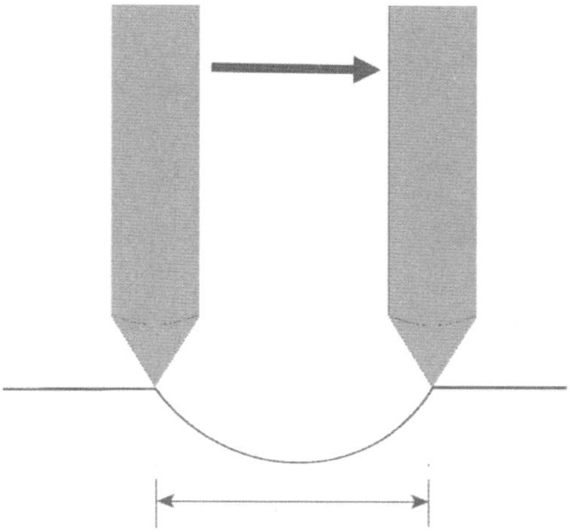

Figure 142: Groove with semi-circular cross section

There are several issues involved in measuring the surface texture in a groove with a semicircular cross section. In addition to the items listed here, see also the discussion about measuring curved surfaces in the section, "**My part is curved. Can I measure it?**" on page **230**.

Stylus damage

If the edges of the groove intersect the adjacent surface at an angle approaching 90°, there is a high probability that the stylus tip will be damaged as it attempts to exit the groove at the end of the measurement.

Instrument range

Many instruments do not have sufficient vertical range to measure a groove. The range can be increased by using a stylus on a longer arm, but this results in a proportional decrease in the vertical resolution so it may not be advisable.

Feature distortion

The amplitude of the surface features will be exaggerated near the groove edges, thus overestimating the actual surface roughness in those regions. This is illustrated in **Figure 143**, showing a roughness trace from a flat surface and a simulation of what that trace would look like if made in a 6 mm radius groove. Note that this simulation does not account for stylus flanking, which will occur when the angle of the surface exceeds one-half the included angle of the stylus tip. For example, flanking would occur when measuring a slope greater than 45° with a 90° stylus.

In some cases, instruments can be developed for particular applications. Optical methods may also be available for challenging measurements such as these.

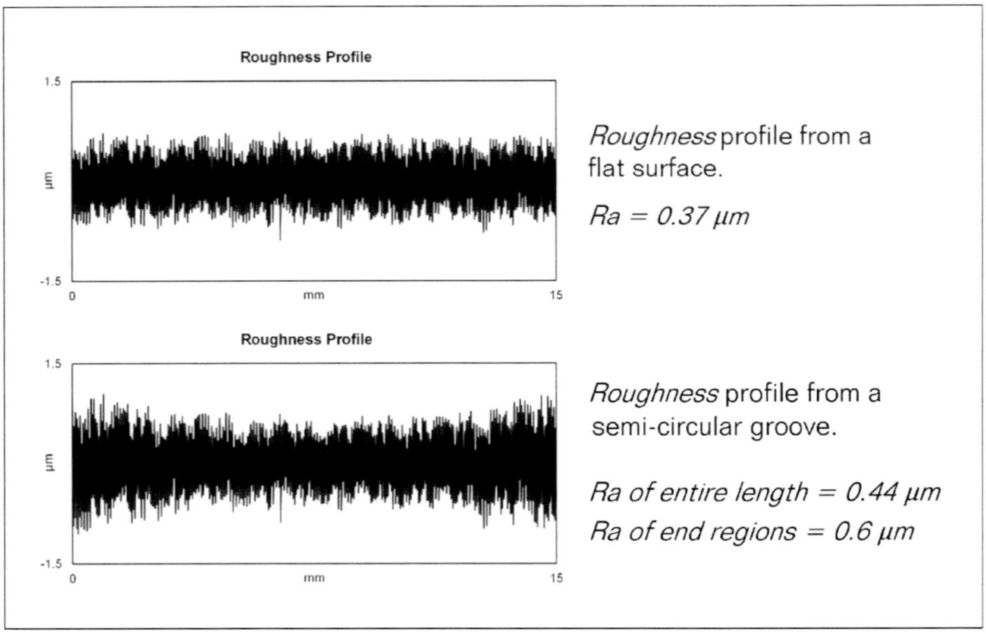

Figure 143: Distortion of the surface roughness when measured in a groove

How can I measure roughness in a fillet radius or corner?

The issues with measuring in a fillet (**Figure 144**) or corner are similar to those with measuring in a groove (see page **214**). The sidewall or other portions of the workpiece can interfere with the travel of the stylus. Special fixtures or instrument configurations are often employed so that the bisector of the feature is perpendicular to the stylus travel. This minimizes the required vertical range and possibly reduces physical interference with the stylus travel. If fillets or corners must be measured routinely, it would be beneficial to invest in this type of equipment.

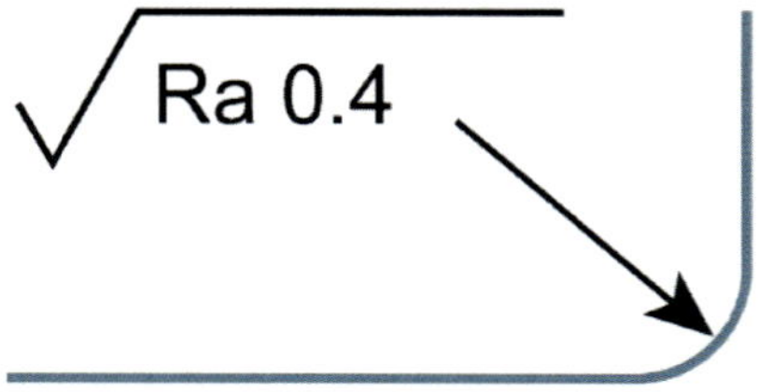

Figure 144: Roughness control of an inside fillet radius

If a fillet isn't accessible, a replica can be used. Additional information on replicas can be found on page **218**.

How do I measure a surface that my instrument cannot reach?

When it is not possible to directly measure an inaccessible surface (without destroying the component), consider measuring a replica. Some replication materials, applied with the proper technique, create a high fidelity negative of the surface with adequate hardness to support a stylus-based measurement. Other replica material may not support a stylus-based measurement but can be measured with optical systems. Internet searches using terms such as "surface replication" or "dental impression material" can identify potential materials and their sources. The various types of replicating materials each have specific preparation, application, and curing procedures, so consult the material source for detailed instructions.

Sometimes these difficult-to-access surfaces are very short. See the section, "**I have a short feature. Do I still need five sampling lengths?**" on page **62** for guidance on short surface measurements.

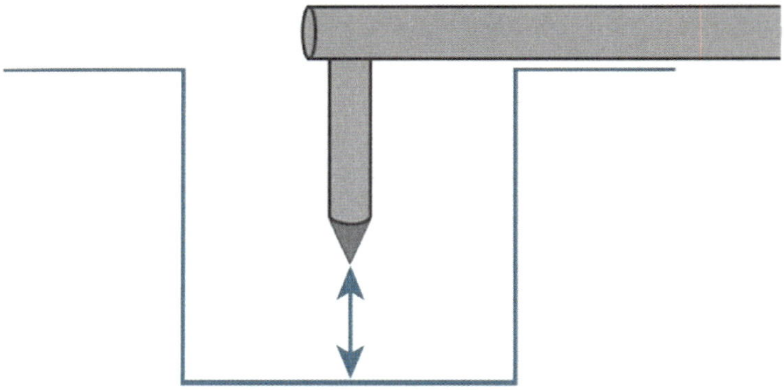

Figure 145: An unreachable groove for the stylus shown

How do I measure lay?

The lay of a surface is the direction of the dominant machining pattern. For example, if a shaft has been turned on a lathe, the lay will be circumferential.

Processes such as turning, boring, and grinding tend to produce a lay in a single direction. A process such as honing can produce a two-directional lay, with an included angle between the two directions of up to 90° (see **Figure 146**). Processes such as tumbling or shot peening produce a random lay. If multiple manufacturing processes are used, the final surface may have a combination of the various lay directions (see **Figure 147**).

The direction of the lay is typically measured optically. Image analysis tools may be used to measure the direction relative to some designated feature or defined axis, or to measure other characteristics such as the included angle on a cross-hatched surface. In this context, an image of the surface and a protractor can be considered as "image analysis tools."

Unless otherwise specified for functional reasons, surface texture measurements should travel across the lay to get a true representation of both the feature heights and the wavelengths. This is indicated on technical drawings by placing the surface texture symbol on a view of the surface where the expected lay is perpendicular to the plane of the drawing (see page **194** for more information on drafting methodology). The symbols given in **Table 17** on page **196** are used to specify the type of lay.

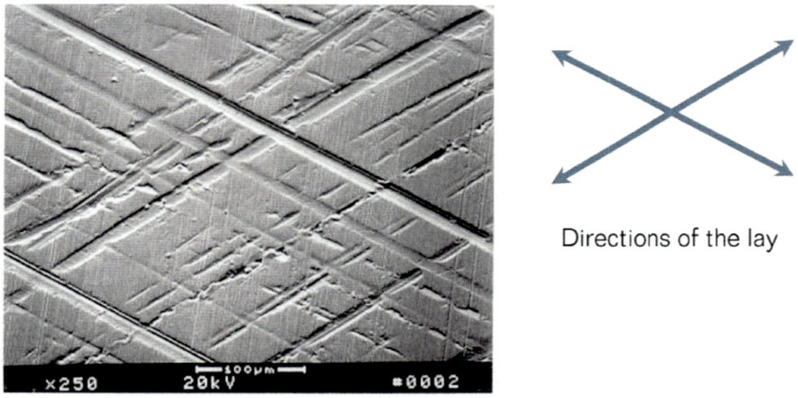

Figure 146: SEM photomicrograph of plateau honed surface with a two-directional lay

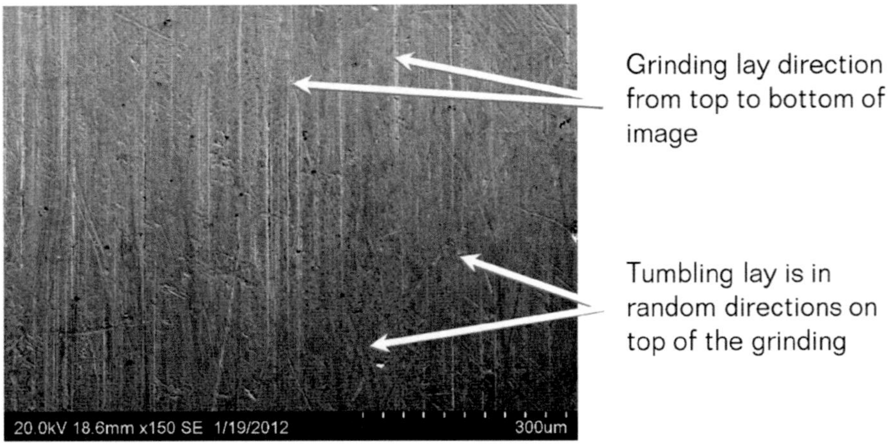

Figure 147: SEM photomicrograph of a ground surface that was subsequently tumbled

How do I control peaks on a surface?

Roughness and waviness peaks can negatively affect the performance of surfaces in many types of applications. Waviness peaks can act as stress concentrators, especially in concentrated (Hertzian) contacts. High spots can abrasively or adhesively wear a mating surface. Waviness and/or roughness peaks can adversely affect the performance of lubricated interfaces by disrupting the lubricant film. Wear of the high spots generates unwanted debris and, in some critical applications, can adversely affect clearance. The height and density of peaks influences a surface's visual appearance. Profile peaks affect coating adhesion. In all of these situations, the specification and control of peak heights are important aspects of a product design.

Numerous parameters include information about peak heights. Various parameters can target the extreme peaks, either as an individual peak (Rp ASME, Rp1 ISO) or as the average of a small number of peaks (Rpm ASME, Rp ISO). Rmr, Rpk, Rpq—parameters derived from the material ratio curve—include peak heights in their calculations. Parameters such as Rpc and Rhsc report peak density. Rsk differentiates profiles with dominant peaks from those with dominant valleys. Peak-to-valley parameters such as Rt, Rz, R3z, and Rmax include peak height extremes in their values. Several of these parameters have analogous "W" parameters for the *waviness* profile. Wcvx (waviness convexity) provides a peak height measurement that includes aspect ratio information. Each of these can be useful, but each must also be used appropriately. The following discussion provides positive and negative aspects of some of these parameters.

Using parameters to describe peaks can be challenging because some of the peaks seen in a measured profile may not be real. It is difficult to have a truly "clean" surface—even the tiniest contaminants can appear as large peaks on a smooth surface. Vibration from the measurement instrument or the environment can create "false" high spots. The challenge with any specification is therefore to use parameters that control the surface without being overly sensitive to these common artifacts.

Another consideration is the actual importance of the higher peaks. In many applications these peaks will be altered or worn away early in the component's life without doing any lasting damage. How much effort should be expended in specifying and controlling them?

Rp and Rpm

Depending on the national or international standard being used, Rp is either the single highest roughness peak in a profile (ASME) or, like Rpm, the average of the highest peaks from each sampling length (ISO)[49]. The single highest peak height is an extremely unstable and non-reproducible value because of the dirt and vibration issues noted above, plus the unlikelihood of two measurements sampling exactly the same features. Rpm is preferable because of its averaging

---

[49] Further information about Rp and Rpm is available in the section, "**What is the difference between Rp and Rpm?**" on page **350**. This discussion will use Rp to refer to the single highest peak and Rpm to refer to the average.

approach, but Rpm is still problematic when surface cleanliness is poor. It requires careful surface preparation prior to measurement and a visual assessment of whether a measurement "makes sense" for the surface being evaluated (see **Figure 148**).

Rmr1, Rpk, Rpq

These three parameters are derived from the material ratio curve and provide some information about profile peaks, although it is much more of an average view rather than a description of the tallest features (especially Rpk and Rpq). More information about these parameters is provided in sections, "**What is material ratio?**" and "**What is the material probability curve and what is it used for?**" on pages **340** and **346**.

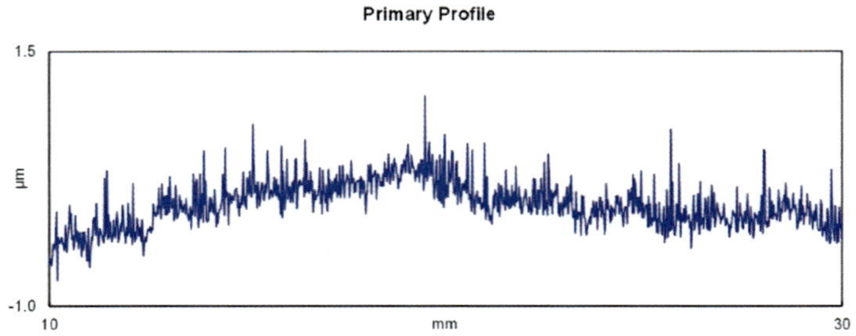

A ground and lapped surface that was poorly cleaned prior to the measurement. The high peaks, much higher than the valley depths, are unlikely for a lapped surface, where the peaks should be similar to or smaller than the valleys.

Figure 148: Debris on a ground and lapped surface

Waviness parameters are much less sensitive to the debris and vibration which can make roughness parameters unreliable. Several parameters that capture *waviness* profile extremes can therefore be quite useful. Note that some of the parameters discussed below are not standardized and may not be available in all commercial software.

Wp

Wp is the highest peak in the *waviness* profile. When using Wp be aware that the highest peak map be at the end of the profile. If this is the case, a longer or shorter trace length could change the value, which makes Wp unstable. For example, measuring into the upward curvature of a fillet can significantly change the Wp value.

Wcvx

Wcvx (waviness convexity) provides a peak measurement that is also dependent on the feature's width or aspect ratio. Wcvx is the height relative to an *opening* profile, rather than the height relative to the mean line used for Wp. The morphological filter used to create the *opening* profile separates "sharp" peaks from those with a more gradual shape, thus identifying potential stress concentration issues. More information is given in the section, "**My part is curved. How do I detect bumps within the curvature?**" on page **239**.

Wseg

A local waviness peak can be detrimental if it lies adjacent to a large drop in the *waviness* profile. This is the basis for Wseg (segment waviness), the largest peak-to-valley distance in a user-defined segment length. The waviness peak's significance is due to this large local change in the surface, which can disrupt a lubricant film. Additional comments about Wseg are included in the section, "**Which parameters relate to lubricated interfaces?**" on page **254**.

Note that these waviness peak measurements are sensitive to the choice of roughness filter cutoff wavelength (lambda-c, or λc). Smaller λc values will often generate higher waviness peak heights because they do a better job of following the profile shape. Further discussion about waviness and λc is available in the section on morphological filters on page **330**.

Robust filter

Surface with extreme features or outliers (either upward or downward) present additional challenges when calculating peak heights. Deep valleys have a large effect on the reliability of *roughness* profile peak height data when using a Gaussian or spline roughness filter (see **Figure 149**). For this reason, a robust filter should be used for surfaces such as these. The section, "**When should I use the robust filter?**" on page **134** provides additional information about the robust filter.

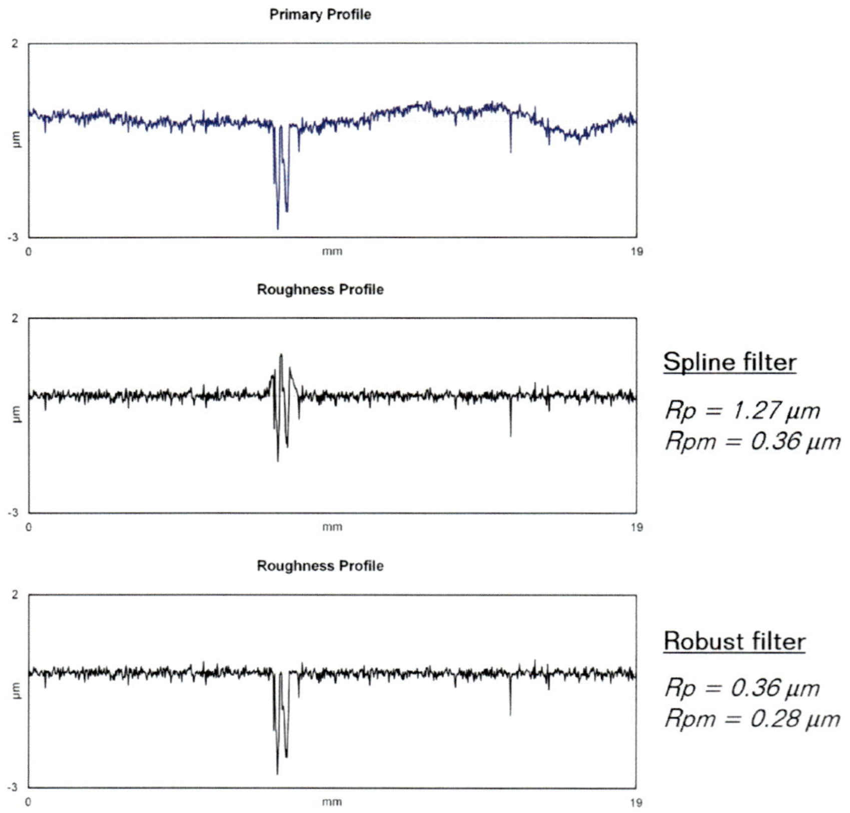

The *primary* profile at the top shows a deep valley in an otherwise smooth surface. Analysis with a spline filter generates a *roughness* profile (center trace) with "false" high spots and large values for the peak parameters. Analysis with a robust filter provides a *roughness* profile (bottom trace) and peak parameters more representative of the actual surface.

Figure 149: Peak parameter values with spline and robust filters

Can I measure around a cylinder instead of along its axis?

Note that this topic is discussed in greater detail in the section, "**My part is curved. Can I measure it?**" on page **230**.

The left side of **Figure 150** shows an accurate measurement made perpendicular to the lay. As the right side of **Figure 150** shows, measuring at any angle other than perpendicular to the lay will make the perceived wavelength of the surface features longer and increase the value of any spacing parameters. If the altered wavelengths are near a filter cutoff value the features may move from the roughness to the waviness domain or, with bandpass waviness, from waviness into form. For more information, see the section, "**How important is sample alignment when I make a measurement?**" on page **79** and **Figure 51** on page **82**.

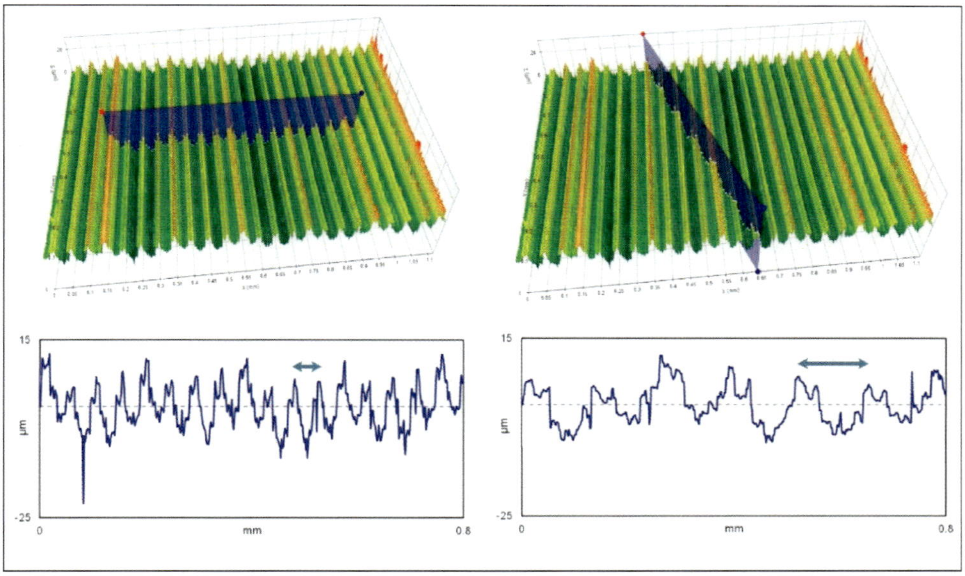

Figure 150: Profile distortion from tracing at angles other than 90° to the lay

It is also important to note that, on a surface with pronounced peaks and valleys, the stylus may be constrained to stay in a valley until the lateral force is sufficient to make it jump over a peak and into an adjacent valley. This can further distort the apparent wavelength content of the surface.

Stylus based traces around smaller cylinders will be typically fairly short because the instrument will quickly go out of range. For example, on an instrument with a 0.8 mm range, the longest possible circumferential trace on a 50 mm diameter cylinder is approximately 12.7 mm (≈ 6.3 mm on either side of the top).

How do I measure the amount of wear on my surface?

The depth of a wear scar can be measured if some of the original surface is available to act as a datum. Ideally, a datum will be available at both sides of the wear so that any curvature within the wear scar can be identified. A trace is made at the appropriate location on the worn sample, starting and ending on the datum surface, and the data is analyzed. An example of this type of measurement is shown in **Figure 151**. The vertical distance between the reference line (aligned with the datum) and the bottom of the wear scar can be measured using cursors or by printing the trace and using a ruler. This example has a maximum wear depth at the ends of the scar of about 6.1 µm. For a worn surface such as this, with curvature at the bottom of the wear scar, a decision must be made concerning what is considered the depth of wear: is it the maximum depth, the depth in the middle, the average depth, or some other value? This choice might be made based on system function, historical data, subsequent calculations (e.g., wear volume), etc.

On a sample with non-uniform wear due to, for example, sample curvature or uneven loading, the location of the measurement must be chosen carefully. If necessary, an approximate determination of the wear volume can be made by making multiple parallel measurements using a precision x-y table to move the workpiece. The wear volume is then determined by integrating the wear area on individual traces (along the x-axis) and then integrating again over the full extent of the worn feature (the y-axis).

If the wear is superficial enough to be less than the total depth of the original surface texture, the wear depth can be determined by comparing the pre- and post-test *roughness* profiles. An illustration of this is shown in **Figure 152**. This procedure assumes a consistent surface finish and/or pre- and post-test traces very close to the same location.

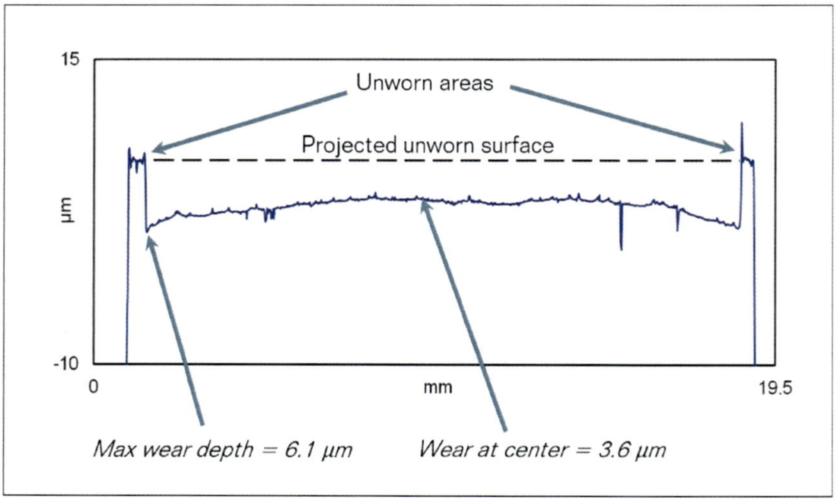

Figure 151: Example of a measurement through a wear scar

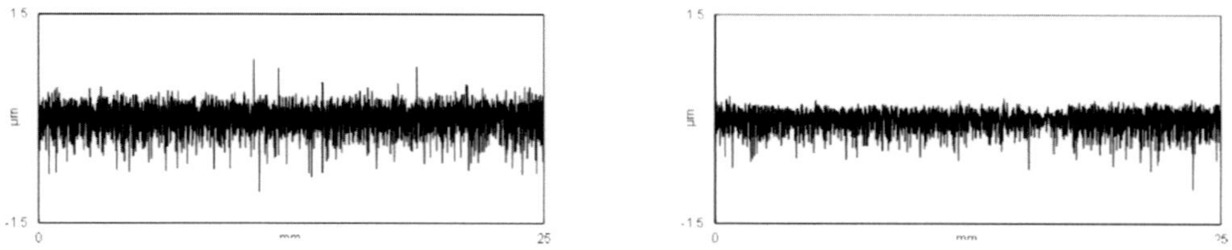

Pre-test *roughness* profile

Post-test *roughness* profile

These traces illustrate how extremely superficial wear may be quantified by comparing the *roughness* profiles pre- and post-test. This test generated approximately 0.4 µm of wear.

Figure 152: Illustration of superficial wear quantified by comparing the *roughness* profiles

The **Figure 152** profiles may lead a person to think that traditional roughness parameters can measure wear. For example, the **Figure 152** pre-test profile appears rougher that the post-test profile. This could lead one to attempt to use the difference in Ra as the "amount of wear." This is very misleading, as the two surfaces have different mean lines, and the common roughness parameters (Ra, Rz, Rpm, etc.) are calculated relative to this mean line. The worn surface sits "lower" than the original surface and the amount of wear is much greater than the simple difference in Ra values[50]. Similar problems exist with attempting to use other roughness parameters as a measure of wear depth. For example, the Rk parameter family is often used to describe surfaces in applications subject to wear (e.g., sliding or bearing surfaces). Rvo, a member of this family, characterizes the volume of the valleys (sometimes called the "oil holding volume" or "crevice volume"). Due to the mathematics involved in calculating the Rk parameters, Rvo can actually increase with wear because the valleys appear to grow as the peaks are worn.[51]

A more quantitative approach for measuring the depth of within-the-texture wear like that in **Figure 152** is based on probability analysis. This was first proposed in 1965 by Williamson[52] as he studied the progression of wear on a surface. At that time, the analysis was performed manually with pencil and paper. **Figure 153** is adapted from Williamson's original data.

---

[50] The Ra difference for the two profiles in **Figure 152** is about 0.03 µm (Ra = 0.13 µm per-test and 0.10 µm post-test). The 0.4 µm wear estimate mentioned in the figure was determined by comparing the two roughness *profiles*, not the Ra *values*. As another example, the two surfaces in **Figure 154** have Ra values of 0.59 µm (unworn, blue) and 0.17 µm (worn, black), a difference of 0.42 µm. However, the probability analysis presented here indicates that the actual wear is closer to 2.2 µm.

[51] Further discussion is available at **https://www.digitalmetrology.com/Papers/Digital_Metrology-Advanced_Wear_Analysis_with_OmniSurf_and_OmniSurf3D.pdf**

[52] J.B.P. Williamson, "Microtopography of Surfaces," *Proceedings of the Institution of Mechanical Engineers*, Vol. 182, Part 3K(1967-68): 21-30

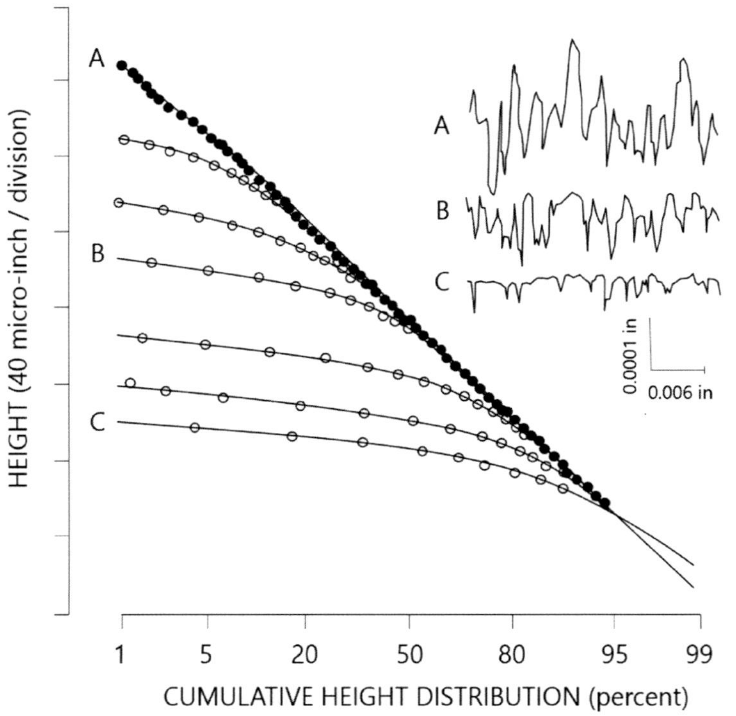

The surface texture traces (upper right) represent successive measurements of a surface that is undergoing wear. The probability curves all converge in the portions representing the original unworn surface (some of which is still present in all traces). The slope of the dark line is the Rq (RMS roughness) of the original surface; the extrapolation of the line to the y-axis provides a prediction of the height of the unworn surface.

Figure 153: Example of probability analysis of wear from Williamson's original paper[52]

With today's software and the analysis method used for the Rpq, Rvq, and Rmq parameters we can easily construct the curves for the analysis of wear depth—see **Figure 154**. All of this assumes, of course, that the wear depth is shallow enough that some of the valley portion of the original surface still exists.

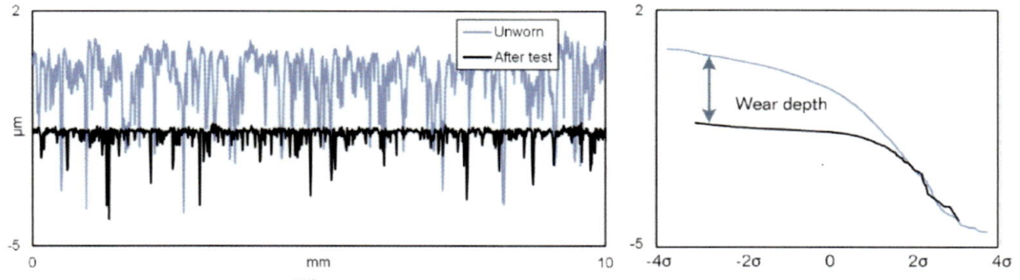

By aligning the valley regions on the material probability curve (right) we can position the *roughness* profiles (left) to visualize the depth of wear. The actual wear depth can be determined by the upper regions of the material probability curves. The x-axis scale of the probability curve is in units of standard deviations.

Figure 154: Wear depth analysis curves based on probability analysis

My part is curved. Can I measure it?

Many critical surfaces have an arc or arc-like geometry with a relatively large vertical height. Examples include ball bearings, bearing races, balls and sockets, and optical lenses. Such components require well-controlled surface roughness and waviness, but it can be challenging to obtain accurate measurement data with stylus-based instruments. This is due to several fundamental issues resulting from the characteristics of the measuring equipment and the analysis of the data. These issues include surface texture lay, vertical resolution, arcuate motion errors, and the filtering of data that still contains the nominal shape.

Errors due to surface texture lay

Accurate measurements of surface roughness require tracing across the lay, not parallel to it. However, the machining of a curved surface often generates a lay that follows that curvature (turning or grinding, for example) so a trace of the curvature will be parallel, or nearly parallel, to that lay. Data from that trace can be misleading. If the stylus dwells in a trough, the roughness features will be hidden. More commonly, the trace will cross only a few peaks and troughs and the distance between features will be distorted, as shown in **Figure 155**. This will obviously impact any feature-spacing parameters, but it may also affect the magnitude of height-based parameters if the measured spacing (wavelength) results in those features being moved from the roughness to the waviness domain. Roughness parameters will then be underestimated, and waviness parameters can be impacted as well.

Solution options:

- Evaluate roughness by tracing across the lay. If the part is a shaft, you could measure the form by making a circumferential trace and measure the roughness with an axial trace.

- Use a 3D (areal) measurement.

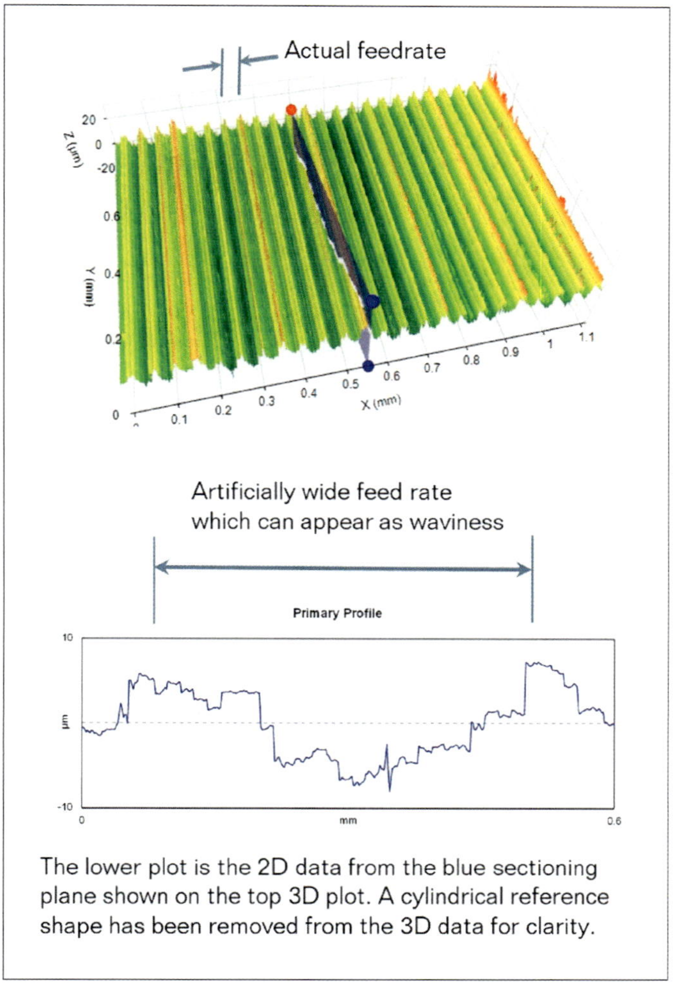

Figure 155: Distortion of feature spacing due to measuring approximately parallel to the lay

Errors due to resolution

Tracing a curved surface may require a fairly large vertical range to obtain data over an adequate length (see **Figure 156**. However, a typical surface texture instrument has a limited number of vertical resolution "steps" within a given range, so a large vertical range will result in larger steps and poorer vertical resolution[53]. This can result in a profile with "digital steps" (**Figure 157**). The unleveled *measured* profile (top plot) shows these steps to be horizontal; however, once we level the profile (bottom) these steps become tilted.

[53] More information about range and resolution is provided on page **371**.

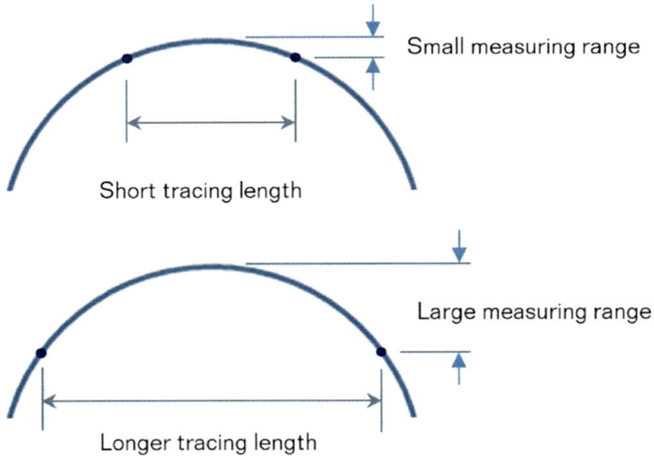

Figure 156: Relationship between tracing length and vertical range on a curved surface

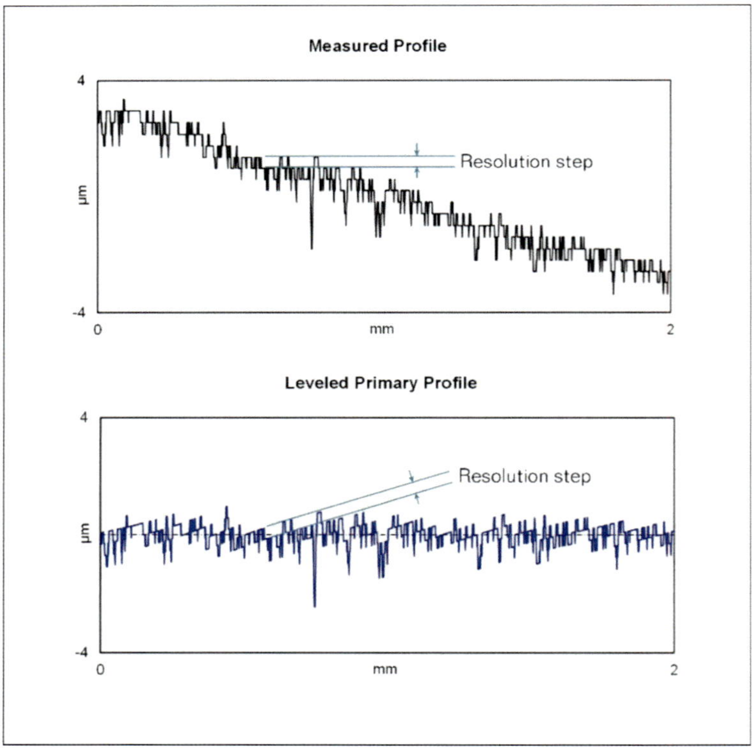

Figure 157: Digital steps in a profile with poor resolution data

Figure 158 (a) illustrates a measurement of a convex surface with this poor resolution. When this curved data is leveled and filtered, these "resolution steps" may appear as triangular roughness

features, as in **Figure 158** (b). For some instruments, this poor resolution enters their calculations and can affect the parameter values.

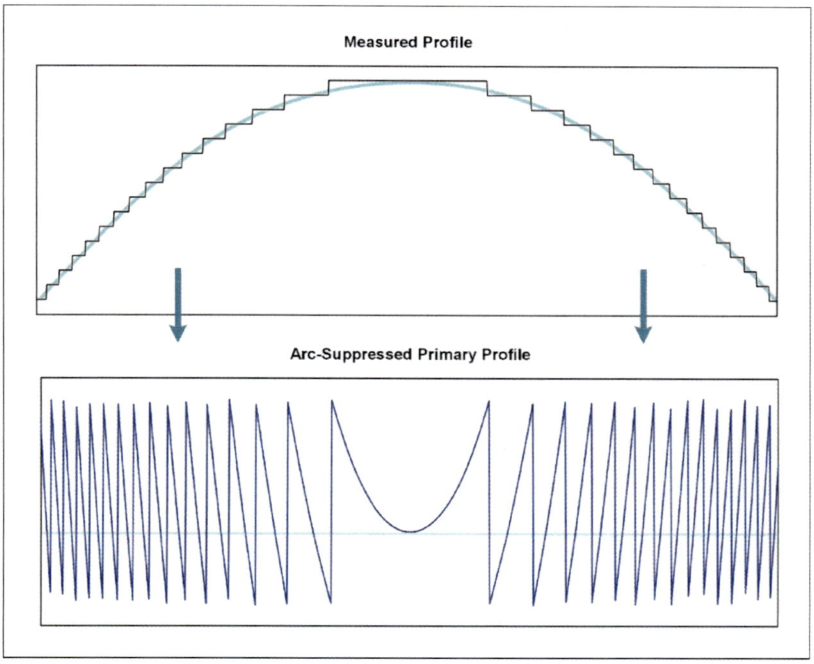

Figure 158: Vertical steps appearing as roughness after form removal

Solution options:

- Use an instrument with a larger range-to-resolution.
- Measure roughness using multiple, short traces, with trace lengths short enough to cover a smaller vertical range. These short traces will very possibly not provide an accurate assessment of waviness. If both roughness and waviness are important, it may be necessary to use multiple short traces to assess roughness and a separate longer trace to assess waviness. Waviness parameters are not as sensitive to the poorer vertical resolution.

Errors due to arcuate motion

The stylus of a profiling instrument does not move vertically but rather along an arc whose radius is the distance between the stylus tip and the pivot point. The actual stylus travel is based on this "arcuate motion" whereby the tip moves along the horizontal (x) axis as a function of its vertical (z) position, as in **Figure 159**. This motion produces a distortion in the measured shape so that it is not truly representative of the actual surface.

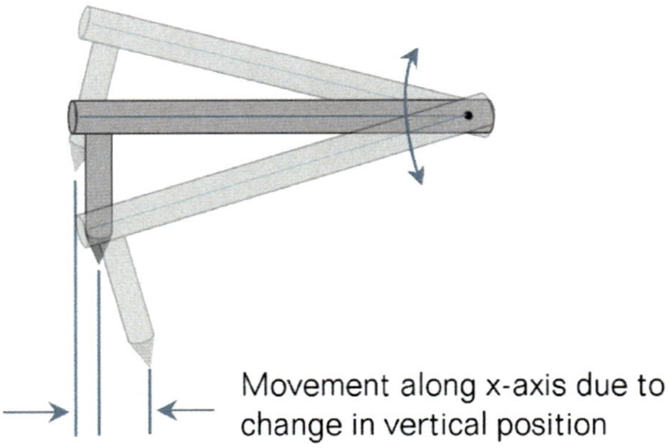

Figure 159: Arcuate motion

Consider a convex shape, as below in **Figure 160** (a). As the stylus moves from left to right, the stylus shifts to the left as it climbs, as in **Figure 160** (a). This has the tendency of lagging behind and taking longer to climb up the radius. Similarly, as the stylus falls down the right side of a convex shape, as in the right side of **Figure 160** (a), it shifts to the right, again resulting in distortion in the measurement. The same error occurs on a concave surface, but in the opposite direction, as in **Figure 160** (b). The general form of the distortion is shown in **Figure 161**, assuming the stylus is moving left to right.

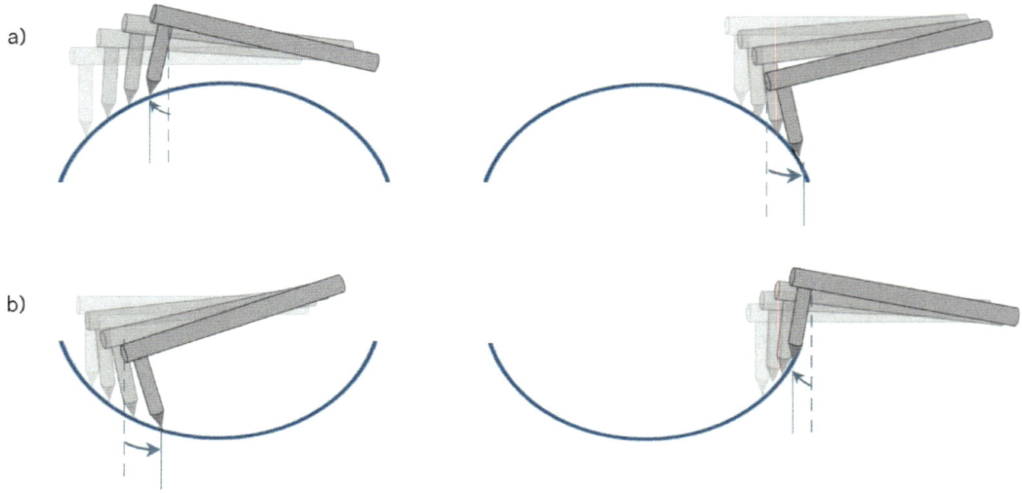

Figure 160: Arcuate motion of a stylus tracing a curved surface

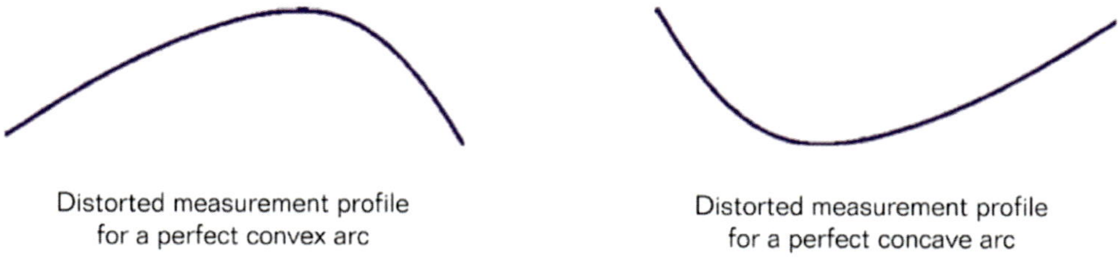

Figure 161: Measured profile distortion due to arcuate motion

The size of the error from this arcuate motion depends on the shape of the surface, the length of the measurement, and the dimensions of the stylus (distance from the pivot point to the stylus tip). **Figure 162** is a mathematically simulated example in which the measured profile drop has an error of over 2 mm on a 10 mm radius arc.

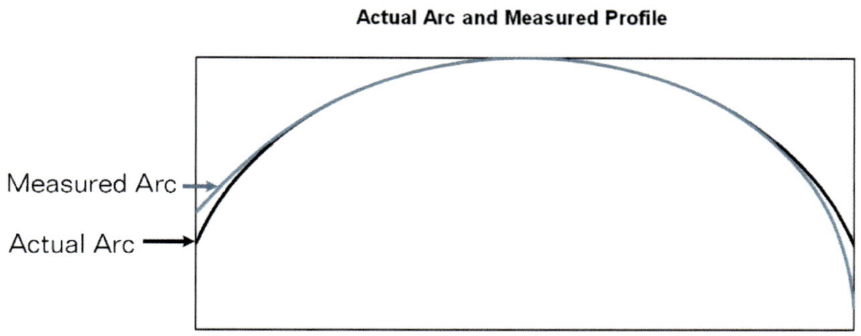

The black line is the actual shape, a perfect semicircle with a 10 mm radius. The blue line is what the measured profile would be on an instrument with no arcuate error correction using a stylus that is 50 mm long (the distance from the pivot point) and 10 mm high. Over the 19 mm long measurement simulated here, the maximum error in measured drop (position in z) is over 2 mm.

Figure 162: Simulation of arcuate motion induced error

One way to determine if arcuate error is affecting your results is to make a measurement, rotate the measured component 180°, and re-measure. If the shape remains constant (for example, the right side is always steep), then the instrument is causing the error. If the shape moves with the component, then the shape is likely attributable to the component geometry.

Solution options:

- The only way that arcuate error can be fully addressed is to use an instrument that calibrates and corrects for the arcuate motion of the stylus. Most common roughness instruments set a single gain value by using a roughness patch. These instruments cannot correct for arcuate motion as that correction requires more than a single gain number. Instruments that can correct for arcuate motion are calibrated with precision spheres/hemispheres or precision angles/prisms.

- The amount of arcuate error can be reduced by using a longer and/or taller stylus (larger distance from pivot point to tip). For the example in **Figure 162**, the amount of measurement error (measured drop) is reduced by over 60% by using a stylus with a length of 100 mm rather than the 50 mm that was originally used. However, with many instruments this longer stylus will negatively impact the vertical resolution, with the resultant roughness errors discussed above.

- Note that the error is also significantly reduced if a shorter trace can be used. The example in **Figure 162** simulated a 19 mm trace over the 10 mm radius curve. If, for example, a 10 mm trace would provide adequate information, the amount of error in the measured drop is reduced from over 2 mm to only 10 µm.

Errors due to filtering a curve

When measuring a curved profile it is important to remove the curvature before filtering for roughness and waviness. Some instruments do not have the ability to suppress advanced geometries such as arcs or polynomials, and this can result in a shift in the roughness profile when using a traditional Gaussian filter.

To understand this shift, consider the process used to create the *roughness* profile. A filter is applied to the *primary* profile to create the *waviness* profile. This *waviness* profile is then subtracted from the primary to provide roughness (see page **114** for more information on filtering). This filter is typically a Gaussian-weighted moving average. However, as shown in **Figure 163**, passing a moving average over a convex surface will produce an average that is biased downward (it will be biased upwards with a concave surface). Subtracting this biased *waviness* profile from the primary will produce a *roughness* profile that lies above the zero line (for a convex surface), with unrealistic values for the roughness parameters. A real-world example is shown in **Figure 164**.

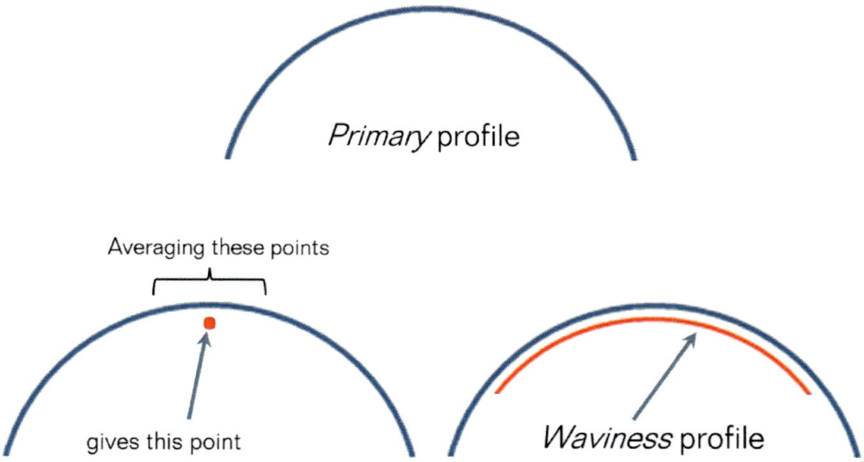

Figure 163: Illustration of the error produced by filtering a curve

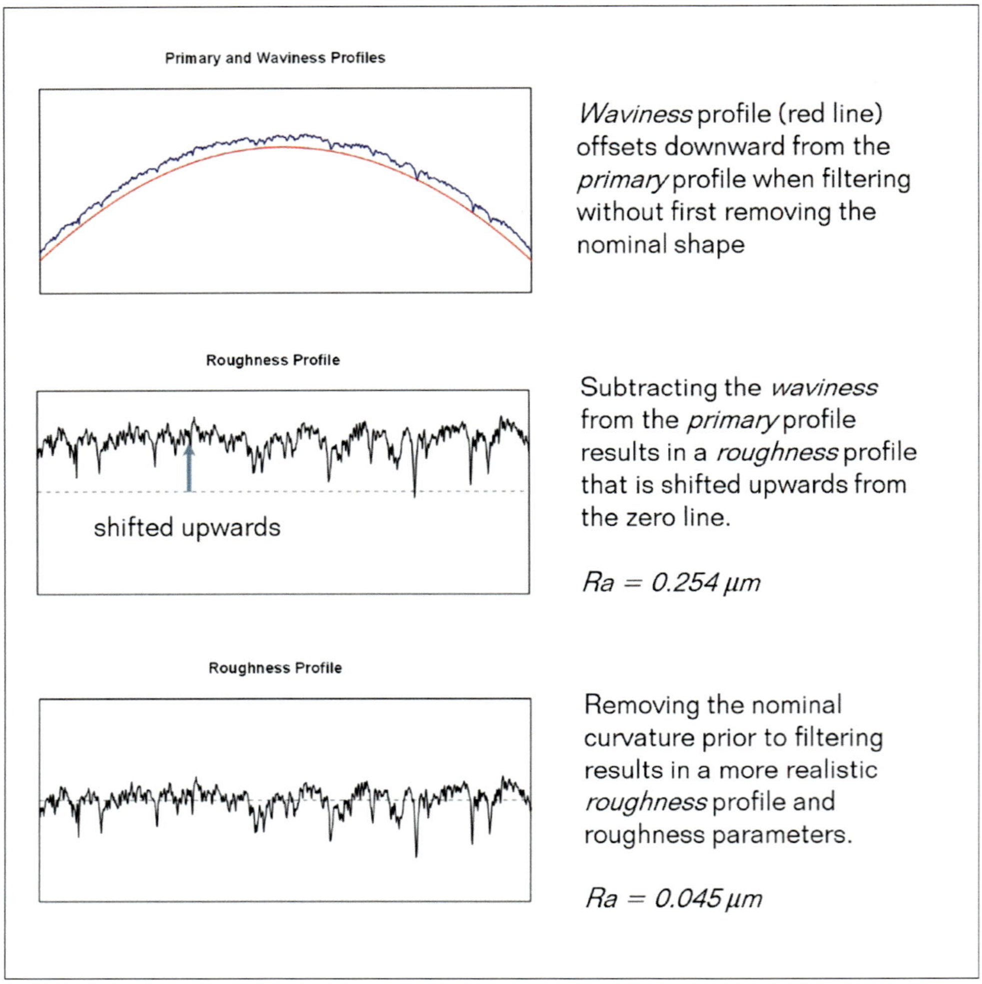

Figure 164: Error induced by filtering a curve—actual data

Solution options:

- The best option is to remove the nominal geometry prior to filtering. If your measuring instrument does not have this capability you should consider using third-party software that does.
- Consider using a "second order filter," which is currently being standardized by ISO, in a future edition of ISO 16610. These filters suppress curvature as part of the filtering process.

My part is curved. How do I detect bumps within the curvature?

Local high spots or bumps can act as stress risers, interrupt an oil film, or in other ways be detrimental to a component's function. Identifying and quantifying those bumps can be challenging, especially if they occur on a surface that has some overall curvature by design (e.g., gear teeth, rolling contact components). In these cases, the height of the detrimental "bumps" may fall completely inside the overall peak-to-valley height of the surface (see **Figure 165**). A bandpass waviness analysis may be able to find them, but often this requires prior knowledge of their sizes and shapes. Furthermore, a bandpass waviness analysis often places the reference (*form* profile) line through the middle of the bumps and doesn't retain their full "outward" appearance, their true magnitude (**Figure 166**).

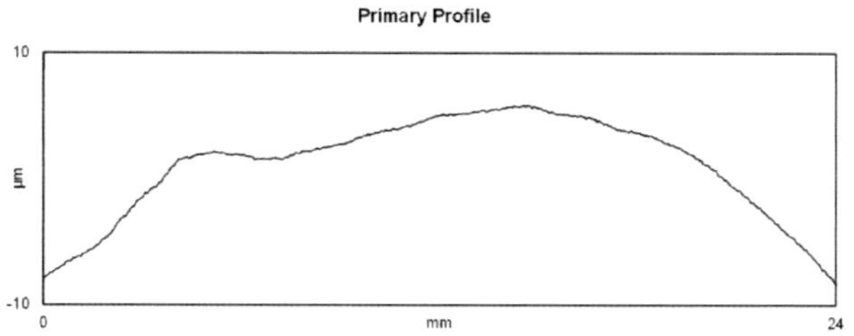

The bump on the left side of this profile is totally contained within the overall profile height and is therefore invisible to a normal one-sided waviness analysis (e.g., Wt).

Figure 165: Profile of a nominally-convex surface with a local bump

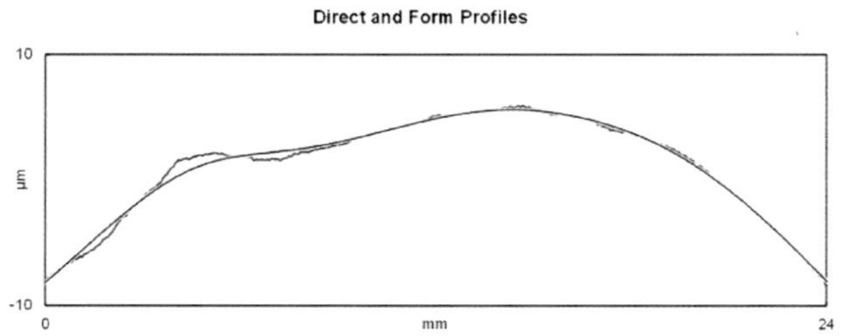

The form profile (black line) that would be used for a bandpass waviness analysis goes through the middle of the bump, reducing its "outward" extent.

Figure 166: The *form* profile is inadequate for "bump" analysis

A better option is to use a parameter such as Wcvx (waviness convexity). Wcvx is analyzed using a morphological opening filter: a virtual ball or disk is passed under the *waviness* profile, with the upper boundary of the ball creating the *opening* profile (**Figure 167**). The maximum vertical distance between the *opening* profile and the *waviness* profile is Wcvx.

Proper selection of the radius of the virtual ball allows the analysis to detect "sharper" bumps that would be detrimental while ignoring the wider features, which are less likely to induce excessive stress. A complete Wcvx callout therefore includes the radius of the virtual ball and the limit on the Wcvx parameter.

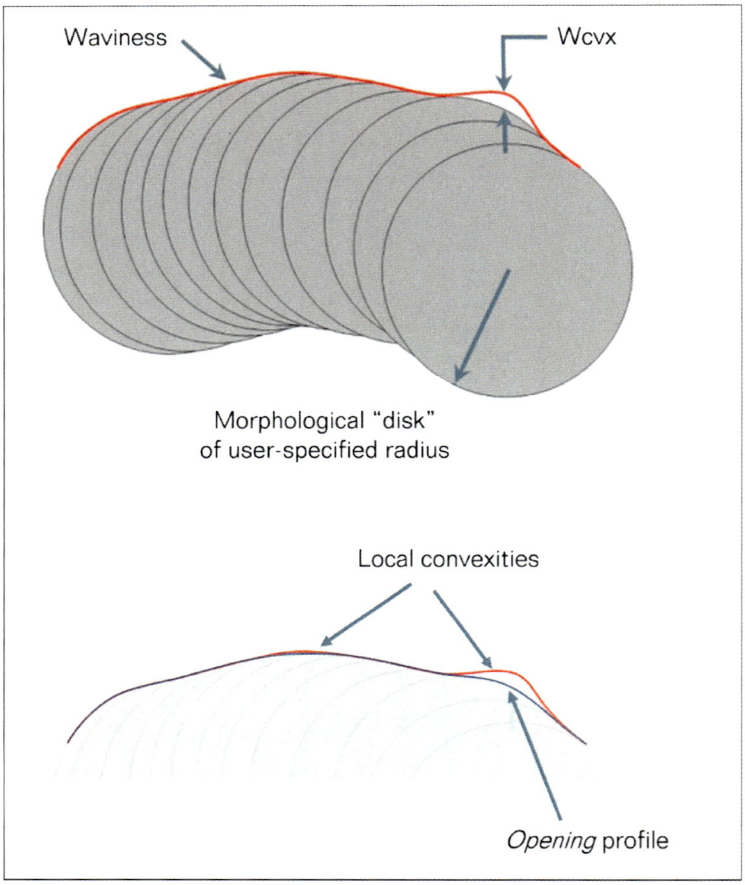

Figure 167: Schematic of a morphological opening filter for evaluating Wcvx

The *opening* profile for the surface in **Figure 165** is shown in **Figure 168** with the *waviness* profile, the typical starting point for this type of analysis.

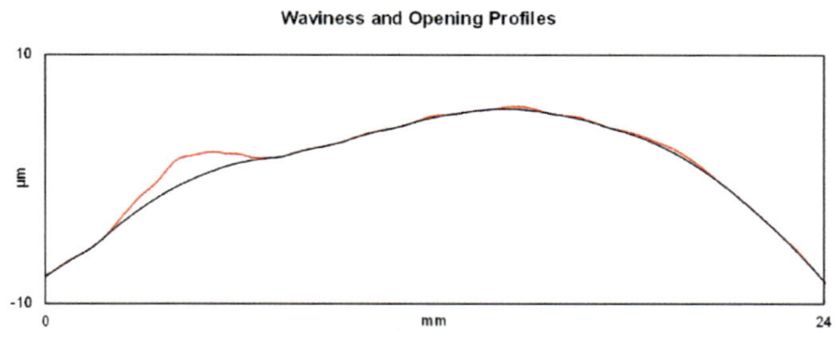

The *opening* profile (black line) calculated using a 2500 mm radius ball, applied to the *waviness* profile (red). This captures the true extent of the protrusion of the bump above the nominal convex profile. *Wcvx = 2.2 µm*

Figure 168: *Waviness* and *opening* profiles illustrating the Wcvx analysis

More information on opening profiles and morphological filters is given on page **330**.

Since this Wcvx-based analysis is performed on the *waviness* profile, the appropriate waviness filters must be selected to create a *waviness* profile representative of the actual surface. **Figure 169** illustrates how this choice of the roughness filter cutoff wavelength (lambda-c, or λc) affects how well an *opening* profile characterizes a surface. With λc = 0.8 mm (top) the *waviness* profile does not follow the actual surface shape. Setting λc to 0.25 mm generates a *waviness* profile more representative of the actual shape, resulting in significant waviness peaks which are detectable by Wcvx.

Note that Wcvx is not yet part of a national or international standard, so it may not be available on all commercial instruments.

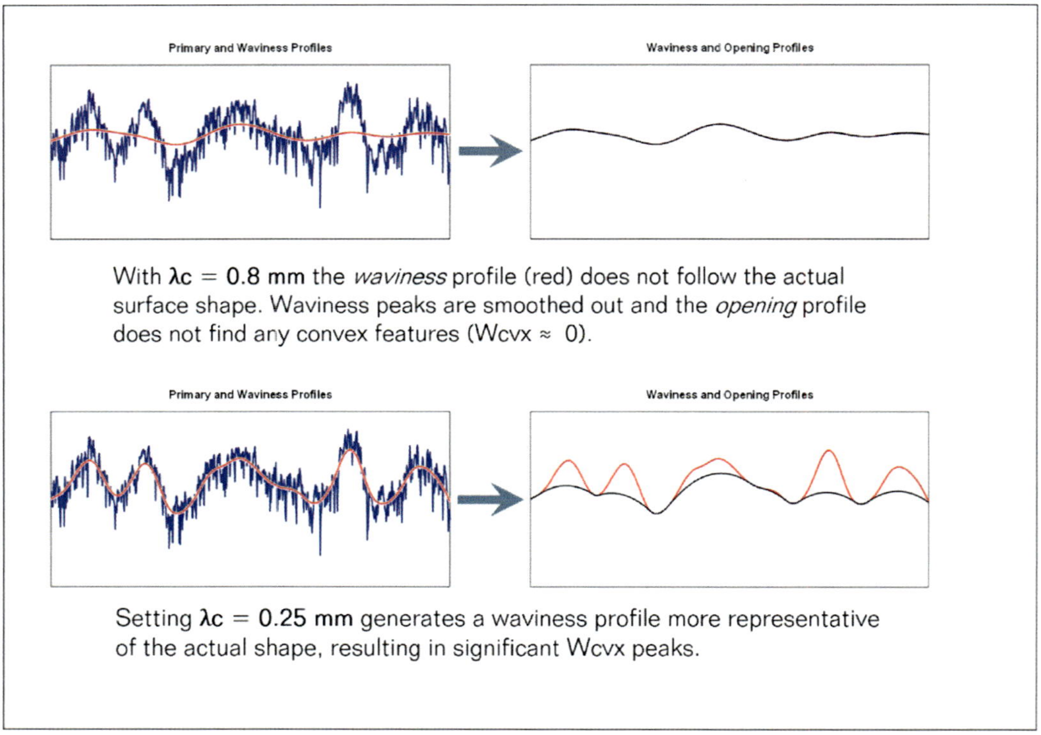

Figure 169: Effect of λc on Wcvx

Why does my gasket leak even though the roughness is correct?

Roughness is not always a good predictor of performance because other surface texture features may have a larger impact on a component's function. Gasketed surfaces and other conformable interfaces are excellent examples of this; they often require a well-controlled waviness. This illustrates the importance of examining the *primary* profile instead of just the *roughness* profile if one wishes to thoroughly understand a surface.

Figure 170 shows an example of two surfaces, one that leaks and one that doesn't. The left-hand *primary* profile shows a periodic feature from the manufacturing process (milling) that might explain the difference in behavior. Unfortunately, Wt, the common waviness parameter, is nearly identical for both surfaces, and other parameters like Wa, Wc, and Ra are actually higher for the surface that seals than for the one that leaks.

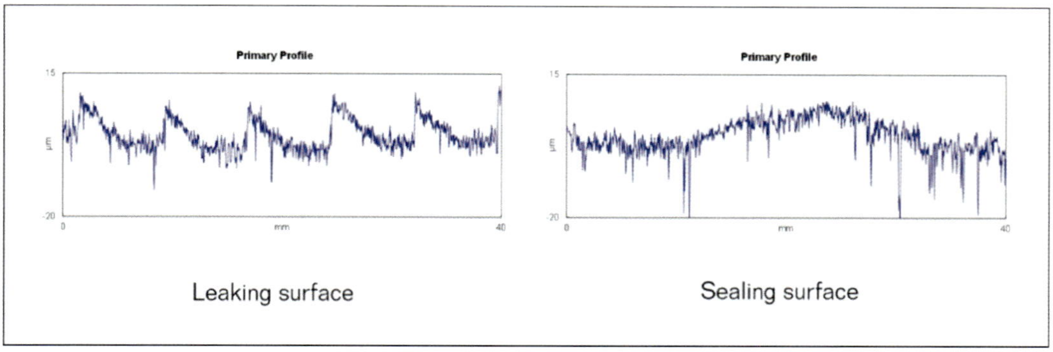

Figure 170: Examples of leaking and sealing surfaces

Figure 171 shows the application of a morphological filter to create a *closing* profile and related waviness parameters that do differentiate between the two surfaces. The morphological filter can be thought of as a ball or disc of prescribed radius that rolls over the surface—in this case, over the *waviness* profile—to create the *closing* profile. The distance between the *waviness* profile and the *closing* profile is analogous to a gap between an insufficiently-conformable mating surface (e.g., a gasket) and the milled workpiece. If this gap is too large, the interface will leak. The amount of leakage will be related to the total gap area over the length of the interface[54], shown as the shaded area in **Figure 171**. The ball radius is related to the conformability of the gasket, with a smaller radius simulating greater conformability of this "virtual gasket."

Since this analysis is performed on the *waviness* profile, the short-wavelength waviness filter (lambda-c, or λc) must be chosen so that the *waviness* profile properly represents the surface. This

[54] The amount of leakage will also be affected by factors such as the pressure differential across the interface, the viscosity of the fluid, the chemical properties of the fluid and the surfaces, and the aspect ratio of the gaps. The gap area does not need to be zero as long as the combination of all these factors creates an acceptable leakage rate (which may or may not be zero).

may require a different λc value than would normally be used for the roughness and waviness analysis. More information on the choice of λc, and on morphological filters, can be found on page **330**.

Wvoid (waviness void area) and Wvdd (waviness void depth) have been proposed[55] as waviness parameters that can be used to quantify gasket behavior and to create a specification. Wvoid is the normalized void (leakage) area, the area per unit length enclosed between the *closing* and *waviness* profiles. Wvdd is the maximum depth of the *waviness* profile relative to the virtual gasket. As seen in **Figure 171**, Wvoid correlates to the behavior of the two sample interfaces. In this analysis the ball radius is 5000 mm, but experimental or analytical work can be envisioned to determine an appropriate ball radius that quantifies the conformability of any type of gasket. The influence of conformability is illustrated in **Figure 172**, where the amount of leakage area (shaded area) is greatly reduced in the lower half of the figure.

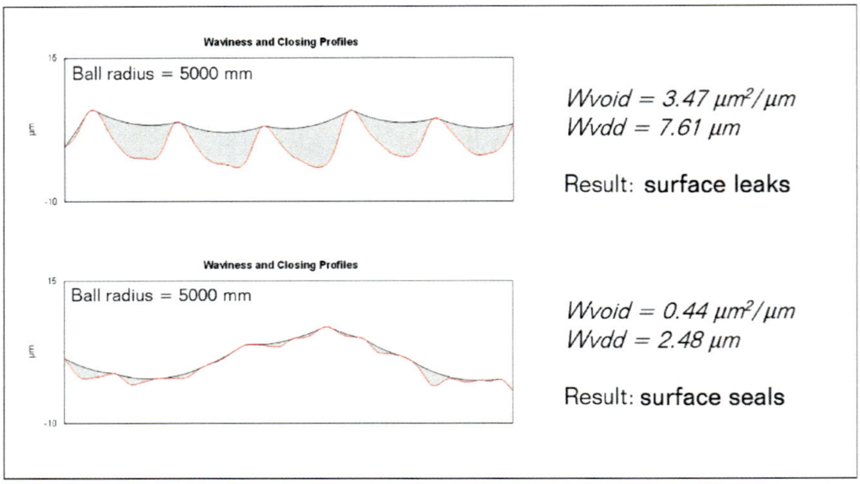

Figure 171: Morphological closing filter analysis of leaky and sealing surface

[55] Mark Malburg, "Surface Profile Analysis for Conformable Interfaces," *J. Manuf. Sci. Eng.* 125(3) (Jul 23, 2003): 624-627

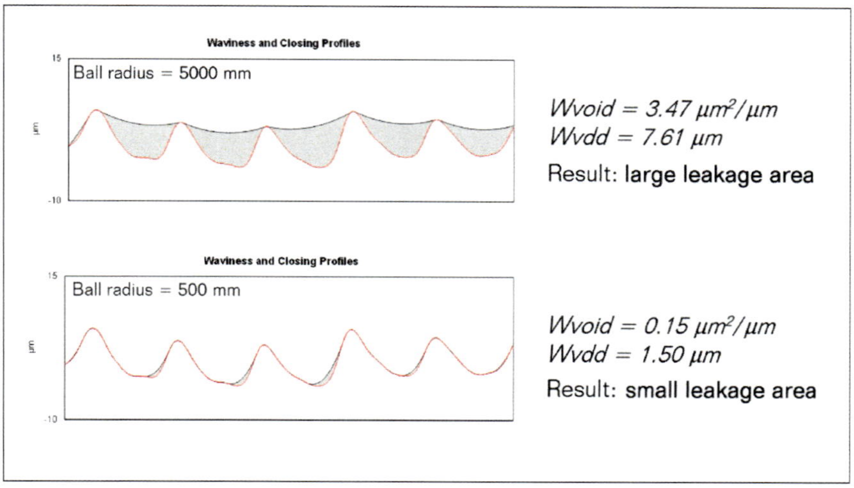

Figure 172: The morphological ball radius simulates gasket comformability

While Wvoid and Wvdd have been shown to help predict gasket behavior, additional characteristics from this analysis of the *waviness* profile may also be helpful. These might include:

- The average void area
- The average void depth
- The number of voids per unit length
- The average void aspect ratio (depth : width)

It should be noted that some interfaces can have leakage in what is traditionally considered the roughness domain. One example would be in the sealing of gases. In these cases, the Wvoid concept can be extended to roughness scale features.

The 2D analysis shown above has been proven to help identify surfaces that will leak. However, for some types of surfaces the voids highlighted by the *closing* profile could be isolated pits rather than leakage paths connecting an "inside" to an "outside" portion of the system. 3D/areal data may be more useful for identifying actual leakage paths in these situations—see **Figure 173**.

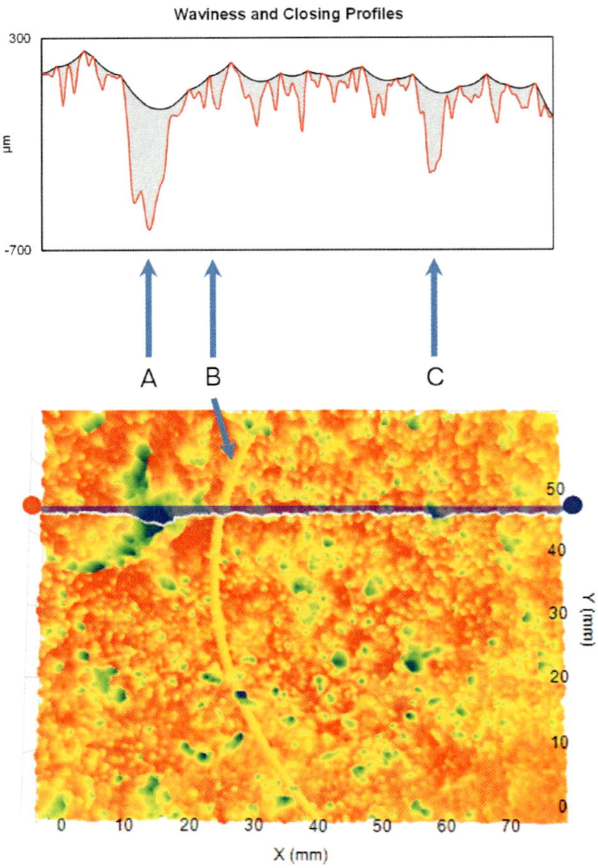

2D data—waviness and closing profiles. Shading indicates the potential leakage areas.

3D data showing the location of the 2D trace. In this image, blue and green are low areas while red indicates high regions.

A 2D measurement of this surface (top) implies that features "A" and "C" could be the worst features for leakage. 3D data, however, reveals that "A" and "C" are isolated pits. Feature "B," even though shallower, could be a potential problem as it spans all the way from the bottom to the top of the surface.

Figure 173: 2D and 3D data from a leaking surface

Which parameters should I use to describe a plateaued surface?

Plateaued surfaces are typically comprised of two textures, so a single parameter cannot adequately describe their shape. Furthermore, the extent of the plateauing will dictate which parameters are most appropriate for describing the surface.

Figure 174 shows three example surfaces with varying degrees of plateauing. Surface (a) is similar to that of "peak honing," or traditional honing, performed in a single machining step. There may be a few deeper valleys but nothing significant. This surface can typically be described using the traditional Ra, Rz, and Rmr parameters.

If surface (a) is used, for example, for an internal combustion engine cylinder bore, the run-in or break-in process during early engine operation will wear the peaks. This will create debris and changes to the surface geometry in a mostly uncontrolled manner. The final surface may not be ideal for the required lubrication, wear, and leakage control. Engine manufacturers developed plateau-honing processes to provide a more functional and durable surface in a more controlled (manufacturing) environment, using a rough-honing operation to create the valleys, followed by a finer hone to create the peaks/plateau. A plateaued surface such as this is shown in profile (b) of **Figure 174.** The traditional parameters become unreliable at distinguishing this surface from the single-honed surface (a). Very different surfaces often have very similar values of Ra, Rz, etc.

Plateaued surfaces such as that shown in **Figure 174** (b) drove the development of the Rk family of parameters. These parameters are based on the material ratio curve, shown to the right in **Figure 174**. The various parameters (Rpk, Rvk, Rk, etc.) separately quantify the peaks, valleys, and central (kernel) regions, and allow better specification and tighter control of these individual geometries and of the related functional behavior.

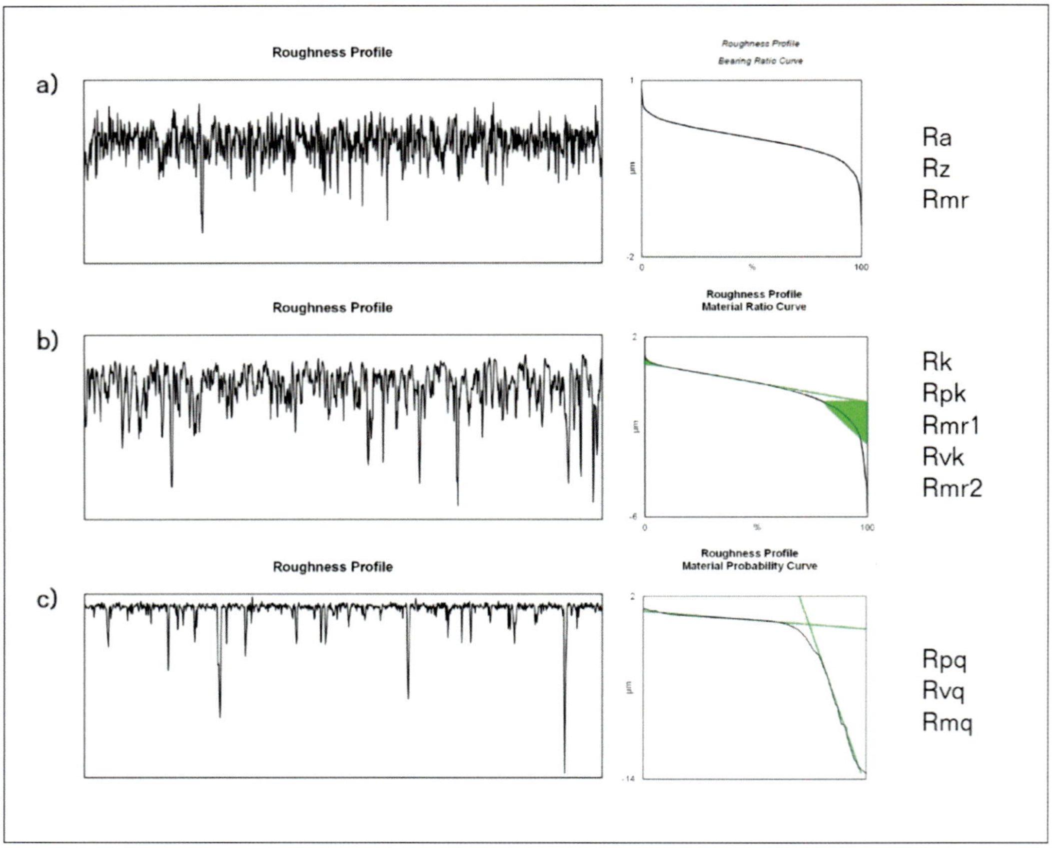

Figure 174: Surfaces with varying degrees of plateauing

Higher performance applications, such as diesel and racing engines, can require even tighter control of the initial surface to minimize dimensional and texture changes during break-in. For these applications, manufacturers create very smooth plateaus with discrete valleys to optimize friction, lubrication, and clearance, often effectively eliminating the break-in period. The primary characteristic of these surface is a strong distinction between the peaks and the valleys, a clear visual indication of where peaks meet valleys, as in **Figure 174** (c).

Unfortunately, the Rk parameters are not adequate for characterizing these surfaces, particularly due to their inability to capture the position on the material ratio curve where the plateau transitions to valleys. The Rmr2 and, subsequently, the Rvk parameter become unreliable. This can be seen in **Figure 175**. The left-hand corner of the green triangle on the material ratio curve represents the location of Rmr2, and it can be seen that the choice of this location is not visually obvious. This difficulty led to the development of the "q" family of parameters.

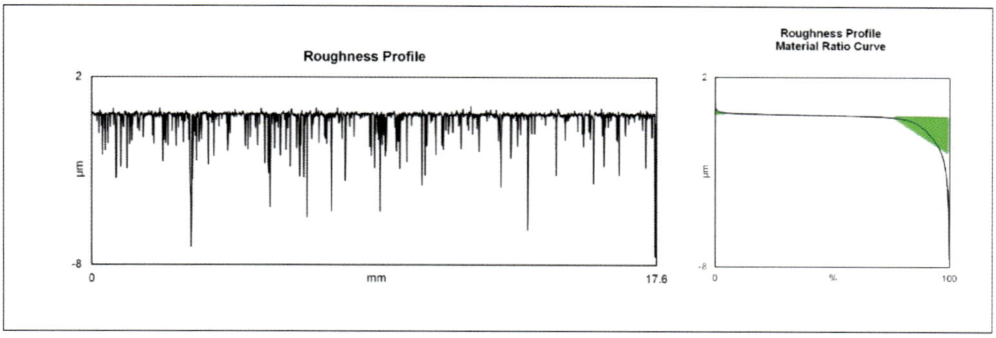

Figure 175: Material ratio curve for extremely plateaued surface

The “q ”parameters—Rpq, Rvq, and Rmq—are derived from the material probability curve, the right-hand graph in **Figure 176** and **Figure 174** (c). The material probability curve is the material ratio curve with the percentages mapped to standard deviations. When represented in this manner, the two portions of the profile—the plateau and the valleys—appear as two clearly linear regions, with a sharp transition or knee in the curve. The "q"parameters derived from this curve can robustly characterize the plateaued surface, and can distinguish the extremely-plateaued surface from the other types of honed surface.

Further discussion about the material probability curve can be found on page **346**. The "q" parameters are described in ISO 13565-3.

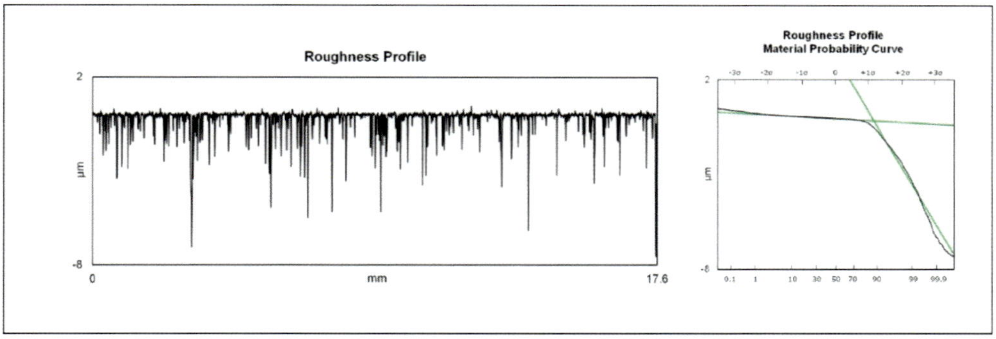

Figure 176: Material probability curve for extremely plateaued surface

To summarize, Ra and Rz can be useful for surfaces with a wide distribution of peaks, valleys, and intermediate heights. The “k” parameters work well when extreme plateaus are not required. The "q" parameters work best when the surface has very distinct plateau and valley regions.

My surface is porous. How can I control the roughness?

When a surface is porous[56], traditional roughness parameters such as Ra and Rz can be artificially high and not representative of the actual machined surface—see the upper profile in **Figure 177**. If the porosity was not present, as in the center section of this profile, we would see much smaller roughness values. These smaller values are more indicative of the actual machining process, as shown in the bottom profile.

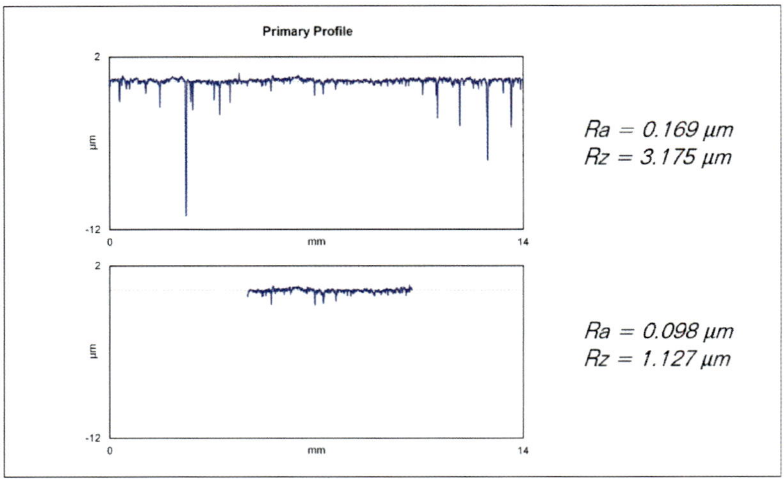

Figure 177: The effect of surface pores on Ra and Rz

There are a few approaches that may be acceptable for dealing with this scenario.

1. Rely on the 16% rule. This will allow some of the traces to be outside the tolerance limit.
2. Remeasure when encountering porosity of a specified depth.
   Note: this remeasurement procedure should be documented in a control plan.
3. Manually exclude porosity regions from the profile. Note: this procedure should be documented in a control plan.
4. Use an alternative parameter. Possibilities that could be considered include:
   - Rk, the "core/kernel" roughness which does not include the porosity
   - R3z, an alternative to Rz that ignores the two deepest valleys in each sampling length
   - Rpq, the plateau roughness.

[56] This porosity may be an inherent characteristic of the material (e.g., casting porosity, porosity in a less-than-theoretically dense powder metal or ceramic) or it may be a consequence of the machining process (e.g., graphite pullout when machining cast iron).

Which parameters specify surfaces with functional requirements?

Surface texture parameters and limits are the language with which one defines the micro-topographical aspects of a functional surface. These parameters and limits are communicated to the people that manufacture and inspect it. A designer must always be mindful of the needs of these other participants in this process. However, a designer's first goal is to specify something that functions, so the selection of parameters and limits should ideally be directly related to the surface's function.

Note that, while a definition of function is key, the number of parameters in a specification should be minimized as much as possible, as it can be difficult to set up manufacturing processes to comply with multiple parameters. Use only as many parameters as required to uniquely define the surface.

In addition to relating to function, parameters should show the appropriate sensitivity to manufacturing process variability. They should not exhibit an extreme or erratic change for normal process variation, but they should show a very noticeable change when the surface approaches its functional limit. For example, controlling the height of the single highest peak (e.g., Rp ASME) can be very difficult for those working in a manufacturing environment and dealing with dirt/debris and vibration.

Creating a design based on a model of the component is becoming more common and is a worthwhile goal. However, our ability to model surfaces and predict functional relationships from first principles is limited, especially the modeling of roughness. Experience may therefore be the best guide for selecting design parameters, even if that selection isn't perfectly logical. When designing a system that will utilize standard components, such as gaskets or o-rings, the recommendations of that supplier should be carefully considered. If time and resources permit, an iterative design process is also helpful, with testing of components that do and do not work. Measurements of "good" and "bad" parts, and of parts before and after test, can highlight the parameters that predict the outcome and the limits between function and failure. This specification development process is discussed further on page 273.

In additional to supplier recommendations, standards and other technical literature may provide guidance. For example, Table J-1 in ASME B46.1-2009 provides parameter suggestions, referencing industry or national standards.

When considering parameter choices, the specific parameter may often be less important than what the parameter represents. For example, it may be that roughness peaks are important and need to be controlled. These features can be controlled by a number of different parameters, such as Rpm (ASME), Rp (ISO), Rpk, or Rpq. The selection of the specific parameter may be influenced by factors such as cost, availability in the measurement system, and experience. But, at the end of the day, what matters most is that the parameter you choose actually controls the desired features.

The following sections discuss parameter options for various functional requirements.

Which parameters relate to load carrying ability?

A key consideration when designing a load-carrying surface is to minimize stress concentrations. The type of loading will determine how this is done. With bending or torsional loading, deep valleys are of concern. Usually, valleys in the *roughness* profile are too small to be significant, and control of larger features such as cracks, notches, and blends is adequate. However, in more critical designs where success depends upon the roughness valleys, those valleys might be controlled by a parameter such as Rvc (valley count), with the valley count threshold related to the severity of the loading. Average values such as Rvm (average maximum valley depth) may be useful. If individual flaws are significant, Rv (maximum valley depth) or Rt (maximum peak-to-valley height) might be appropriate, although it must be understood that extreme value parameters such as these will show significant measurement variability, and are therefore difficult to control.

The depth of larger valleys can be controlled with a morphological closing filter and a waviness parameter such as Wvdd (waviness void depth). Specify a small ball radius for the morphological filter, typically on the order of two or three millimeters, depending on the shapes of the valleys you're concerned about. Select a roughness filter cutoff wavelength (lambda-c, or λc) that generates a *waviness* profile which accurately captures the shape of the valleys of interest, noting that this λc may be smaller than what is commonly used to quantify roughness or waviness. More information about morphological filters and the selection of λc can be found on page **330**.

If loads are applied by surface-to-surface contact, high spots will concentrate the stresses. Roughness peaks are typically not of concern (except in lubricated contacts—see below), but waviness peaks should be considered. Since the amount of stress concentration depends on both the waviness height and width (peak curvature), a parameter such as Wcvx (waviness convexity) can be useful (see **Figure 178**). If Wcvx isn't available it might be possible to use some combination of wavelength and amplitude waviness parameters, perhaps Wsm (waviness mean spacing) and Wc (waviness mean profile height), with appropriate discrimination levels). However, this approach has not been proven.

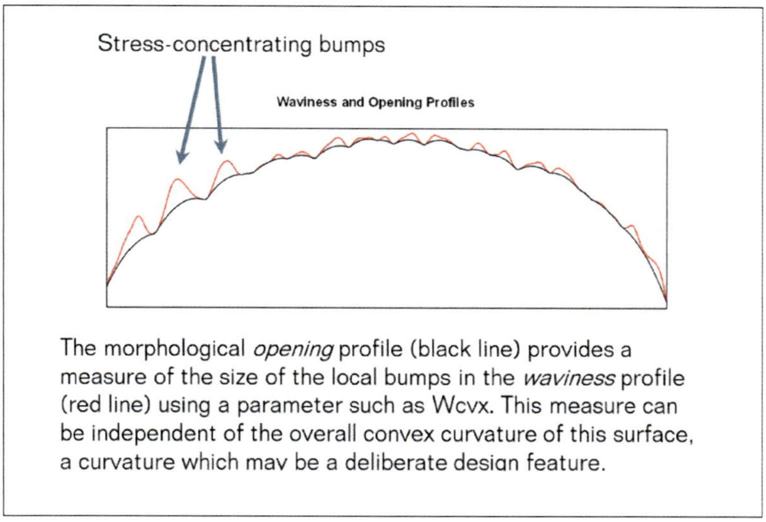

Figure 178: Controlling stress-concentrating peaks on a contact surface

Another design aspect of some load-carrying surfaces is their ability to resist wear. A small amount of run-in wear may be acceptable, but the wear rate will likely need to approach zero after limited material loss. Material ratios (Pmr, the primary material ratio, or Rmr, the roughness material ratio) are good parameters for these situations.

Load-bearing components are often more likely to fail due to manufacturing or handling defects than from features created by a well-designed manufacturing process. It is not appropriate to use surface texture parameters as a control of these defects or flaws, since their random nature makes it unlikely that a surface measurement will find their worst-case size, or, in fact, even find them at all. Further discussion can be found in the section, **"Which parameters should I use to detect scratches and dents?"** on page **184**.

Which parameters relate to lubricated interfaces?

A key consideration with most lubricated interfaces is minimizing metal-to-metal contact by keeping the asperity heights smaller than the minimum oil film thickness[57]. Rpm (ASME) is one option for controlling this, although Rpm is not always very repeatable.

From the longer list of averaging parameters, Rq can be preferable to Ra. Like Ra, Rq doesn't differentiate between peaks and valleys, so it isn't a good *functional* description, at least not by itself. However, Rq is influenced more by higher peaks and deeper valleys (due to the squaring of the values), which means it is sensitive to those features most likely to be detrimental to a lubricated interface. Furthermore, the Rq parameter has more mathematical meaning in that it represents the standard deviation of the data points along the profile. It has therefore historically been used in the lambda ratio[58] calculation, originally used by the rolling element bearing industry to relate the composite roughness of the two mating surfaces to the minimum elastohydrodynamic (EHD) oil film thickness. In a simple way, this lambda ratio combines surface texture with the system's operating characteristics (load, velocity, lubricant properties, material properties, and geometry) to predict performance. Achieving an adequately high lambda value (≈ 3) will maintain separation between the surfaces. Given Rq's lack of discrimination between peaks and valleys, this approach obviously makes assumptions about the type of surface, assuming a Gaussian surface produced by the grinding processes common in the bearing industry.

For reasons that often aren't obvious, non-Gaussian surfaces may be preferred in some partial-EHD systems. Averaging parameters such as Ra, Rq, or Rz are inadequate for defining these negatively-skewed surfaces (those with valley depths greater than peak heights) as in **Figure 179** (c). It is tempting to use Rsk, but it tends to vary greatly between measurements—even within the same surface (see, "**Should I use Rsk or Rku to specify my surface roughness?**" on page **338**). A combination of parameters such as Rpm and Rz, with upper and lower limits on Rz, may provide the desired control.

---

[57] This is only possible when there is relative motion between the interfaces; except in hydrostatic systems, there will always be surface contact (no sustainable oil film) when the system is stationary or operating at low speeds. It is commonly assumed that this condition constitutes an insignificant portion of the system's duty cycle, and that other design factors such as material selection protect the surfaces during this time.

[58] $\Lambda = \frac{h_{min}}{\sqrt{(Rq1)^2+(Rq2)^2}}$, where h_{min} is the minimum oil film thickness and Rq1 and Rq2 are the Rq roughnesses of the two mating surfaces. Numerous articles about the lambda ratio can be found in the technical literature.

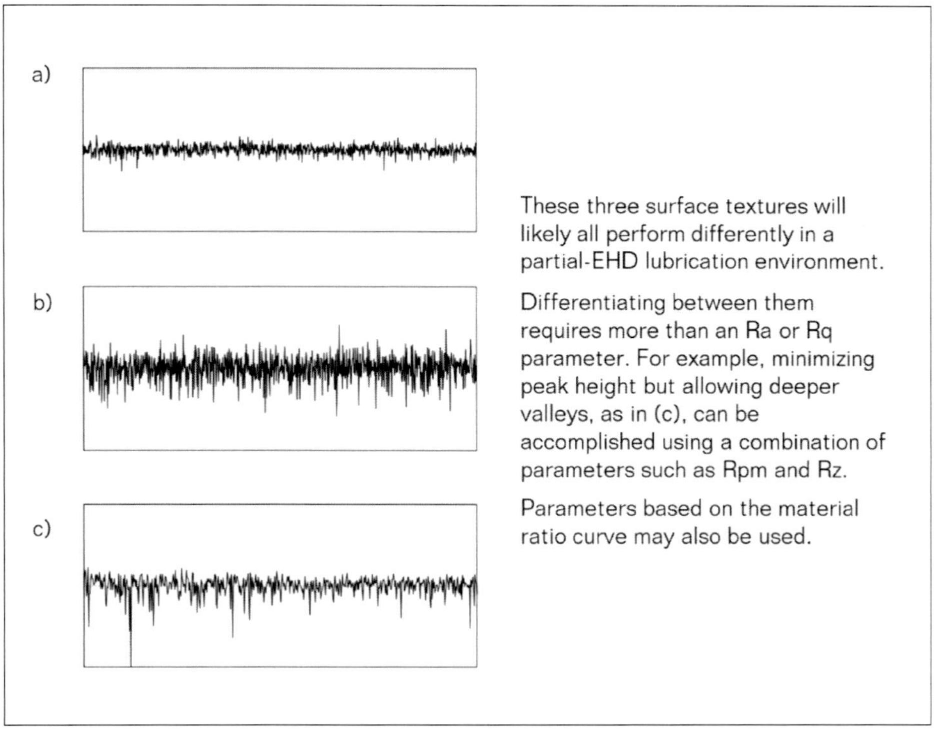

Figure 179: Differentiating *roughness* profiles with a combination of parameters

Parameters based on the material ratio curve, such as Rmr or the Rk family of parameters (Rk, Rpk, Rvk, Rmr1, Rmr2—see below), can provide a limit for the heights of the peaks above some amount of load-bearing surface, without unduly restricting the depths of the valleys.

Under hydrodynamic lubrication conditions waviness can be disruptive of the oil film. This is most commonly experienced as circumferential waviness (i.e., out-of-roundness), but axial waviness can also be influential (see the discussion on waviness starting on page **173**). It is not only the maximum waviness (Wt) that can be disruptive, but also the average waviness (Wa or Wc) and local features (steps) controlled by a parameter such as segment waviness (Wseg)—see **Figure 180**.

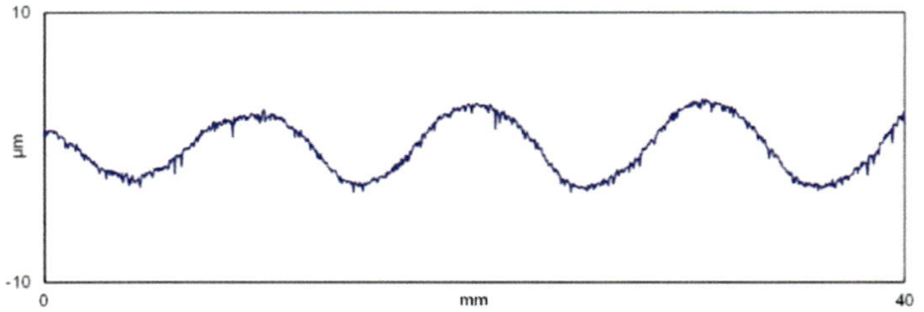

A pattern of repeated waviness can be controlled with Wa or Wc.

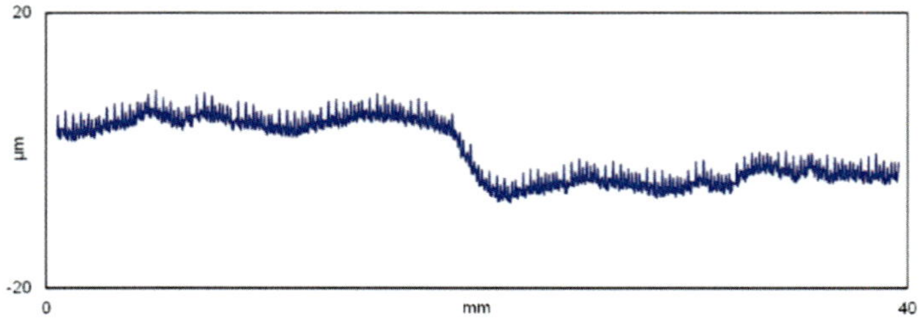

The overall straightness might be acceptable, but the step can be detrimental. This can be controlled with a parameters such as Wseg (segment waviness).

Figure 180: Waviness detrimental to plain bearing systems

Hydrodynamic bearings may be sensitive to short wavelength waviness but relatively insensitive to the long wavelength geometry or form—see **Figure 181**. This is due to the system's ability to conform to the longer wavelengths but not to the short ones. Bandpass waviness parameters are often appropriate in these situations. They allow the designer to impose tighter limits on the features that matter, with less stringent controls on the long wavelength form.

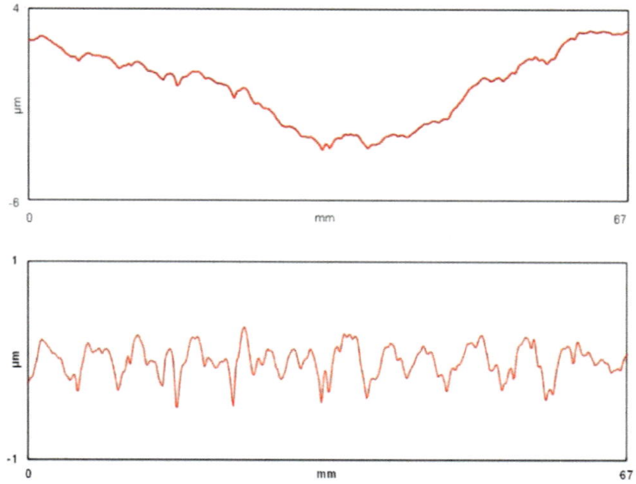

A hydrodynamic bearing system was relatively insensitive to the long wavelength concavity shown in the top *waviness* profile, but developed bands of distress corresponding to the shorter wavelength features shown in the bottom plot. Bandpass waviness analysis is the best tool for separating these features. Bandpass waviness parameters can provide tighter control of the shorter wavelengths while allowing more latitude in the overall geometry or form. Note that the bottom plot has a much smaller y-axis range than the top one.

Figure 181: Bandpass waviness analysis for plain bearing systems

Sliding lubricated interfaces have their own set of parameters, especially reciprocating systems for which the oil supply is limited and oil flow must be controlled (e.g., cylinder bore/piston ring interfaces in internal combustion engines). These types of surfaces typically require very small peaks to resist wear (adhesive and abrasive) but deep valleys to act as oil reservoirs and oil channels. These are commonly created using two honing processes—one coarse, to create the deep valleys, and one fine, to create the small peaks—so the specification system must apply to both. The overall process is typically called plateau honing (**Figure 182**). Rpm and Rz can be used in combination for a plateaued surface, with a small Rpm and a large Rz; the Rz/Rpm ratio quantifies the asymmetry. However, two other sets of parameters provide a more complete description of this type of surface:

- Rk, Rpk, Rvk, Rmr1, and Rmr2 are directly derived from the material ratio curve, as shown in **Figure 262** on page **345**. These are not as specific to a plateau-honed surface as Rpq, Rvq and Rmq but they are commonly used for these types of systems as well as other types of lubricated interfaces. Rk describes the central ("kernel") part of the material ratio curve, Rpk relates to the peaks, Rvk relates to the valleys, and Rmr1 and Rmr2 identify the intersection points.

- Rpq, Rvq, and Rmq are probability parameters related to the material ratio curve but using a probability analysis. They are defined in ISO 13565-3 and illustrated in **Figure 264** on page **347**. Rpq describes the surface peaks, Rvq describes the valleys, and Rmq is the point of intersection between the two.

See page **247** for further comments on specifying plateaued surfaces.

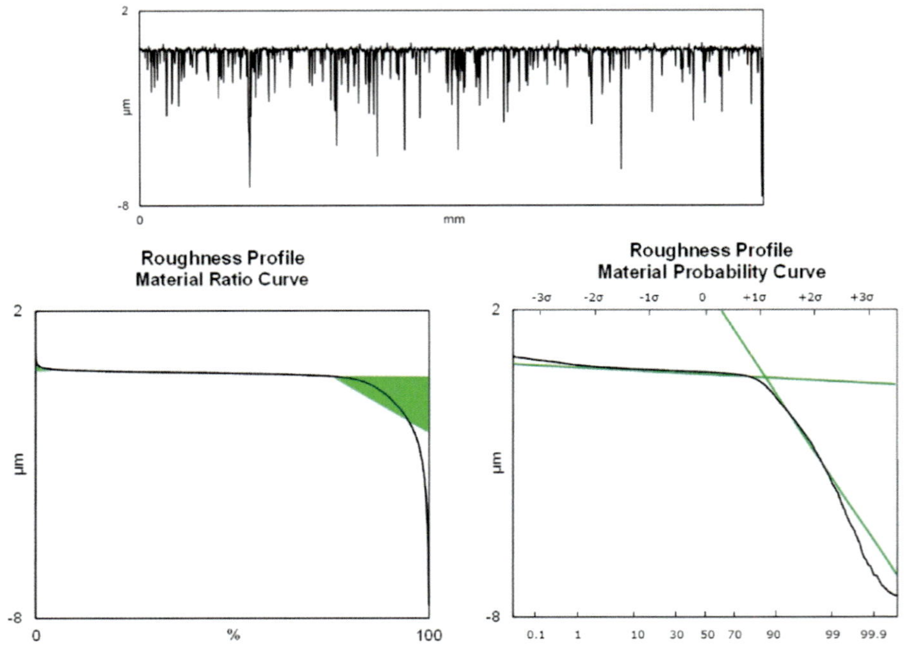

A plateau-honed surface roughness profile (top), with its bearing ratio curve (bottom left) and material probability curve (bottom right). For this highly plateaud surface, the material probability curve, with parameters Rpq, Rvq, and Rmq, provides a better representation. For other honed surfaces, the Rk parameters, or the basic Rpm and Rz parameters, might be better suited.

Figure 182: A plateau-honed surface with its Rk and probability analyses

The modeling of lubricated interfaces has historically been limited by its computational demands, especially in an elastohydrodynamic or mixed lubrication regime where surface roughness is especially important. Models have therefore tried to incorporate texture by using parameters that consider the statistical nature of the surface rather than looking specifically at peaks and valley. The spectral moment parameters Rm0, Rm2, and Rm4 are examples of these (not to be confused with Rmr0 and Rmr2). Other parameters such as Rda, or Rcl in combination with Rq, have also been used. Despite their utility for modeling, these parameters should typically not be used in a specification because, with the possible exception of Rq, the relationship between the parameters and the manufacturing process is not well understood.

Which parameters relate to sealing?

A seal interface often involves two materials with significant difference in hardness—for example, a hardened steel shaft and a polymeric O-ring. If there is relative motion between the two (intentional or unintentional), the hard surface's roughness must be low enough to avoid excessive abrasion of the soft component. Product literature commonly recommends using parameters such as Ra, Rz, Rp, Rmax, and Rmr (tp). Some of these parameters are not ideal for defining functionality, or are not particularly repeatable; however, the suppliers' experience makes them worth considering. R3z is also sometimes used because it disregards the most extreme peaks and valleys and thus improves measurement repeatability.

Sealing (rotating shaft seals)

The limiting values for a shaft's surface texture parameters depend on the surface material and hardness, the seal material, and the expected life. Seals can polish lower hardness surfaces, so a higher initial surface roughness may be acceptable. However, this polishing-induced wear may become excessive over a longer lifetime, so long durability systems may need a harder, and smoother, shaft surface. Hard chrome plated surfaces tend to experience more in-service polishing than HVOF (High Velocity Oxygen Fuel) coated surfaces, especially HVOF coatings containing tungsten carbide particles. HVOF coated surfaces may therefore need to be smoother to avoid damaging the seal. Elastomeric seals experience less wear and lower friction with a shaft surface that promotes boundary layer lubrication. PTFE (Polytetrafluoroethylene) seals, on the other hand, are self-lubricating, so the boundary lubrication-promoting characteristics of the shaft surface are less important.

The lay is a critical characteristic for rotating shaft seals since it must not "pump" the fluid in either direction: through the seal, creating leakage, or away from the seal so as to dry it out. Normal surface texture measurements don't characterize lay well, so tests often involve methods such as strings and weights to determine the lead angle. Some seal suppliers recommend avoiding these directional surfaces in favor of a random lay, but manufacturing process options and cost may make a randomized lay impractical.

The roughness filter cutoff should be selected to approximate the scale of the seal's contact width. If the cutoff is much less than the contact width, waviness parameters may need to be included (e.g., Wa, or possibly Wt). **Figure 183** is an illustration of how the relationship between the seal contact width and the surface profile wavelengths determines whether or not waviness parameters are required.

Another very important factor in the performance and lifespan of softer sealing materials is the presence of torn and folded material on the hard side of the interface. This often goes unnoticed as it cannot be easily detected by traditional stylus instruments (see the section, "**Does my surface contain tears, burrs, or folded material?**" on page **281**). Nonetheless, small microscopic tears and/or folds in a surface can quickly abrade the seal as it passes over these sharp edges.

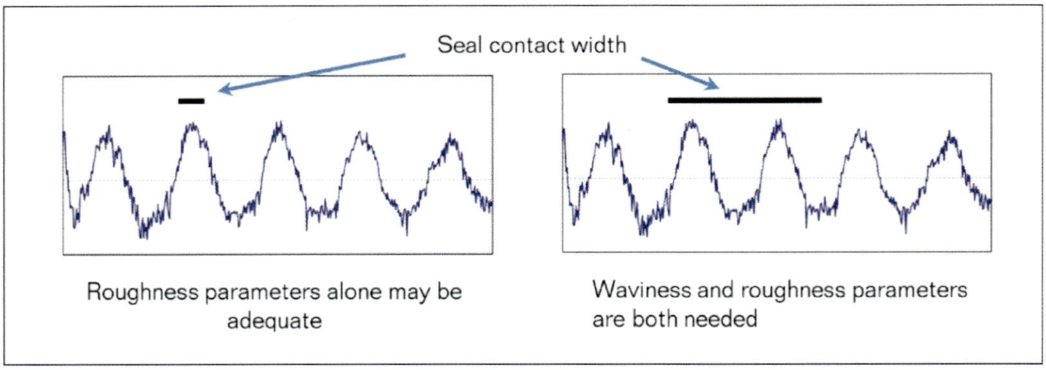

Figure 183: Waviness parameters for rotating shaft seals

Industry and international standards provide parameter and parameter limit recommendations, and these should be consulted for specific applications; e.g., RMA (Rubber Manufacturers Association) OS-1-21, SAE J946, and ISO 6194. Seal suppliers often have recommendations for their products. As noted above, it is essential to treat your sealing interface as a system, taking into consideration shaft and seal materials, hardness, operating environment, and life expectations. Seal installation must also be taken into account during the design process. To avoid installation damage, surface texture must be controlled on any portion of the shaft over which a seal must pass during assembly.

Sealing (stationary gaskets)

In basic terms, gaskets seal by filling in the gaps between two surfaces. If the gasket cannot conform to all the potential leakage paths, the system will leak. The gasket industry often uses parameters such as Rz or Rmax for roughness control. However, waviness may be even more important because of the potential for larger cross-sectional area leakage paths; fluid viscosity and surface tension may prevent leaks through roughness features but not through waviness[59]. Standard waviness parameters such as Wt or Wa may be useful, especially if calculated using bandpass analysis to separate waviness from the longer wavelength geometry or form[60]. More functionally-relevant parameters such as Wvoid (waviness void area) and Wvdd (waviness void depth), calculated using a morphological closing filter (a virtual gasket), are even better at predicting performance (**Figure 184**), especially if the morphological filter radius is tuned to the conformability of the specific gasket. More information about Wvoid and Wvdd can be found on page **243**.

Note that this discussion only applies to the roughness and waviness aspects of surface texture. The long wavelength form or flatness of both surfaces is obviously also important, but is not addressed here. Note, too, that the parameter values shown in **Figure 184** only apply to these specific surfaces

---

[59] For these same reasons, roughness may be important when sealing gases. In that situation, the concepts described here for waviness can be applied to roughness.

[60] A gasket may be able to conform to a long wavelength of a given amplitude, but unable to conform when that amplitude occurs over a shorter distance.

in a specific application, and are for illustration purposes only. They should not be considered as functional limits for any other situation. Those limits must be determined empirically, taking into account the actual gasket and fluid properties and the system's operating conditions.

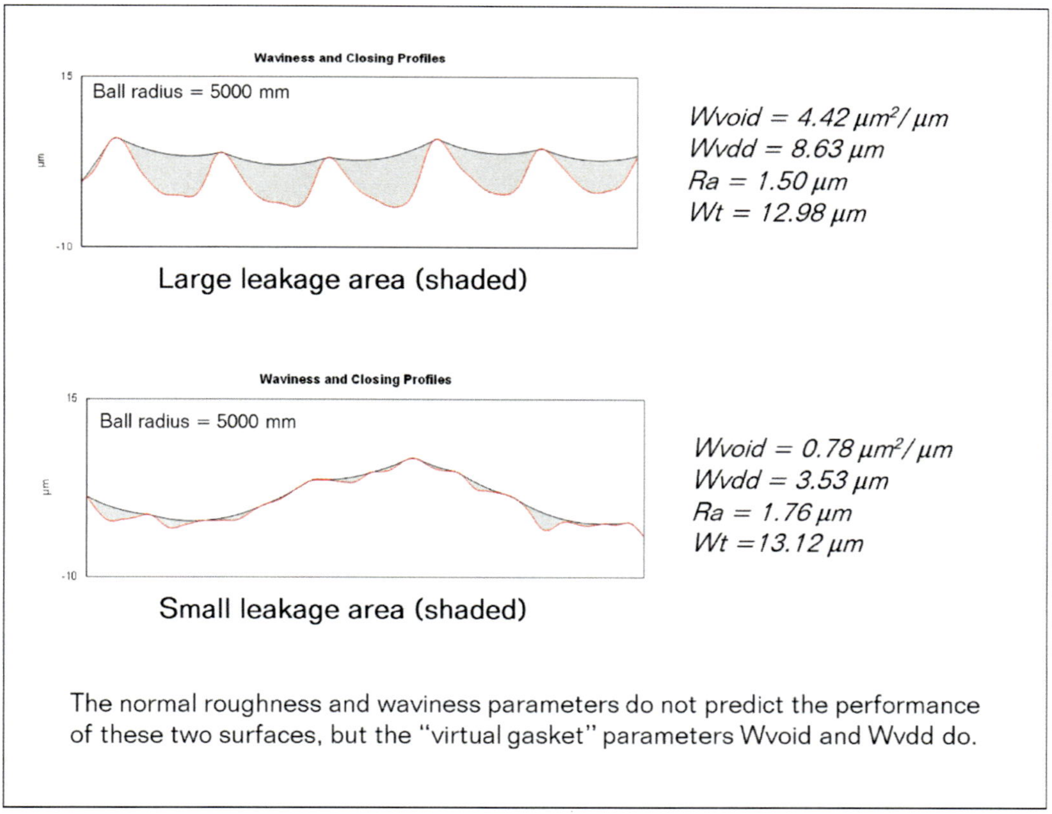

Figure 184: Morphological closing filter analysis of leaking and sealing surfaces

Which parameters relate to noise and vibration?

Vibration is commonly caused by a periodic waviness on one or more of the mating surfaces. If the waviness features combine with the operating speed to create vibrations in the audible range, this is perceived as noise. The waviness spacing as well as the amplitude are therefore important. This spacing is most commonly encountered in a circularity measurement of a shaft as undulations per revolution (upr), but could also be characterized by Wsm (waviness mean spacing) if evaluated with a linear measurement. Waviness amplitude can be characterized with Wa (mean height of *waviness* profile) or possibly Wc (mean height of waviness elements).

Note that a surface may "run-in" and become quieter with use due to wear of the roughness asperities (see **Figure 185**). A surface should therefore be rough enough to allow this wear to "mask" the underlying waviness; smoother may not necessarily be better. The surface roughness can be controlled with a material ratio, perhaps an upper limit at a given section depth. Rdc (*roughness* profile section height difference) or Rpk (reduced peak height) may also be useful parameters.

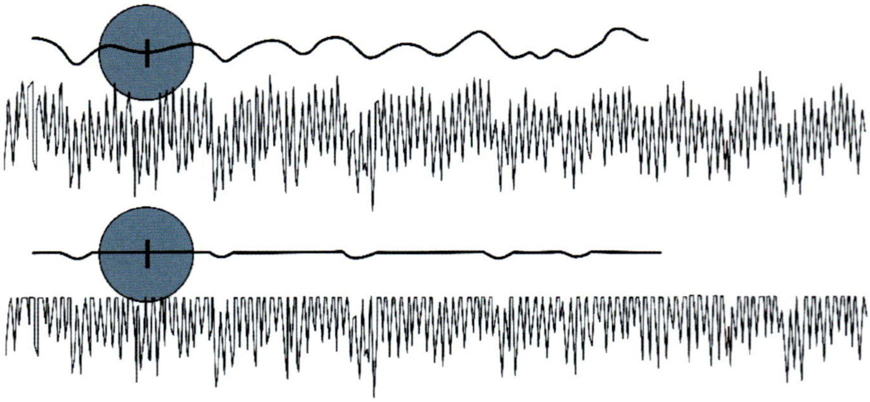

Figure 185: The running-in of the roughness diminishes the impact of the underlying waviness

Which parameters relate to appearance?

Surface appearance is directly related to the reflection of light, which can be affected by various characteristics of that surface. The following are comments about several surface texture parameters that have been found to correlate to appearance in specific situations.

The surface lay can be important. A strong dominant lay can look rough, even if the texture is relatively smooth. If the lay is a random pattern a surface will have a matte appearance, even if it is relatively rough.

The local slopes may be influential as they relate to the scattering of light. Thus, a slope parameter such as Rdq may be more important (within reason) than an amplitude parameter such as Ra; equivalent Ra values can visually appear very different. In one actual example, new ground surfaces had an Ra of 0.085 μm and a Rdq of 3.10°. After use, some of these surfaces were very shiny, and were found to have an Ra of 0.076 μm (a 10% change) and an Rdq of 1.59° (a 50% change). Other samples, tested under slightly different operating conditions, were dull after use; these had an Ra of 0.10 μm and an Rdq of 4.30°. The appearance therefore correlated much better with Rdq than with Ra. One possible explanation is that as the small, high frequency features which raise the Rdq value wear away, the surface becomes shiner. This is illustrated in **Figure 186**.

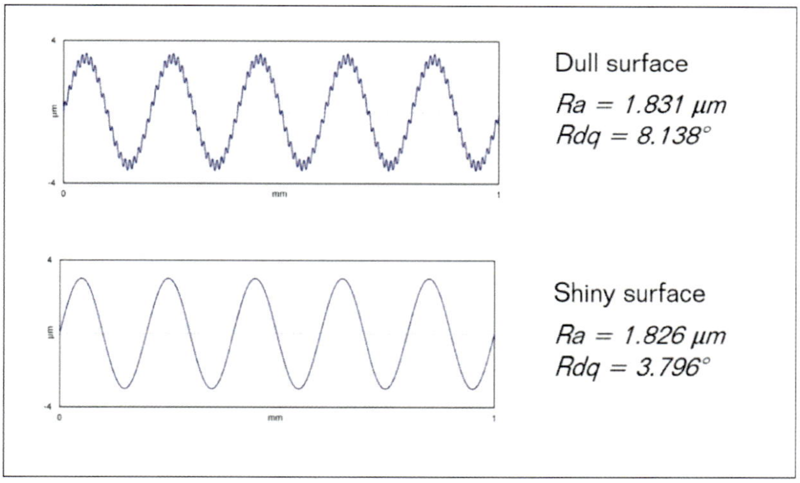

Figure 186: Illustration of the effect of local slopes (Rdq) on surface appearance

The correlation length, Rcl, is also sometimes correlated with appearance, with shorter lengths tending to appear more matte (think fine-grained sand), while longer lengths can give an "orange peel" appearance. However, the definitions of "short" and "long" depend on the application; a mirror may have a very short correlation length, but it is also very smooth.

The aesthetics of coated or painted surfaces such as car body panels is affected by the number of roughness peaks. The peak counting parameters Rpc or Rhsc[61] may be appropriate for controlling these surfaces. Surface waviness can create an orange peel appearance so this may need to be specified and controlled to a low amplitude.

The eye is sensitive to changes in the shape of the surface, especially when the change is sudden or "sharp." The eye may not detect changes that are gentle, even when the features are relatively high. The Wcvx (waviness convexity) parameter, which simultaneously assesses height and sharpness, has been shown to differentiate visible, undesirable, surface blemishes from other features that are not perceived as blemishes. **Figure 187** shows one example, in which the waviness peak near the center of the profile was visually undesirable, but the peak was not detectable with traditional parameters. The *opening* profile (in black) highlighted this peak. A Wcvx numerical limit and the associated morphological opening filter radius can therefore be used as part of the quality control system. More information on Wcvx is provided on page **239**, and a further discussion of morphological filters is given on page **330**.

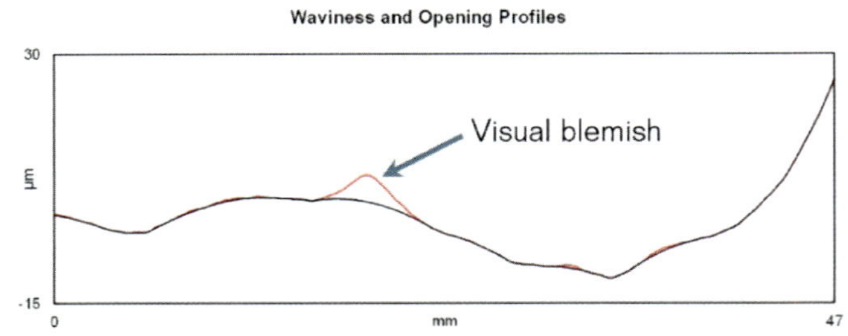

The *opening* profile (black) highlights the peak in the *waviness* profile (red) that was seen as a blemish on this cosmetic surface.

Figure 187: Morphological opening profile applied to a blemish on a cosmetic surface

Aluminum Foil Study

Common aluminum foil has two significantly different appearances: one shiny side and one dull side. **Figure 188** is a study of a piece of aluminum foil using both 2D and 3D data to illustrate some of the factors discussed here. It is not possible to separate out the influence of the various factors or to know which is more influential, so **Figure 188** should not be used as guidance for design decisions. It is included here for illustration purposes only.

---

[61] Rpc is the number of peaks per unit length above a given threshold, with valleys on either side which drop below a second threshold. Rhsc, the high spot count, is the total number of peaks higher than an upper threshold. See the respective entries in the ***Definitions*** section for more information.

The 3D data covers an area 0.78 x 0.58 mm, while the 2D traces are 8 mm long. The 3D spacing is 1.218 µm in both directions, while the 2D spacing is 0.26 µm. The 3D data was filtered using a least squares plane reference geometry, a 2.5 µm short filter cutoff wavelength (lambda-s, or λs), and a 0.25 mm roughness filter cutoff wavelength (lambda-c, or λc); this λc was chosen to remove some wavelengths that appeared to be more like waviness than roughness. The 2D data was processed using a least-squares line reference geometry, a 2.5 µm λs, and a 0.25 mm λc, with this λc being chosen for consistency with the 3D analysis. The peak count (Rpc) and high spot count (Rhsc) analyses used a 0.5 µm threshold. Note that Sal is the shortest autocorrelation length and may therefore not be in the same direction as Rcl.

Comparing the 3D maps in **Figure 188** (a) and (b) shows the difference in lay. The shiny side has a unidirectional lay. The dull side has a remnant of that lay, but it has a strong random lay superimposed over it. This is consistent with the earlier statement that a random lay tends to be dull or matte.

The 3D parameters all differ by a factor of 1.5 to 2 between the two surfaces. This difference can be even greater if a targeted, narrow band of wavelengths is considered.

Figure 188 (c) and (d) are the *roughness* profiles from 8 mm traces taken using a 2D measurement system. The measurement direction is believed to be perpendicular to the lay. The roughness profiles are somewhat similar, with identical Ra values, although the dull surface is more positively skewed. The difference in Rcl corresponds to the general rule of thumb noted above: shorter = duller. The peak counting parameters show opposing trends, but with Rpc showing the largest relative difference.

It is difficult to draw many conclusions from this study. The value of 3D versus 2D data is clear. Trends in a specific parameter do not necessarily determine the appearance; the change in Rdq in the example noted above does not apply to the aluminum foil. It may be that parameters other than those mentioned here are more powerful for predicting appearance.

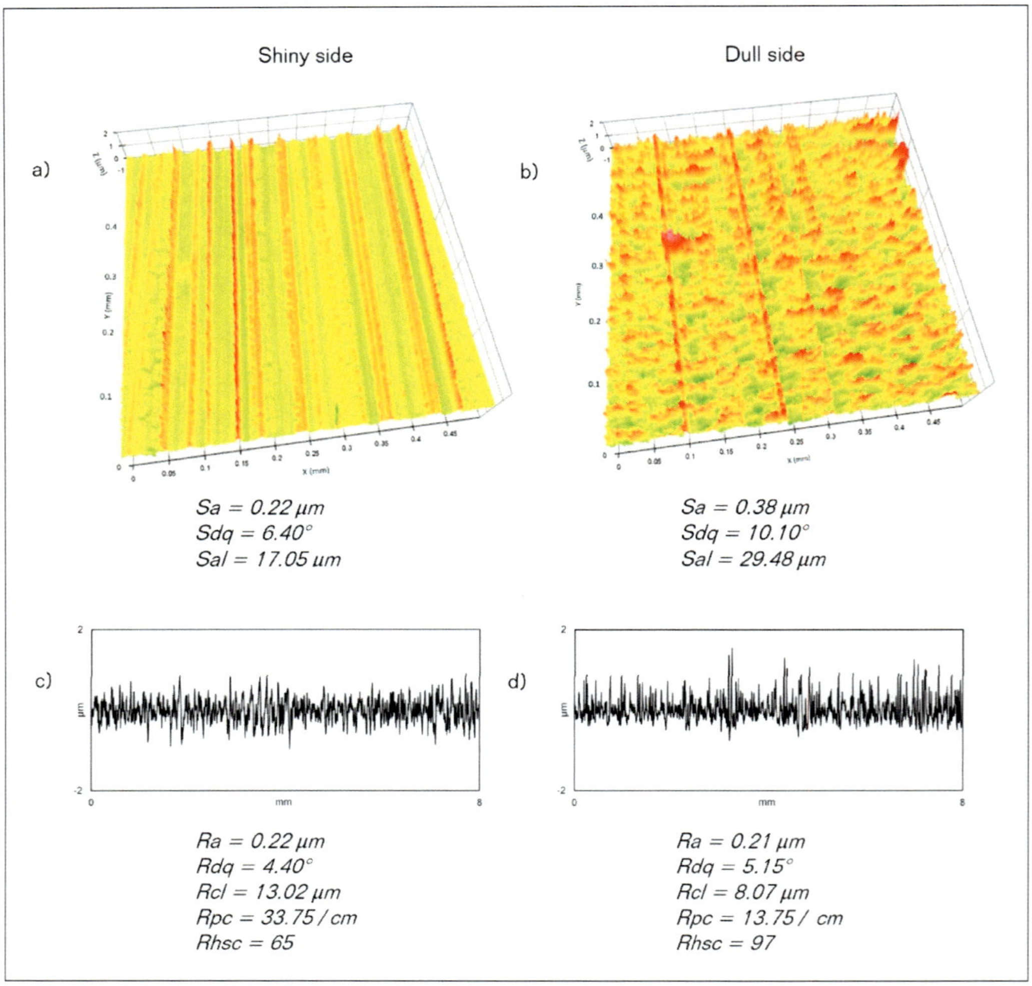

Figure 188: Surface texture analysis of aluminum foil

In many industries, a multiband approach is used to control various aspects of appearance. For example, short wavelengths can affect aspects such as haze, gloss and clarity. Medium wavelengths are related to aspects such as "orange peel." The longer wavelengths of a surface can influence overall distortions in a reflection. One particular example of multiband analysis is shown in **Figure 189**. The numerical range above each graph indicates the range of wavelengths included in that graph.

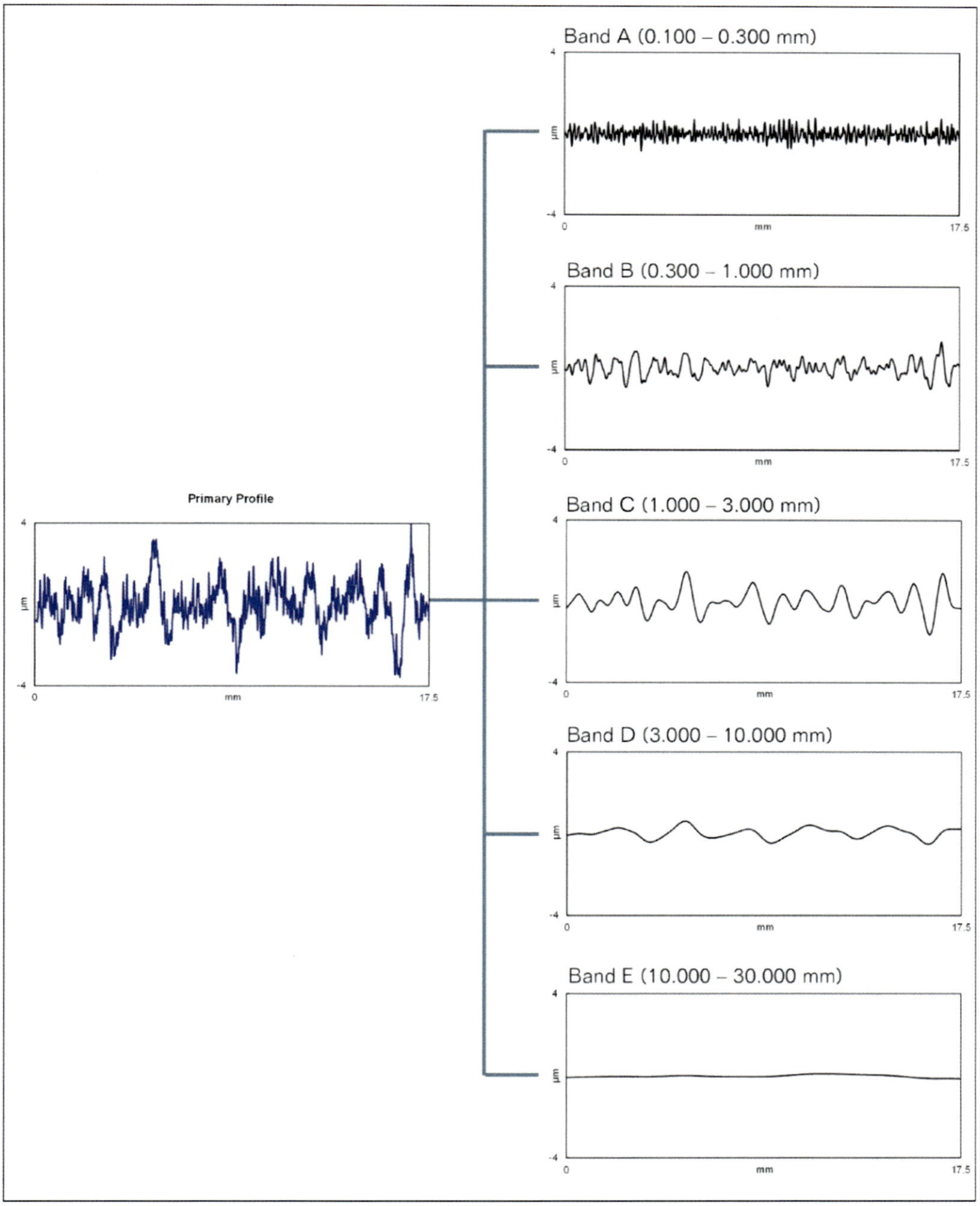

Figure 189: Multiband surface texture analysis

Which parameters relate to coating adhesion[62]?

Before discussing 2D surface texture parameters and coatings, it should be noted that fractal or multiscale analysis has shown promise in correlating adhesion to surface texture (see the section, "**What are fractal methods in the context of surface analysis?**" on page 359). Numerous references are available, including a recent CIRP article[63] which discusses the analysis technique and multiple potential applications.

In general, a rougher surface has better adhesion because of the greater surface area. It is also believed that peak count in addition to peak height is important because, as the peak number and height increase, there is a shift in debonding forces from tension toward shear. The coating industry often uses Rt and Rmax to assess peak height, and Rpc (peak count) or Rhsc (high spot count) to assess peak count. The actual target values for these parameters will vary based on the type of coating being applied.

Given the non-repeatable nature of Rt (roughness peak-to-valley height) and Rmax (roughness maximum height), a potentially better choice would be a parameter such as Rz (10-point roughness height). Rpm would also be a possibility if the peaks are most important. However, the coating industry's history of using Rt and Rmax supports their continued use.

Peak height and peak count are related to the actual length of the surface (or surface area in 3D), so a measure of the actual length, or of the ratio of the actual length to the evaluation length, might be useful. This approach is similar to the fractal analysis mentioned above, but with a single parameter, Rlo (the point-to-point length of the *roughness* profile). Rlo is a measure of this actual length, but it does not appear to be widely used. One reason may be that the ratio of Rlo to the evaluation length is typically very close to 1. Secondly, Rlo is dependent on the data spacing, both because that spacing is used in the Rlo calculation[64], and because the more closely one "examines" a surface (the smaller the spacing), the larger Rlo becomes. Therefore comparisons between surfaces can only be made with a consistent spacing.

The concept of adhesion, and its relationship to surface texture, can extend beyond coatings. For example, studies by Clarke et. al.[65] looked at the "adhesion" of tennis shoes on various tennis court surfaces, specifically studying the belief that players have a greater risk of injury on acrylic hard

---

[62] Some material taken from H. J. Roper et. al., "The Effect of Peak Count of Surface Roughness on Coating Performance," *Journal of Protective Coatings and Linings* (June 2005). (Available at www.paintsquare.com)

[63] Christopher Brown et. al., "Multiscale analyses and characterizations of surface topographies," *CIRP Annals – Manufacturing Technology*, 67(2018), 839-862

[64] $Rlo = \sum_{i=1}^{N-1} \sqrt{\Delta x^2 + (z_{i+1} - z_i)^2}$, where N is the number of data points, Δx is the data spacing, and z_i are the heights of the *roughness* profile.

[65] James Clarke et. al., "The effect of normal load force and roughness on the dynamic traction developed at the shoe-surface interface in tennis," *Sports Engineering* (2013) 165-171.

court surfaces than on clay courts. While their work focused on the Ra of the surface and found a varying traction—Ra relationship, their raw data suggests that other surface texture parameters might be more powerful in predicting behavior. **Figure 190** shows how the interaction of contact length and other aspects such as conformability and wetting combine to affect adhesion.

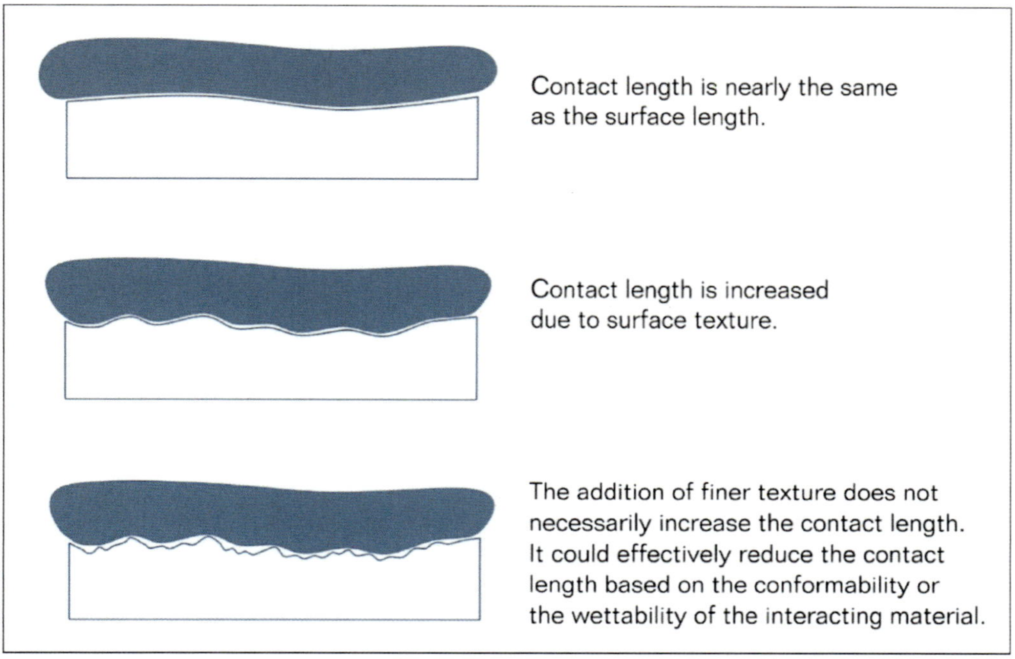

Figure 190: Adhesion is influenced by the contact length and conformability of the materials

Which parameters should I use to control my processes?

A surface texture specification is the means by which a designer communicates a surface's micro-topographical functional requirements to the manufacturing facility. Ideally, manufacturing will use that same specification—the analysis details and the parameters—to set up and monitor their process. There are situations, however, when surrogate parameters or a different analysis may be required:

- Historical data or operator experience. Sometimes companies have internal databases and knowledge of parameters that may not be on the component drawing. This local knowledge may be useful for process control.
- Shop floor measurement capability. The specification may require sophisticated measurement systems to evaluate a surface. It may not be convenient or timely to use that equipment for real-time monitoring of a process, so simpler instruments and analyses may be employed.

Surrogate parameters are only acceptable when there is a demonstrated correlation between them and the drawing callouts. This is not necessarily a mathematical correlation; directional indications (is the surface getting better or worse) and limiting values may be adequate. Final surface quality decisions must still be based on the specified parameters.

The following are a few examples of process control parameters. Each of these parameters is also discussed in the **Definitions and Terminology** section.

Ra

Given the ubiquitous nature of Ra (average roughness), there is a lot of experience using it for process monitoring. While it is poor at defining functionality, once a process is developed and stable, Ra can identify when the process has drifted. Following are some of its advantages and disadvantages:

- Ra has the advantages of being statistically stable and repeatable. It can be useful for random surfaces such as those from grinding, and when process settings remain constant (grinding wheel grade, coolant, tool tip geometry, speeds and feeds, etc.). Ra is also available on even the lowest cost surface texture instruments.
- A disadvantage of Ra is that it is unable to discriminate peaks from valleys. Its averaging can mask excessively rough or smooth areas. It is relatively insensitive to outliers and is not often useful for describing heavily skewed surfaces (those with Rsk outside ± 2).

Rz

Rz (10-point roughness height) is the second most common surface texture parameter so, like Ra, it comes with a great deal of experience and historical value. It has many of the same advantages and limitations as Ra. It differs primarily because of its increased sensitivity to the higher peaks and valleys, making it more responsive to some process changes.

Rq

The RMS nature of Rq makes it more sensitive to extreme peaks and valleys than Ra, so it is good for critical surfaces where those features matter.

Rz and Rmax

Using Rz and Rmax together provides a measure of process consistency. Similar values indicate a consistent surface, while a significant difference indicates the presence of defects or rough areas on an otherwise consistent surface. Be aware, however, of the general non-repeatability of Rmax.

Rz and Rpm

Used together, Rz and Rpm can help monitor processes generating skewed surfaces without as much instability as the Rsk parameter.

Rpc, Rhsc, Rp

Industries involved in sheet metal forming and coating industries often use one or more of these for applications where appearance is important.

Rpq, Rvq, Rmq

These are often used for specifying plateau-honed or other two-process surfaces where the two processes are required for functional reasons. Whenever possible, they should also be used for monitoring these types of processes because they provide good feedback on the individual process steps—the roughness generated by each and the amount of stock removal. More information is available on page **346**.

Rk, Rpk, Rvk, Rmr1, Rmr2

Like Rpq, Rvq, and Rmq, the Rk parameter set can provide somewhat independent feedback on multiple aspects of a manufacturing process or set of processes. More information is available on page **340**.

Another tool for monitoring and controlling manufacturing processes is the wavelength content graph (**Figure 191**). This plots amplitude versus wavelength[66] for all the features that make up the surface texture, with peaks at the wavelengths of any periodic or repeating features. Some of these will be a normal part of the process, such as feed rate in a turning or boring operation, while others may be an unwanted characteristic such as chatter caused by vibration. The wavelength content plot

---

[66] The wavelength content plot is generated by performing a Fourier transform of the data. The y-axis represents the mean-to-peak amplitude at each wavelength.

provides a quick visual assessment of which wavelengths are dominant so that appropriate corrective action can be initiated.

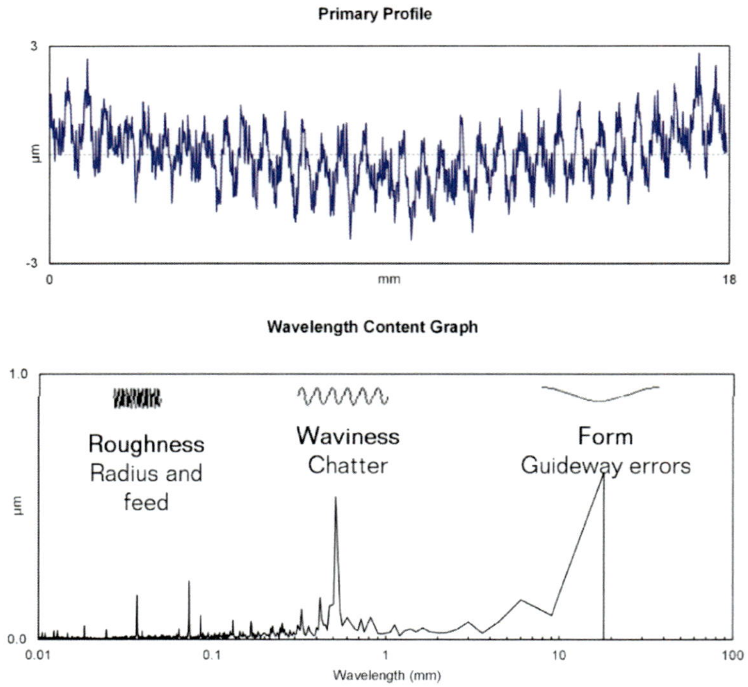

The wavelength content plot for the profile shown at the top reveals several prominent wavelengths that provide information about the machining process and the machining environment. This can be used to identify the source(s) of problems (e.g., chatter) or to confirm that the process parameters are as desired (e.g., feed rate).

Figure 191: Wavelength content plot useful for process monitoring

How do I create a surface texture specification?

Your company is developing a new or improved component and you are responsible for the surface texture part of the design. How do you decide upon the right parameters and numerical limits?

Mathematical modeling and computer analysis are routinely used to determine many design features—stiffness, strength, hydrodynamic behavior, system dynamic behavior, etc. Unfortunately, modeling techniques are not yet sophisticated enough to reliably predict the behavior of surface texture, especially of roughness. Experience and experimentation are therefore often the best guides. The modelling of form and, to some extent, waviness is much more progressed and may be sufficient to base a design on, at least for the initial prototypes.

A surface texture design consists of a list of parameters, their limiting values, and the wavelength ranges to which those parameters apply. As a starting point, the sections starting on page **251**, "**Which parameters specify surfaces with functional requirements?**," provide suggestions for parameters for various applications. Be certain your choices truly reflect the function of the surface. Do roughness peaks matter? Roughness valleys? The number of features? Waviness? In any case, the design and specification should consider all the possible wavelengths. For example, simply controlling "roughness only" may lead to future problems in the uncontrolled waviness domain. When possible, choose parameters that have manufacturing process "levers," ones your supplier knows how to control. Your suppliers' level of sophistication in terms of measurement instrumentation is also a consideration.

Following are some approaches that can lead to an acceptable specification. None of these are foolproof—coming up with a reliable and cost efficient surface texture callout can be difficult. They are simply a good place to start. They are listed in approximately the order of the amount of effort required—least to most.

1. Does your company already make a similar component? Does it operate under similar conditions? Does it have acceptable durability and reliability? If so, copy its specification. This is an easy approach, but there are a few things to avoid.

    a. First, does the manufacturer of the existing component use the whole surface texture specification range, or do they provide parts that only use a small portion of that range, possibly because they've had many years to fine-tune their process? Do you know that parts at the extreme limits will still function adequately?

    b. Is the existing specification adequate, based on all the things you've learned from reading this book? Does it address the component's function or is it simply an Ra or Rz value? Does it define the wavelength ranges? If not, it would be wise to revise or expand it.

    c. Do you really understand your customers, current and future, well enough to be certain the operating environment won't be different for the new design? Relatively small changes in, say, operating temperature or lubricant formulation may result in a previously adequate design becoming unacceptable.

2. As an offshoot of number 1, measure a reasonable sample of your current supplier's product and establish your specification from that data. This is especially useful if you're wanting to include new parameters on your drawing. Be careful not to make your limits too tight, since this can unnecessarily increase the manufacturing cost. This can happen if your sample isn't random enough, or if point 1a applies.

3. If you will be purchasing a component that mates with yours, such as an o-ring or gasket, use that supplier's recommendations. Their parameters may not always be logical, but if you follow their suggestions for the parameter and for the manufacturing process, the result may be acceptable. Their experience is valuable.

4. Measure parts that are known to work well and others that don't, then set your limits between the two. This may be an option if your existing design does not have an impeccable service history or if your supply base has been less than stellar. An example of this approach is shown in **Figure 192**, where the cause of a bearing problem was evident after tracing a good and a bad part. In this case, the addition of a waviness control solved the problem.

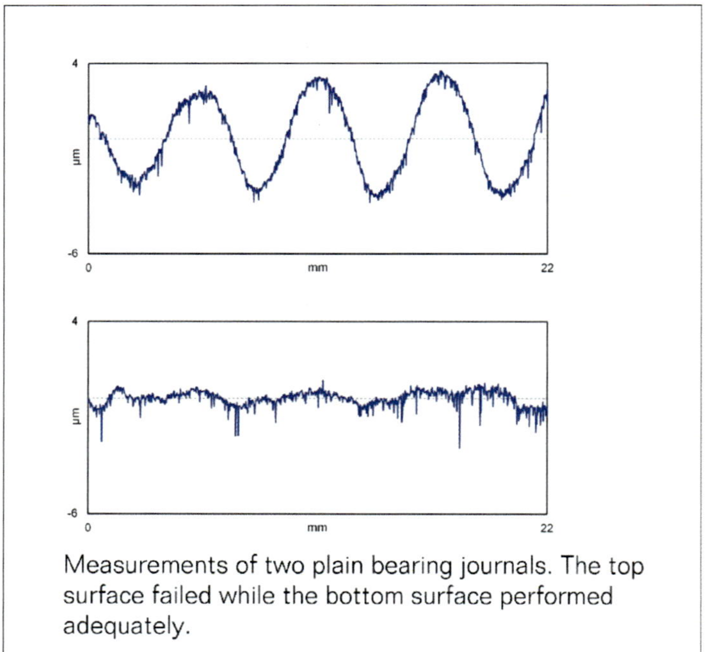

Figure 192: Example of measurement data revealing the cause of failure

5. Measure parts before and after testing. While some change in texture may occur as part of the run-in process, large changes in a surface may indicate potential problems. Changes in

the surface mean the mating components are doing the final machining of your component, often in an uncontrolled manner. **Figure 193** shows an example of a Hertzian contact where the surface roughness showed a significant change after running, especially in the height of the peaks. Adding a finishing process and controlling the peaks with a parameter such as Rpm significantly extended this product's life. See the section, "**How do I measure the amount of wear on my surface?**" on page **226** for more information about measuring wear.

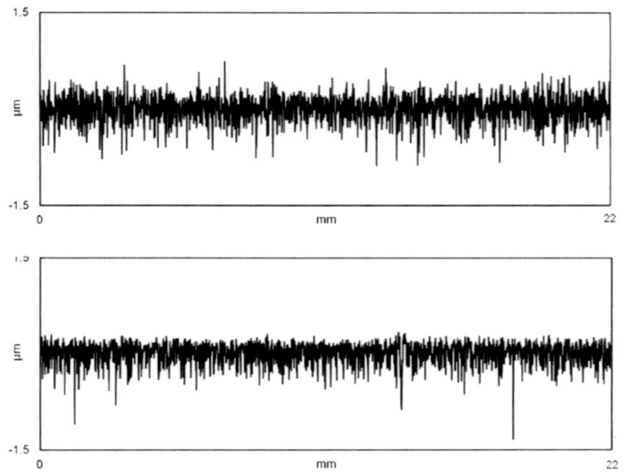

A Hertzian contact surface before and after test showing the removal of the peaks. Adding an Rpm limit and removing these peaks using a controlled polishing process extended the life of this surface.

Figure 193: Changes in surface features with running

6. Make an educated guess. Then make lots of measurements, do lots of testing, and refine that guess into a workable design. Careful consideration of the function of your component using the information in this book will allow your guess to be an educated one, even if it isn't perfect.

If your design dictates the manufacturing process that must be used (e.g., die casting), your choice of parameter limits may be determined by that process's capability (see **Figure 195**).

The selection of the transmission bands for your parameters can usually be based on the default settings (see, e.g., **Table 13–Table 15** on page **147**). There may, however, be situations where this gives an undesirable result, such as the example shown in **Figure 194**. This surface has controls on both roughness and waviness, and the roughness is smooth enough that the default roughness filter cutoff wavelength (lambda-c, or λc) is 0.25 mm. However, the manufacturing process creates occasional deep valleys, and, with a small λc, the *waviness* profile is "pulled" down into these valleys, creating large waviness parameter values (middle graph in **Figure 194**). With a larger λc value the valleys have a smaller effect on the *waviness* profile, yielding waviness values more representative of the actual surface (bottom graph in **Figure 194**). In these cases, robust filtering should also be considered[67].

Some of the specialized waviness parameters, such as Wcvx (waviness convexity) and Wvoid (waviness void area)[68], can also be very sensitive to the transmission band. Limiting values should therefore be closely tied to the wavelength ranges, and analytical or experimental validation is essential for confirming that the specification is functional.

Sometimes none of the standard transmission band limits provide what you need. An example of using a non-standard filter for roughness control is discussed at the end of the section, "**If a cutoff wavelength is not specified on my drawing, what should I use?**" starting on page **145**.

Creating a specification is often an iterative process, especially if it applies to a component or a set of operating conditions with which you have limited or no experience. Inputs to that process can come from every stage of a product life cycle, starting with the initial concept and continuing through the analysis of end-of-life components. Examples of specific inputs from each stage in a product's life are given in **Table 20** below.

---

[67] The robust filter de-weights the influence of outliers such as this valley. Unfortunately, the robust filter is not always available. This filter is discussed further in the section, "***When should I use the robust filter?***" on page **138**.

[68] Wcvx is typically used to control the local stress caused by bumps on a surface (see the section, "**My part is curved. How do I detect bumps within the curvature?**" on page **247**). Wvoid helps control leakage on a gasketed surface (see page **251**).

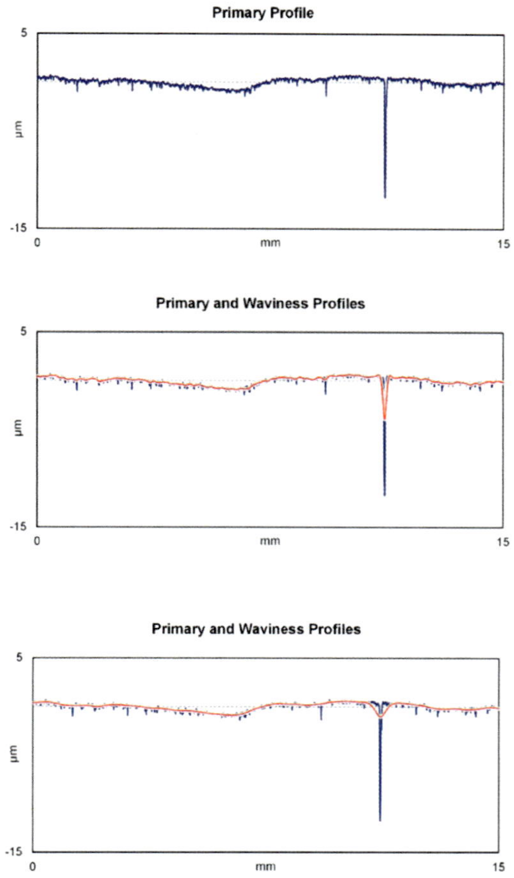

A *primary* profile showing a relatively smooth surface with one deep valley

The *waviness* profile, with λc = 0.25 mm, superimposed over the *primary* profile. The *waviness* profile is "pulled" into the valley, resulting in a large total waviness that is primarily due to the valley.

Wt = 4.556 µm

The *waviness* profile, with λc = 0.8 mm, superimposed over the *primary* profile. The *waviness* profile is less affected by the valley, resulting in a total waviness more representative of the surface.

Wt = 1.649 µm

Figure 194: The effect of filter cutoff selection on a waviness parameter

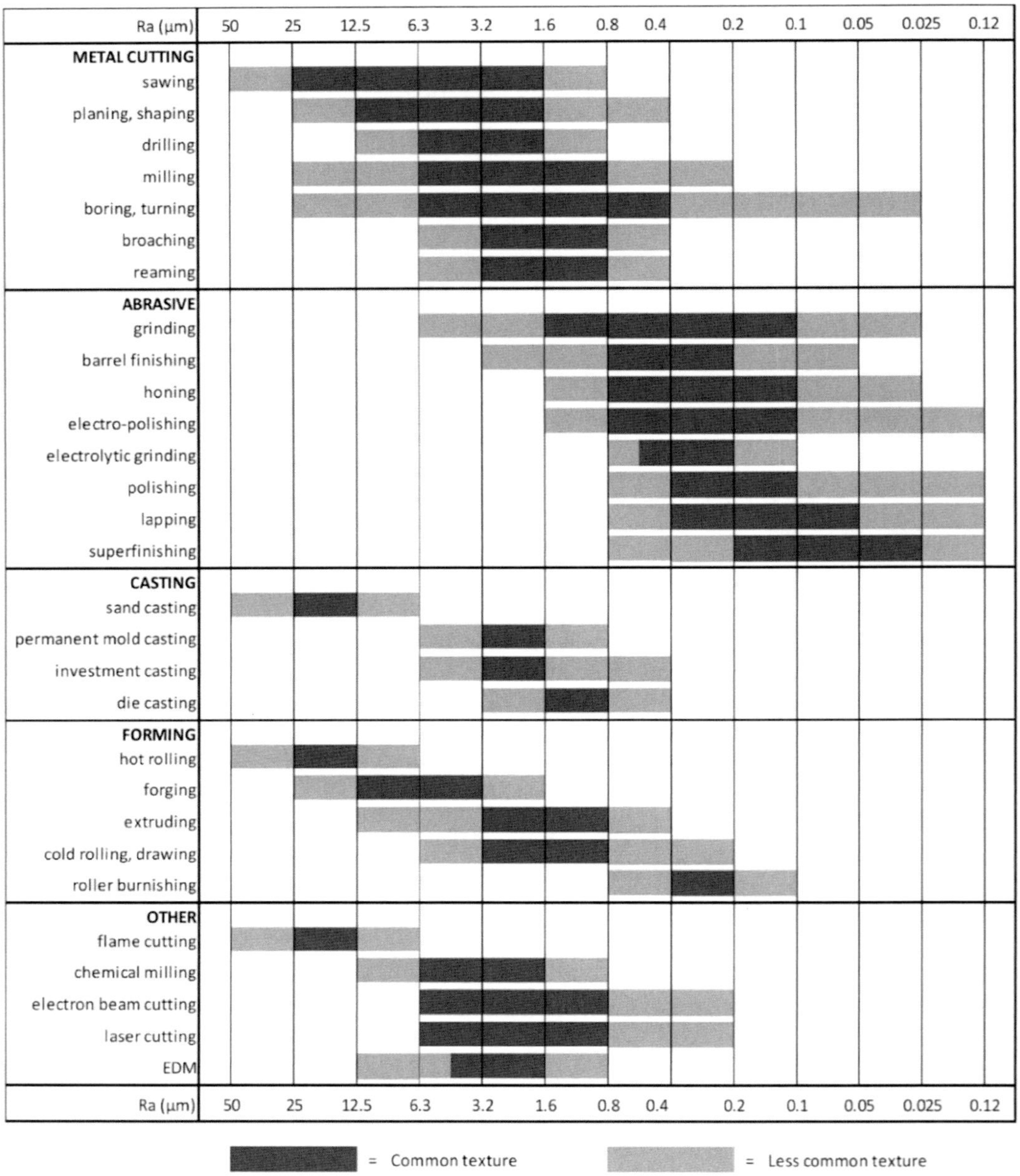

Figure 195: Typical surface textures (Ra) from common manufacturing processes

Table 20: Surface texture considerations during the product development process

Development Step	Surface Texture Considerations and Activities Examples
Concept	What type(s) of interaction occur with mating component(s)? ▸ Sliding ▸ Sealing ▸ Lubricated What stresses will the surface experience? ▸ Bending ▸ Tangential ▸ Hertzian contact What will be the operating conditions/environment (temperature, speed, etc.)? Which surface features might be needed for proper function? ▸ Valleys ▸ Bearing area ▸ Surface area Which surface features might be detrimental to proper function? ▸ Peaks ▸ Deep and narrow valleys ▸ "Bumps" ▸ Wide valleys
Design	Which parameters can control the required features? Which parameters can limit the undesirable features? Which "sizes" (wavelength ranges) are important? Which parameters are appropriately sensitive to process drift (both the normal and the unanticipated changes)? Which parameters are familiar to and understood by your supplier? Which parameter limits separate functional from non-functional?
Document	Is the engineering drawing callout clear and unambiguous? (use ISO 1302 if possible) Are all wavelengths covered (roughness, waviness, form)? Are any supporting inspection documents required? Are they available and up-to-date? Does the supplier understand what is specified, and can they measure it?

Table 20: Surface texture considerations during the product development process (continued)

Development Step	Surface Texture Considerations and Activities Examples
Product development	What are new components like? (measure as many test parts as possible) How do surfaces change with time? (pre- and post-test measurements) Do test results correlate with initial surface texture data? ▸ Which parameter(s) predict performance? (these may be parameters not included in the original design) ▸ What parameter values separate "good" from "bad"? ▸ What needs to change? (iterate with design step and with manufacturer)
Manufacturing development	Which processes, tooling options, process parameters provide the desired surface? What frequency of tool change/wheel dressing is required? What inspection frequency is required? What is the control plan? What is required if parts go out of print? What is the short-term process capability? Is the proper measurement system available? Is in-line gaging adequate or is laboratory equipment required? Communicate with product development organization
Production	What is the long-term process capability? Do routine material changes, sub-supplier changes, etc. affect the process?
Failure analysis	Could the failure have been surface-initiated? Are there unworn areas available for surface texture measurements? Do these meet specifications? Do used but undamaged areas show greater than anticipated changes from the original state? Communicate with product development and, if appropriate, with design and concept groups

Does my surface contain tears, burrs, or folded material?

Machining defects such as burrss and folded metal can be detrimental to a component's function, especially in applications with tight clearances or sensitivity to loose debris. These defects can also cause excessive wear when running against soft materials such as O-rings.

In some cases simply rubbing a cotton ball over the surface can reveal local "snags" along the surface. However, the best detection tool is an optical or scanning electron microscope; an example photomicrograph is shown in **Figure 196**. While some of this equipment is expensive, simple light microscopes capable of over 300x magnification are available for less than $100. With a trained eye, these less expensive microscopes are very capable of detecting the features of interest. This should be the first choice for surface inspection and process control.

In situations where visual inspection isn't adequate or practical, it may be possible to detect burrs or folded metal using standard surface texture equipment. One procedure has been published elsewhere[69] and is summarized here.

Defect detection involves the phenomenon called flanking, where a surface feature is contacted by the side (flank) of the stylus rather than by the spherical tip. An illustration is given in **Figure 197**, where a stylus of tip radius (r) and included angle (α) traces over a step on a surface. This behavior is experienced when the local slope is greater than the stylus' included half-angle, or with a step or a reentrant feature with a height greater than the radiused portion of the stylus. This height is easily calculated[70], and is often less than one micron. Calculation of the slope at each data point in the *primary* or *roughness* profile allows a comparison with this stylus half-angle. Since the slopes for a normal machined surface are typically much less than this half-angle (45° for the common 90° stylus), the higher slopes will be strongly correlated with burrs or folded material.

In order for this procedure to work, the stylus included angle must be known reasonably accurately, and the sample must be as level as possible during the measurement. Further information can be obtained from the referenced journal article[69]. A sample data set from that article is shown in **Figure 199**; it is derived from **Figure 198**, a trace of the surface shown in **Figure 196**.

---

[69] Malburg, Chetwynd, and Raja, "Local Slope Analysis in the Stylus Based Assessment of Surface Integrity," Tribology International, (1997), 30(7), 527-532 (July 1997)

[70] If α is the tip angle, R is the tip radius, and $\beta = 90° - \alpha/2$, then the height h of the radiused portion of the stylus is equal to $R(1 - \cos\beta)$. For a 2.0 μm tip radius and a 90° included angle, h = 0.95 μm.

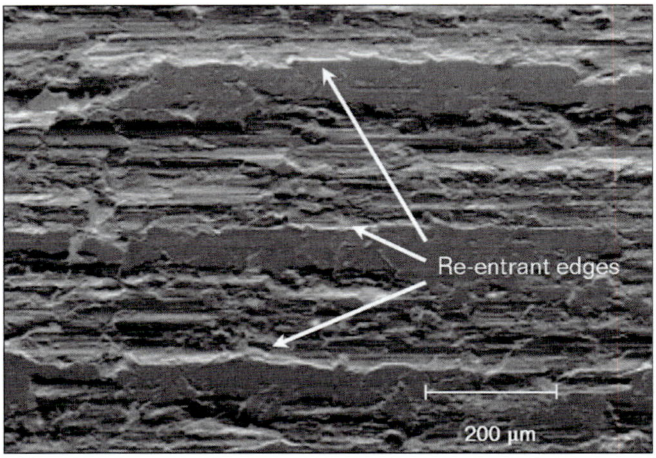

Figure 196: SEM photomicrograph showing smeared and folded metal

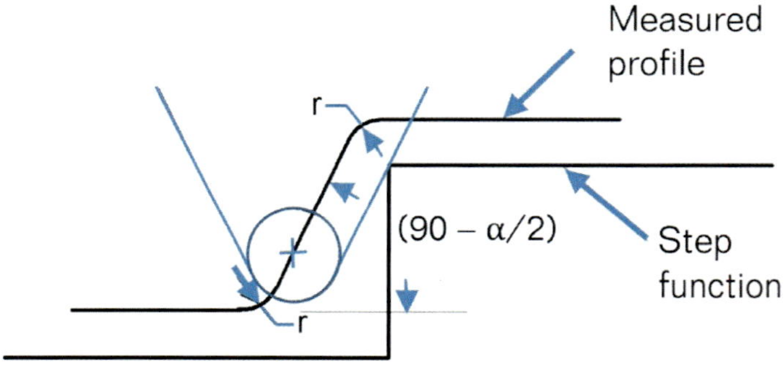

Figure 197: Illustration of stylus flanking

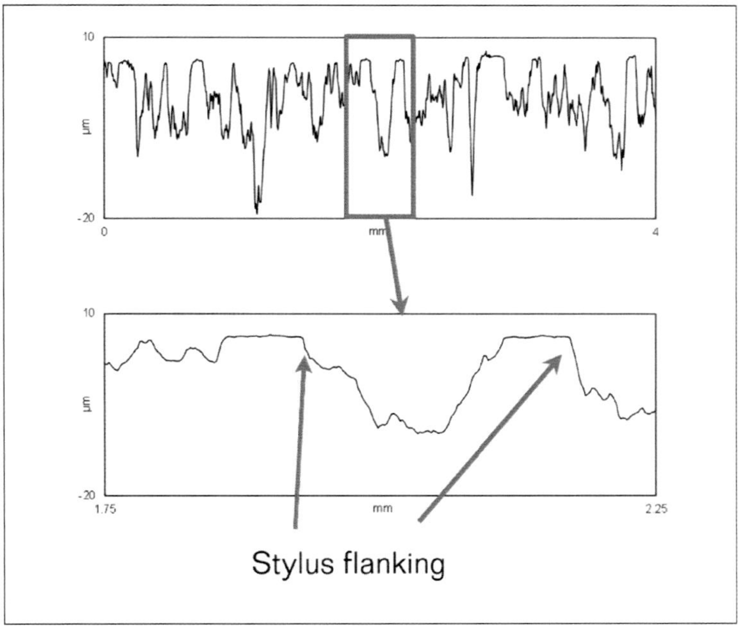

Figure 198: *Roughness* profile from the surface in **Figure 196** showing where flanking occurs

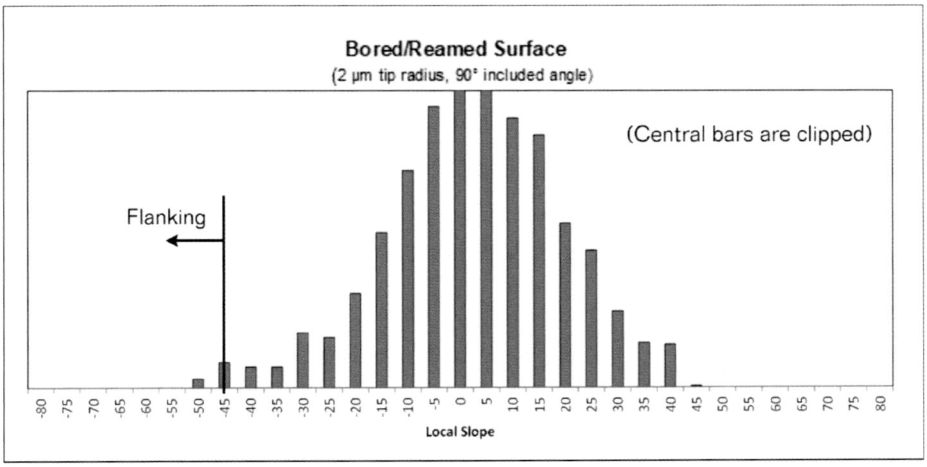

Figure 199: Sample data set showing flanking indicative of folded material (see footnote **69**)

Can texture distinguish material properties from machining properties?

Some materials have holes, pores, cracks, pull-outs, etc. in their surface due to the type of material or the forming process. Examples are cast iron components that have had the graphite flakes or nodules "pulled out" during machining, cast components with porosity, and powder metal components with less than theoretical density. The holes can be quite deep and can have a large influence on parameters such as Rz (see the section, "**My surface is porous. How can I control the roughness?**" on page **250**). For surfaces with stringent surface texture specifications it may be desirable to determine if high measured Rz values are due to machining issues that need to be addressed, or to features that are inherent to the material and cannot be eliminated[71].

Typical machining processes yield relatively shallow local slopes. On the other hand, material properties, such as porosity or cracking, will exhibit very steep local slopes. Thus, a study of the local slopes can facilitate such a determination.

Figure 200 shows two traces from a machined cast iron surface. The top trace is from the surface shown in **Figure 201**, with extensive graphite pullout (the dark spots in the photograph). The bottom trace is from a region with minimal pullout. The holes left by the pull-out can be differentiated analytically from the machined features using the distribution of local slopes, as shown in the histograms in **Figure 200**. This is a similar procedure to that used to identify torn or folded metal—see page **281**. The two histograms clearly show a difference in the distribution of slopes that correlates with the visual appearance of the surface and of the traces.

[71] Note that features such as graphite pullout may actually be machining issues. However, distinguishing between them and deep valleys from machining may still be important since the corrective actions may be different.

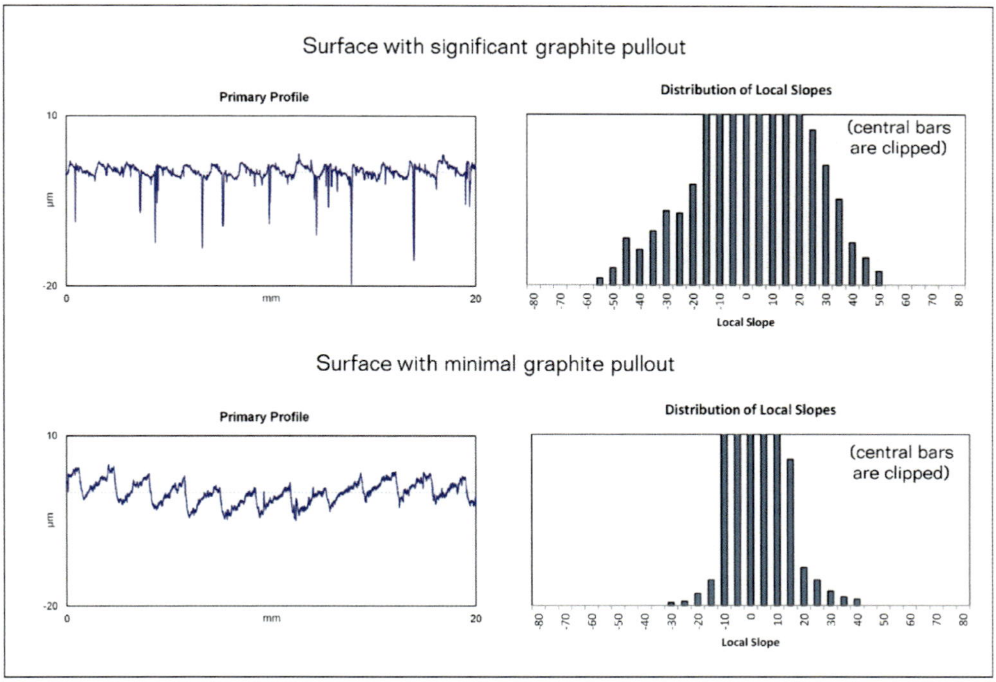

Figure 200: Differentiating holes from machining valleys on a cast iron surface

The histograms can be easily created in a spreadsheet using the x-z data from a *primary* profile. Slopes can be calculated using two points ($\Delta z/\Delta x$) rather than the seven-point formula used for Rda or Rdq (equation 3 or 4 on page 376).

Note that this analysis does not address whether or not the pullouts are detrimental to the performance of the component. That must be determined by someone with materials expertise or with prior knowledge of the functional requirements. This analysis only helps differentiate between properties of the material and results of the machining.

Figure 201: Photograph of machined cast iron surface showing graphite pullout (dark spots)

Can I measure grinding damage via surface texture analysis?

There is no reliable way of finding grinding damage (grinder burn) with surface texture tools.

Grinding damage is a change in the metallurgical structure, mechanical properties, and residual stress state of the near-surface material of a workpiece. This change is caused by excessive heat generated at some point during the grinding process, sometimes combined with localized plastic deformation. (Note that similar damage can occur in other material removal processes, but grinding is the most common source.)

Grinding damage can happen at any time during the grind cycle. It can occur early in the process when, for example, there is an excessive amount of grind stock, or when an irregularly-shaped part isn't fixtured or oriented properly. It can occur in the middle of the process if, for example, there is a temporary interruption of coolant. It can happen late in the process if the grinding wheel isn't adequately dressed. Damaged material from early in the process may or may not be removed by the subsequent material removal.

The final surface texture is created during the final steps of the processing sequence and may or may not be associated with the part of the cycle that creates grinding damage. Since the contributors to the two characteristics aren't necessarily related, you cannot depend on one to be related to the other.

Having said all that, there are no doubt some situations where a surface texture measurement may reveal grinding damage. One example would be if the burn is severe enough that it generates surface cracking, and these cracks are detected in a surface trace. The risk of burn tends to increase as the ground surface texture gets smoother (all other things being equal), so finding a very smooth ground surface may prompt further checks of the surface integrity. Severely burned regions may be softer than the surrounding non-damaged material and could experience higher material removal rates, thus distorting the surface profile. These relationships, however, are tenuous at best, and cannot be relied upon. It should therefore never be assumed that surface texture measurements will reveal grinding damage.

DEFINITIONS AND TERMINOLOGY

DEFINITIONS AND TERMINOLOGY

Any discussion of surface texture uses numerous specialized terms and definitions. This section contains short descriptions of many of the more common terms and concepts. Longer discussions are provided for terms that can be confusing or that warrant an extended treatment. Some parameters are included that are not yet recognized by the national and international standards organizations but that may be available in some software.

Standardized parameter definitions and measurement terminology can be found in numerous locations:

- National and international standards such as ASME B46.1 and ISO 4287.
- Instrument documentation
- The help system in many software packages
- Wall charts supplied by some of the instrument manufacturers
- Web sites
- Various reference books (see the section, "**Background reading material — books and journal articles**" for a partial list)

In case of dispute, the official definitions found in the applicable national or international standard shall take precedence over what is presented here.

Definitions

Parameter name equivalence

Many surface texture parameters are represented by multiple symbols. This can be for historical reasons, for ease of communication, or because of differences between ASME and ISO. **Table 21** lists some of these parameters, with column two giving the modern designations used in this book. This table only lists parameters that differ in name but are equivalent in value. It does not include those parameters that are calculated differently due to, for example, being based on sampling length instead of evaluation length.

Table 21: Alternate but equivalent parameter names[72]

Historical	Modern	Description
AA	Ra	"Arithmetic average" roughness height
CLA	Ra	"Center line average" — arithmetic average roughness height
Δa, RΔa	Rda	Average asperity slope, calculated using the absolute values
Δq, RΔq	Rdq	Standard deviation of the asperity slopes; the RMS slope
Mr1	Rmr1	Material ratio at the peak-to-core transition
Mr2	Rmr2	Material ratio at the core-to-valley transition
Pc	Rpc	Peak count; the number of peaks per unit length
Htp, Rδc, RHtp	Rdc	Vertical distance between the section levels corresponding to two material ratios
RMS	Rq	Standard deviation of the asperity heights; the RMS roughness
Rtm	RzDIN or Rz	Average over the evaluation length of the height between the single highest peak and deepest valley in each sampling length
Ry	Rmax	Maximum peak-to-valley distance in a single sampling length
Sm	Rsm	Average spacing in the *roughness* profile
tp	Rmr or Pmr	Material ratio (sometimes called bearing ratio or bearing area)
tpa	Pmr	Material ratio of the *primary* profile
tpi	Rmr	Material ratio of the *roughness* profile

[72] When *primary* and *waviness* profile parameters exist that are analogous to the roughness parameters listed in the left column, this name equivalence applies to them as well.

Definitions

Definitions of the more common surface texture terms and parameters are given below. The list includes selected parameters from both the ASME and ISO systems, a few parameters from other national systems, and some useful parameters that are not yet implemented in national standards. Note that similar parameters may exist with different names in other standards or software. Official definitions can be found in ASME B46.1 and ISO 4287. Selected parameter mathematical equations are given in the section, "**Equations**" starting on page **375**. It should be noted that standards are continually evolving. At the time of this writing, the ISO 4287 standard will soon be embodied in an upcoming ISO 21290 series of standards.

Note that parameter names that are common to both ASME and ISO may in fact represent different values because ASME parameters are evaluated over the evaluation length while many ISO parameters are based on the sampling length. Where applicable, this difference will be noted in the definition.

Definitions will typically be given for the roughness parameter. For many of these, analogous parameters exist for the *primary* and *waviness* profiles.

A1: The reduced peak volume from the material ratio curve. A1 is a member of the Rk set of parameters (see the section starting on page **340**). A1 is the area bounded by the material ratio curve, the peak-to-core transition height, and the 0% axis. For surfaces subject to run-in wear, A1 may represent the material that is removed during running in. Because it is calculated from a two-dimensional graph with one axis being percent, A1 has units of length.

A2: The reduced valley volume from the material ratio curve. A2 is a member of the Rk set of parameters (see the section starting on page **340**). A2 is the area bounded by the material ratio curve, the core-to-valley transition height, and the 100% axis. On lubricated surfaces this parameter may provide a measure of the oil retention volume. Because it is calculated from a two-dimensional graph with one axis being percent, A2 has units of length. Also known as Rvo.

AA: Arithmetic average, an old designation for Ra.

Amplitude distribution function (ADF): Histogram of data points as a function of profile height (see **Figure 202**), or the probability there will be a surface point at a given height. The shape and symmetry of the ADF are described by the skew and kurtosis parameters. Integration of the ADF provides the material ratio curve.

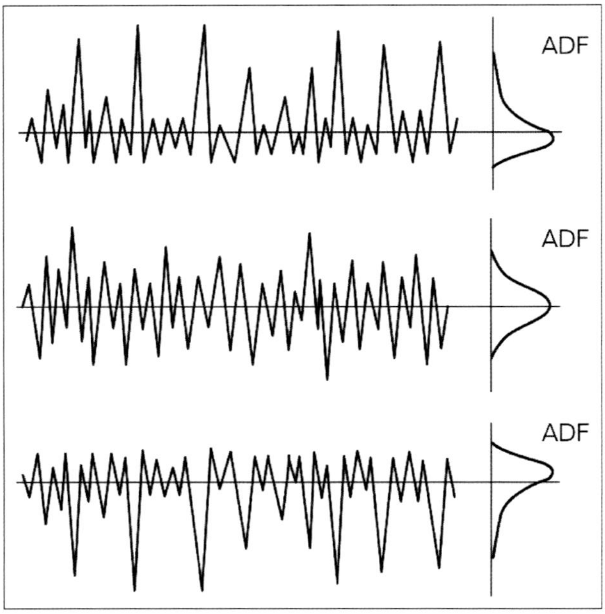

Figure 202: Amplitude distribution function for three different profiles

AR: Average roughness motif width. See page **354** for a discussion of the "motif" parameters.

Assessment length: See evaluation length.

AW: Average waviness motif width. See page **354** for a discussion of the "motif" parameters.

Bandpass: The application of both a short-pass and a long-pass filter to select a specific range, or band, of wavelengths.

Bearing ratio curve: See material ratio curve.

CLA: Center-line average, an old designation for Ra.

Closing profile: The profile obtained by applying a morphological filter to the top of a profile (see **Figure 203**). The morphological filter is most commonly represented by a disk/ball rolling over the surface. The *closing* profile can simulate the behavior of a conformable component as it touches the surface. It can provide an assessment of localized valleys that might, for example, act as leakage paths. It is used to calculate the specialized parameters Wvoid (waviness void area) and Wvdd (waviness void depth). An example of its use can be seen in the section starting on page **243**. More information about morphological filters can be found on page **330**.

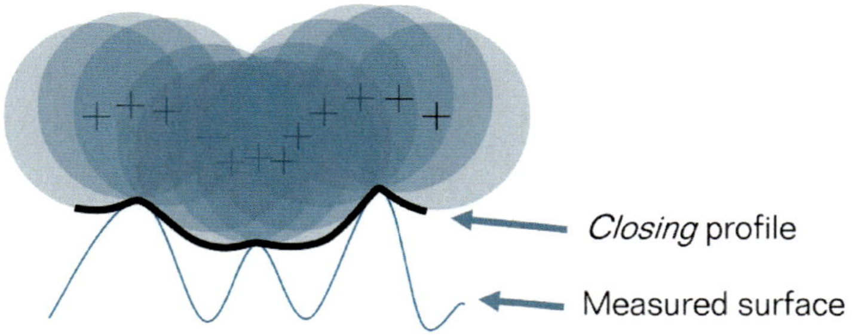

Figure 203: Illustration of a *closing* profile

Cutoff: The wavelength at which, for a given filter, 50% transmission occurs.

Direct profile: See *measured* profile.

Discrimination level: A threshold that must be met or exceeded before a feature is included in the calculation of parameters such as Psm, Wsm, Rsm, Pc, Wc, and Rc. This threshold avoids an undue influence from narrow or small features. The default ASME and ISO discrimination levels for these parameters are a height of 10% of Pz, Wz, or Rz, and a length of 1% of the sampling length. Both conditions must be met. The height discrimination applies to the peaks and valleys individually. The length discrimination applies to the distance between the two mean line crossings that define the peak or valley. Additional information is available on page **352**.

Evaluation length: The distance along the x-axis over which a surface is evaluated (see **Figure 204**). It is also known as the assessment length, and is designated as *ln* ([lowercase L] n). The evaluation length will typically consist of multiple sampling lengths. Its definition is part of the product specification, and it can be specified in three ways:

1. a default length of five sampling lengths (for roughness parameters)
2. an integral number of sampling lengths (e.g., eight sampling lengths for the callout "Ra8")
3. an absolute length, in mm, included in the surface texture callout as ln = xx.x.

Historically, parameters were evaluated over the entire evaluation length; however, ISO 4287 generally defines parameters to be calculated within sampling lengths. See page **119** for more information.

Note that if you are applying a filter (e.g., a Gaussian filter) that discards some amount of data from each end (e.g., the length of one-half or one full cutoff), the evaluation length is the length of the data minus those discards. For example, if you filter 5.6 mm of valid data with a 0.8 mm Gaussian filter and a one-cutoff end discard, the evaluation length is 4 mm.

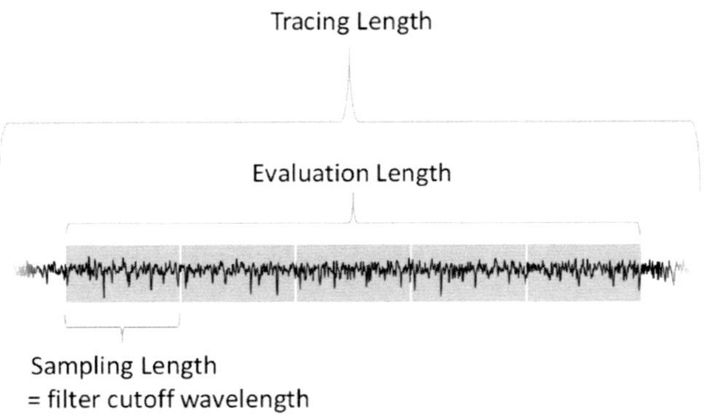

The Evaluation Length is used to determine texture parameters. All instruments discard data from both ends of the Tracing Length in order to achieve the Evaluation Length—but the discards vary greatly between instruments (fading regions in the image above).

Figure 204: Illustration of tracing length, evaluation length, and sampling length

Flanking: Contact of the workpiece surface by the side, or *flank*, of the stylus instead of by the spherical tip. Flanking occurs when the local angle of the surface, relative to the horizontal, equals or exceeds one-half the included angle of the stylus (45° for the common 90° stylus). This can happen when measuring surfaces with very deep, narrow valleys, when measuring features such as outside corners, deep narrow grooves, or small radii, or when the surface has smeared or folded metal.

Flaw: Unintentional surface imperfection. Nicks, dings, and scratches are common types of flaws, but a flaw can be any feature agreed upon by the buyer and seller. If flaws are considered undesirable, the buyer-seller agreement shall include controls on size, shape, number, and location. ASME B46.1 and ISO 8785 provide additional terminology and examples of flaws. Surface texture methods and parameters are typically unreliable for detecting flaws and are therefore not commonly used to determine the acceptability of a flaw.

Form: The "shape" of a surface, with the exact meaning depending on its context. Additional comments are found on pages **179** and **181**.

- When used in a GD&T context, "form" refers to features/characteristics such as straightness, roundness, flatness, etc. Depending on which GD&T standard is being used, "form" can include all surface deviations regardless of wavelength, or it can include only those wavelengths longer than roughness. For the latter definition, the GD&T value is dependent on the surface texture filter selection. (Note that the definition of GD&T "form" is a subject of ongoing debate.)

- In a strict surface metrology context, form is the longest wavelengths on the surface, those longer than waviness. Form is revealed by suppressing all wavelengths shorter than the form cutoff wavelength (lambda-f, or λf), creating the *form* profile.
- "Form" is often (but incorrectly) used to describe the "shape" of a surface that is suppressed prior to analyzing for surface roughness or waviness. The shape suppression is performed using a reference geometry such as a least squares line, a least squares arc, a polynomial, or some other nominal shape. The suppressed shape, such as a convex crown represented by an arc, is often called "form," but it is more appropriately called a *reference geometry* profile.

Form profile: The long wavelength data set obtained after application of the form removal filter—see **Figure 205**. The "F" parameters are based on this profile.

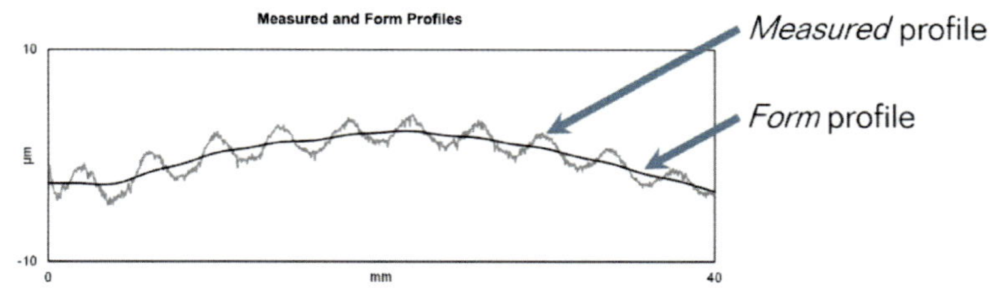

Form profile obtained by applying an 8 mm spline filter (λf) to this *measured* profile

Figure 205: Illustration of a *form* profile

Ft: The vertical distance between the highest peak and the deepest valley in the *form* profile.

λc: The long wavelength limit for roughness data, also called the roughness cutoff. It is sometimes abbreviated as *lc* ([lowercase L]c) and may be written as lambda-c.

λcw: The ASME designation for the long wavelength cutoff for a bandpass waviness analysis. This is equivalent to ISO λf.

λf: The long-wavelength limit for waviness when using a bandpass analysis. It is sometimes abbreviated as *lf* ([lowercase L]f) and may be written as lambda-f.

λs: The short wavelength limit for the *primary* and *roughness* profiles. Features shorter than lambda-s are considered to be noise. It is sometimes abbreviated as *ls* ([lowercase L]s) and may be written as lambda-s.

λsw: The ASME designation for the short wavelength cutoff of waviness. λsw should be set equal to the roughness cutoff, λc, so that roughness and waviness are treated as a continuum.

Lateral resolution: The ability of a system to discriminate between two horizontally separated points. In the context of stylus-based measurement this is dictated by instrument characteristics such as the stylus radius and data point spacing.

Lay: The direction of the predominant pattern left on the surface by the machining or manufacturing process. Typically the lay can be thought of as the direction that the valleys are running. Additional information is given on page **219**.

Long-pass: A filter that lets the long wavelength/low frequency features through while attenuating the short wavelengths. Also referred to as "low pass" in some instruments.

Material probability curve: A version of the material ratio curve in which the material ratio is plotted based on standard deviation rather than percentage. The parameters Rmq, Rpq, and Rvq are derived from the material probability curve, and are often applied to stratified or plateaued surfaces per ISO 13565-3. Additional information is given on page **346**.

Material ratio curve: A graph of the percentage of material (x-axis) at various depths within the profile (z-axis)—see **Figure 206**. The material ratio curve is the basis for the material ratio parameters (e.g., Rmr) and the Rk family of parameters. It is also known as the Abbott-Firestone curve or the bearing ratio (or bearing area) curve. More information on material ratio can be found on page **340**.

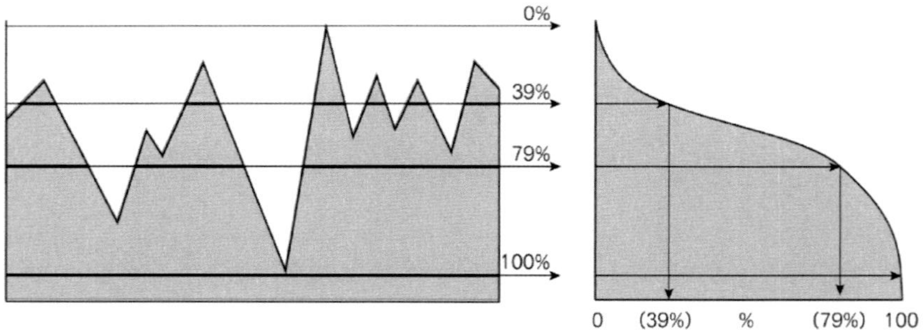

The material ratio curve on the right is created by plotting the percent bearing area on the x-axis as a function of the depth below the surface on the y-axis.

Figure 206: The derivation of a material ratio curve

Mean line: The line about which variations of a profile are evaluated to generate various parameter values. The mean line corresponds to the long wavelength component that is suppressed by the profile filter. The mean line for the *roughness* profile is the *waviness* profile from lambda-c; for the *waviness* profile it is the output of lambda-f (for bandpass waviness) or the reference geometry; and for the *primary* profile it is the reference geometry (commonly a least squares line) or the form profile.

Measured profile: The data from the measurement instrument without any modification. The *measured* profile is the starting point for reference geometry removal, filtering, and the rest of the data analysis process. The *measured* profile is also known as the *direct* profile and, in some cases, is referred to as the "raw data" or *raw* profile.

Morphological filter: A method for assessing the size and shape of local features in a profile. The filter is applied by rolling a virtual shape of user-defined size over or under the profile, creating a *closing* or *opening* profile, respectively. It is most commonly applied to the *waviness* profile. The filter can be applied using any shape; for surface texture measurement purposes, however, a disk or sphere of user-specified radius is most often used. Morphological filters are described in ISO 16610-40. See page **330** for more information.

Opening profile: The profile obtained by applying a morphological filter (most commonly a disk/ball) to the underside of a profile, commonly the *waviness* profile (see **Figure 207**). The *opening* profile can provide an assessment of features like localized "bumps" that might act as stress concentrators. It is used to calculate the specialized parameter Wcvx (waviness convexity). An example of its application can be seen on page **239**. More information about morphological filters can be found on page **330**.

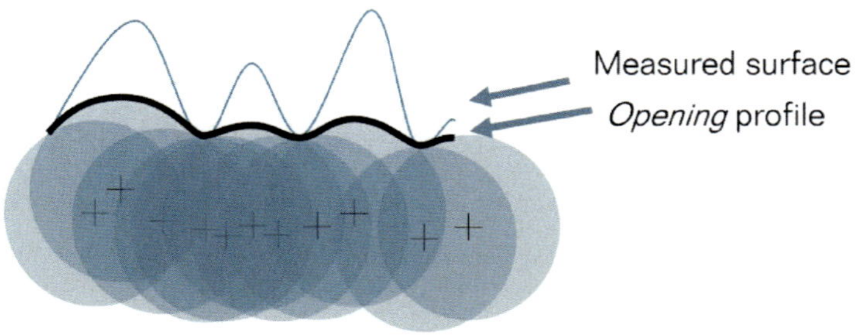

Figure 207: Illustration of an opening profile

Peak: see Profile peak.

Profile: In a 2D measurement, the line created by the intersection of a plane and a surface, creating a two-dimensional representation of that three-dimensional surface. The sectioning plane is typically perpendicular to the surface.

Profile element: A profile peak and its adjacent profile valley, or a profile valley and its adjacent profile peak (see **Figure 208**).

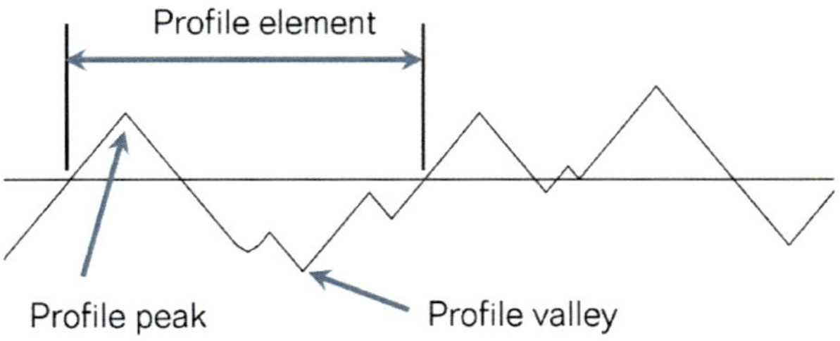

Figure 208: Illustration of profile peak, profile valley, and profile element

Profile peak: The highest point between two mean line crossings in a part of the profile that lies above the mean line (see **Figure 208**).

Profile valley: The lowest point between two mean line crossings in a part of the profile that lies below the mean line (see **Figure 208**).

Primary profile: The dataset obtained after reference geometry or form removal and the optional application of the short filter. The *primary* profile is bounded by the reference geometry or form on the long wavelength end and the short filter on the short wavelength end. The "P" parameters are based on this profile.

R: Average roughness motif depth. See page **354** for a discussion of the "motif" parameters.

R3z: Average height of the third-highest peak-to-valley distance in each sampling length. The third-highest peak and the third-deepest valley are determined in each sampling length. These are combined to determine the third-highest peak-to-valley distance, and these values are averaged to calculate R3z (see **Figure 209**). This parameter attempts to desensitize the analysis to outlier peaks and valleys.

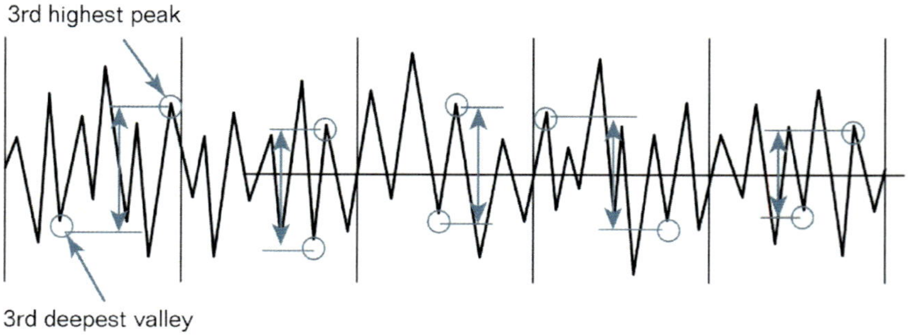

Figure 209: Illustration of R3z

Ra (Wa, Pa): The arithmetic average, computed as the average absolute distance between each point on the *roughness* profile and the mean line (see **Figure 210**). Additional information about Ra is given on page **209**.

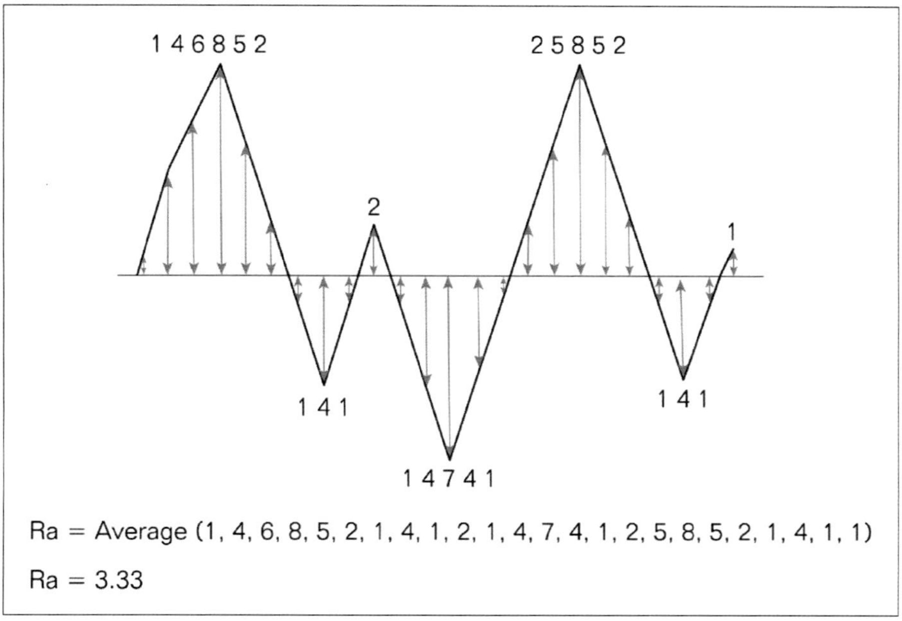

Figure 210: Calculation of Ra

Ra1: The largest Ra value within any window one sampling length wide. This parameter can identify rough (or smooth) regions, regions that may be individually out of specification, embedded in a long feature that is cumulatively within specification. An example is shown in **Figure 211**.

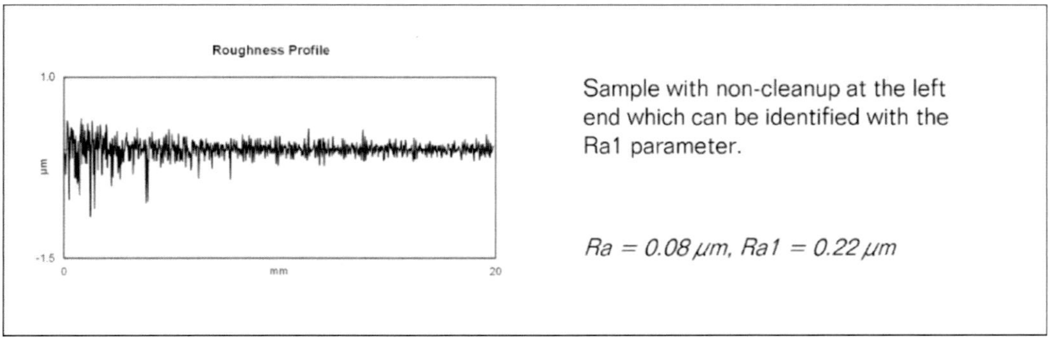

Figure 211: Example of the use of the Ra1 parameter

Ramax: In some instruments, the largest Ra of any sampling length within an evaluation length. The Ramax parameter is not recognized by ISO or ASME, and can be easily confused with the ISO max-rule applied to the Ra parameter, which is also written as "Ramax". See the section, "**What does "max" mean in a surface texture callout?**" on page **187** for information about the max-rule.

Range: The maximum vertical (z-axis) distance over which a surface texture measurement system can collect data. See page **359** for more information.

Rc: See Wc.

Rcl (Pcl): The *roughness* profile correlation length, the distance along the x-axis at which the autocorrelation function decays to (1/e) or 37%. See page **362** for more information.

Rda (Pda, Wda): The arithmetic average slope of the *roughness* profile. The local slope is calculated at each point and the average of the absolute values is reported as Rda. Note that the seven-point Lagrangian equation used to calculate the slope (equation **3** on page **375**) contains the data spacing in the denominator, so Rda is sensitive to that data spacing.

Rdc (Pdc): The vertical height between two user-defined material ratios (see **Figure 212**). Rdc is also sometimes designated as Htp, RHtp, or Rδc.

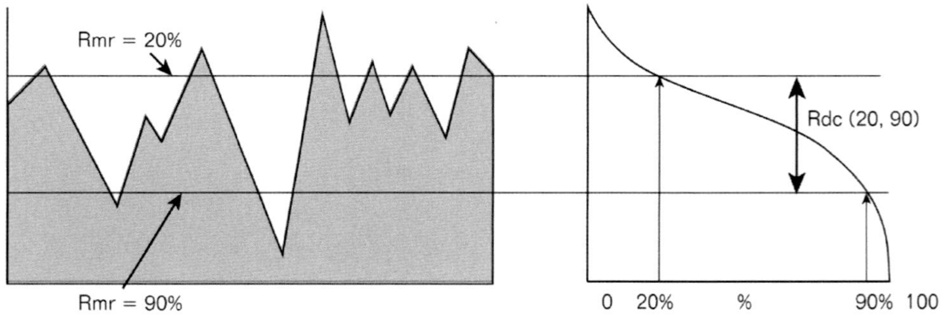

Figure 212: Illustration of Rdc

Rdq (Pdq, Wdq): The RMS slope of the *roughness* profile. The local slope is calculated at each point and the standard deviation of these slopes is reported as Rdq. Note that the seven-point Lagrangian equation used to calculate the slope (equation **4** on page **376**) contains the data spacing in the denominator, so Rdq is sensitive to that data spacing.

Resolution: The ability of the system to discriminate between two points. In the context of stylus-based measurement, resolution refers to the vertical resolution of the measurement system. See **Vertical resolution.**

Rhsc (Phsc, Whsc): The peak (high spot) count based on a user threshold. Rhsc is the number of peaks that exceed a user-specified height, with a peak defined as the highest point between two crossings of the threshold value (see **Figure 213**). Unlike a peak count (Rpc), the high spot count includes any peak that crosses the threshold, regardless of the adjacent valley depths. The high spot count is assessed over the entire evaluation length and is not normalized, so the evaluation length must be clearly stated. A comparison of the peak and valley counting parameters is given in **Table 22**.

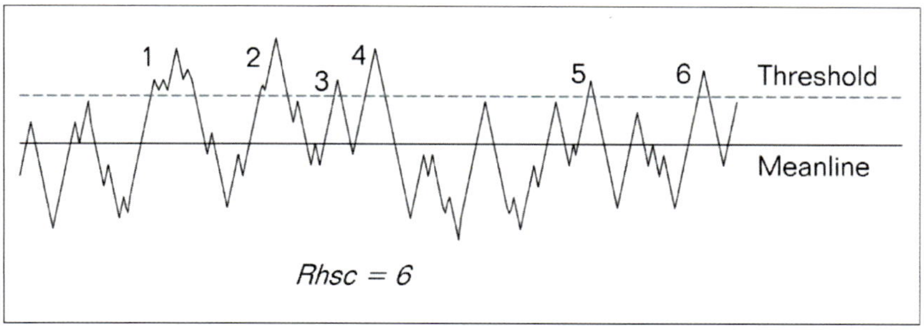

Figure 213: Illustration of Rhsc

Table 22: Comparison of peak and valley counting parameters

Parameter	Thresholds	Number per . . .
Rhsc	One	Evaluation length
Rpc	Two, parallel to and equidistant from the mean line	Unit length
Rvc	One	Evaluation length

Rk (Pk): The core (kernel) roughness, derived from the material ratio curve. See page **340** for a discussion of the Rk parameter family.

Rku (Pku, Wku): Profile kurtosis, describing the randomness of the profile heights and the sharpness of the distribution of profile points about the mean line (see **Figure 214**). A Gaussian distribution of points has a kurtosis of 3.0. Values greater than 3 indicate a profile with extreme peaks and valleys, outside the normal statistical limits of a Gaussian distribution. Values less than 3 indicate a profile without extreme peaks and/or valleys. Rku is strongly influenced by isolated points and therefore tends to be unstable and non-repeatable; see page **338** for more information.

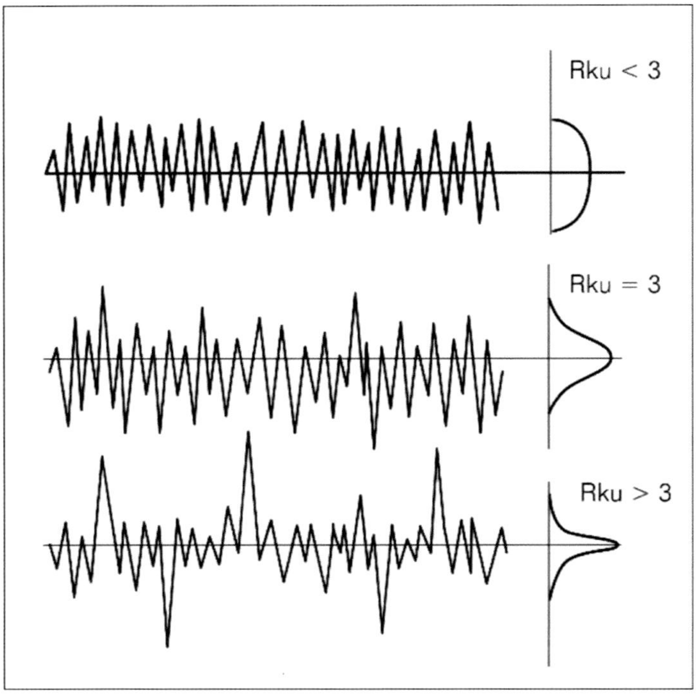

Figure 214: Illustration of kurtosis values corresponding to various profile distributions

Rm0 (Pm0): The zeroth spectral moment. It is equal to the variance of profile heights, calculated as the RMS roughness squared (Rq^2). The spectral moment parameters are sometimes used in the simulation of contact between rough surfaces[73].

Rm2 (Pm2): The second spectral moment. It is equal to the variance of the first derivatives of the *roughness* profile, calculated as the RMS slope squared (Rdq^2).

Rm4 (Pm4): The fourth spectral moment, equal to the variance of the second derivative of the *roughness* profile.

Rmax or **Ry** or **Rz1**: The maximum peak-to-valley distance of any single sampling length within an evaluation length (see **Figure 215**).

[73] See, for example, John I McCool, "Comparison of models for the contact of rough surfaces," *Wear*, 107(1) (Jan 1986): 37-60

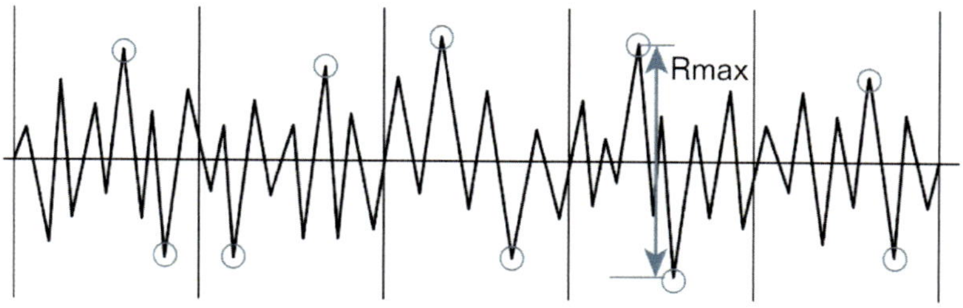

Figure 215: Illustration of Rmax

Rmq (Pmq): Material ratio at the peak-to-valley transition of a stratified (plateaued) surface, derived from the material probability curve. Additional information about the material probability curve and its associated parameters can be found on page **346**.

Rmr (Pmr): Material ratio, the percentage of material at a given height or depth (section or slice level) within the profile (see **Figure 216**). The material ratio curve (aka Abbott-Firestone curve or bearing ratio curve) plots the material ratio (x-axis) as a function of this section level (z-axis).

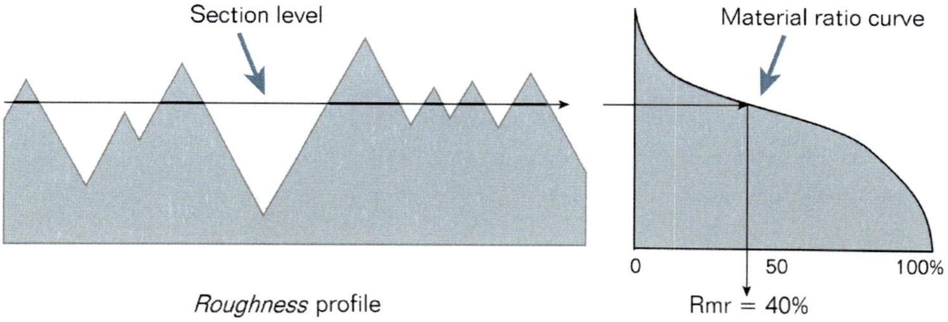

Figure 216: Illustration of Rmr and the material ratio curve

Rmr1 (Pmr1): Material ratio at the peak-to-core transition of the material ratio curve. See page **340** for a discussion of the Rk parameter family.

Rmr2 (Pmr2): Material ratio at the core-to-valley transition of the material ratio curve. See page **340** for a discussion of the Rk parameter family.

Roughness profile: The short wavelength data set obtained after reference geometry removal and the application of the roughness (lambda-c) and short wavelength (lambda-s) filters. The *roughness* profile is bounded by the roughness filter on the long wavelength side and the short filter on the short wavelength side. The "R" parameters are based on this profile.

Rp (Pp, Wp): The maximum peak height. In ASME B46.1 this is the maximum peak height in the evaluation length. In ISO this is assessed over each sampling length and then averaged over the evaluation length. The ISO Rp is therefore equivalent to the ASME Rpm.

Rpc (Ppc, Wpc): The peak count based on an upper and lower user threshold. Rpc is the number of peaks per unit length (#/cm) in a *roughness* profile. In order to be counted, a peak must rise above a threshold level, and the valleys on either side must drop below a second threshold below the mean line (see **Figure 217**). The thresholds are parallel to and equidistant from the mean line. A comparison of the peak and valley counting parameters is given in **Table 22** on page **303**.

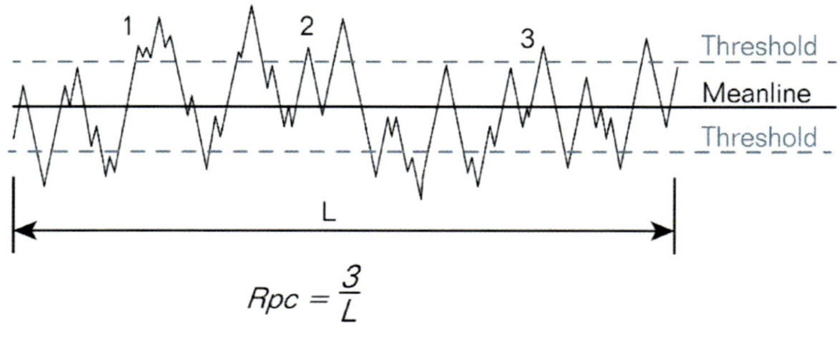

Figure 217: Illustration of Rpc (three detected peaks)

Rpk (Ppk): The reduced peak height, an average of the heights above the core profile. See the section starting on page **340** for a discussion of the Rk parameter set.

Rpm (Ppm): Mean peak height. The average of the highest peak in each sampling length, averaged over the evaluation length (see **Figure 218**). This is an ASME parameter, and is equivalent to the ISO Rp.

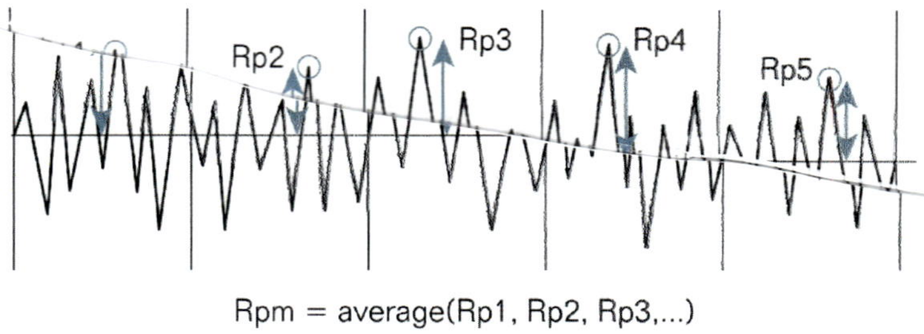

Figure 218: Illustration of Rpm (equivalent to ISO Rp)

Rpq (Ppq): The RMS roughness of the peak component of a stratified (plateaued) surface, derived from the material probability curve as the slope of the plateau region. Additional information about the material probability curve and its associated parameters can be found on page **346**.

Rq (Pq, Wq): The RMS amplitude of the *roughness* profile, equivalent to one standard deviation of the distribution of profile distances from the mean line (see **Figure 219**). Rq is more sensitive to peaks and valleys than Ra because its value is based on the square of the amplitudes (see Equation **9** on page **376**).

Figure 219: Graphical representation of the Rq parameter

Rs (Ps, Ws): The average spacing between peaks, with a peak defined as the highest point bounded by two meanline crossings (see **Figure 220**). Unlike Rsm, Rs does not discriminate based on the feature height or width. Note that, by convention, spacing parameter values are stated in millimeters rather than micrometers, consistent with the axis to which they refer (length rather than height).

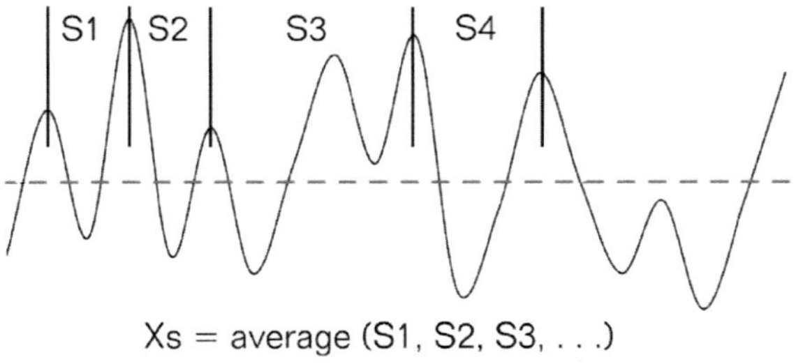

Figure 220: Illustration of the "s" parameter

Rsk (Psk, Wsk): The profile skewness, describing the shape or symmetry of the distribution of profile points. Positive skewness values indicate a profile with dominant, isolated peaks while negative values indicate dominant, isolated valleys (see **Figure 221**). Due to this strong influence of isolated peaks and valleys, Rsk tends to be unstable and non-repeatable; see page **338** for more information.

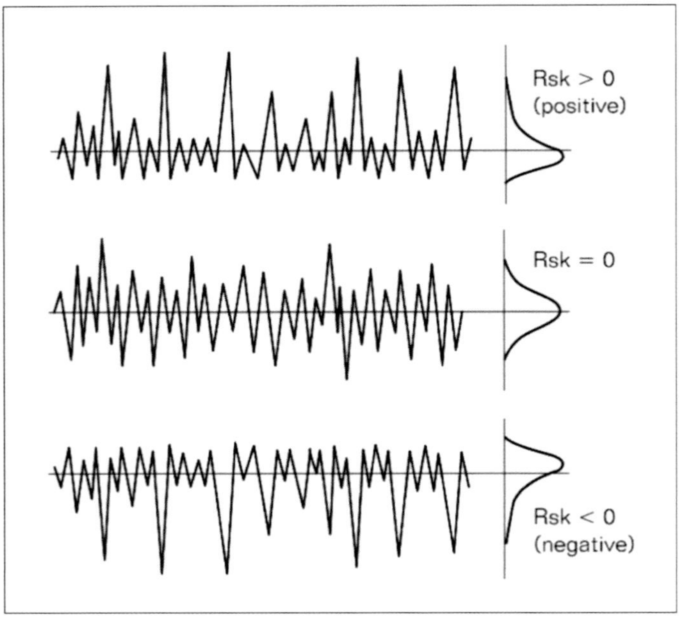

Figure 221: Illustration of skewness values corresponding to various profile distributions

Rsm (Psm, Wsm): The average spacing between profile elements (see **Figure 222**). This can also be thought of as the spacing between all the significant positive or all the significant negative meanline crossings. Spacing and height discrimination is required for the evaluation of these spacing parameters because the spacing value can be significantly impacted by narrow and/or short features that may not be functionally significant; this is especially true for Wsm and Psm. An example of the effect of different threshold or discrimination levels can be seen in **Figure 223** for Wsm (see page **352** for more information). Page **201** describes a proposed method for including these discrimination levels on a technical drawing. Note that, by convention, spacing parameter values are stated in millimeters rather than micrometers, consistent with the axis to which they refer (length rather than height).

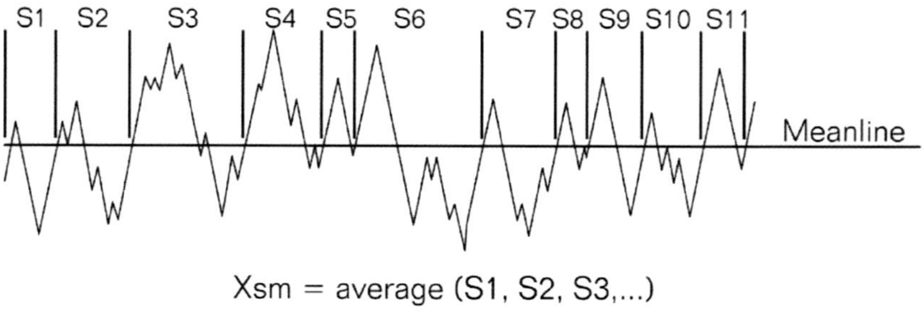

Figure 222: Illustration of the "sm" parameter

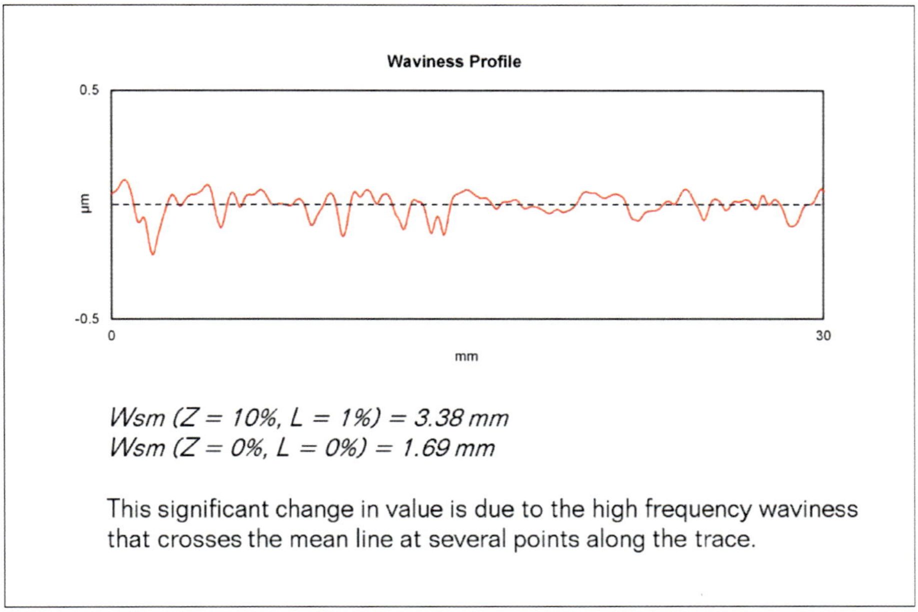

Figure 223: Illustration of the effect of discrimination level on Wsm

Rt (Pt, Wt): The total peak-to-valley height; the distance between the highest peak and the deepest valley in the evaluation length (see **Figure 224**). Note that this definition is consistent between ASME and ISO.

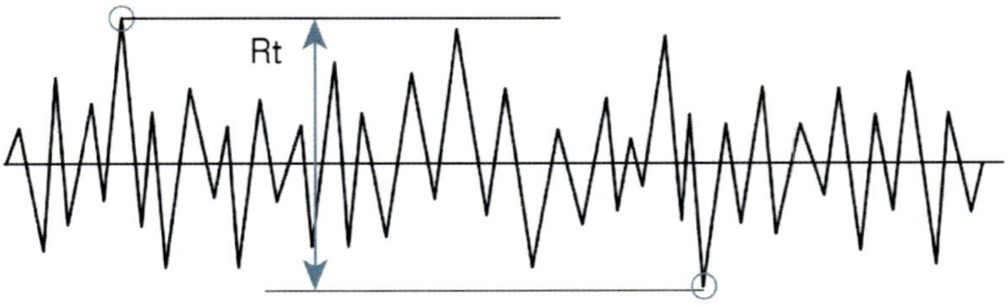

Figure 224: Illustration of Rt

Rv (Pv, Wv): The maximum valley depth, reported as a positive value. In ASME B46.1 this is the maximum valley depth in the evaluation length. In ISO this is assessed over a sampling length and averaged over the evaluation length. The ISO Rv is therefore equivalent to what in ASME terminology would be called Rvm (see Rvm).

Rvc (Pvc): The valley count based on a user threshold. Analogous to Rhsc, Rvc is the number of valleys over the evaluation length that are deeper than a user-specified threshold (see **Figure 225**). A valley is defined as the deepest point between two threshold crossings. The valley count is not normalized so the evaluation length must be clearly stated. A comparison of the peak and valley counting parameters is given in **Table 22** on page **303**. Rvc is not a standardized parameter and may not be available in all commercial software.

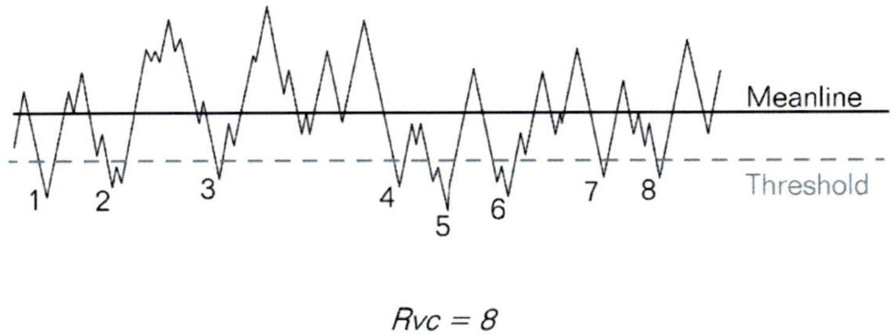

Figure 225: Illustration of Rvc

Rvk (Pvk): The reduced valley depth. See the section starting on page **340** for a discussion of the Rk parameter set.

Rvm (Pvm, Wvm): Mean valley depth. The average of the deepest valley in each sampling length, averaged over the evaluation length (see **Figure 226**). It is equivalent to ISO Rv, and is reported as a positive number. Rvm is analogous to the ASME parameter Rpm, although Rvm is not listed in ASME B46.1.

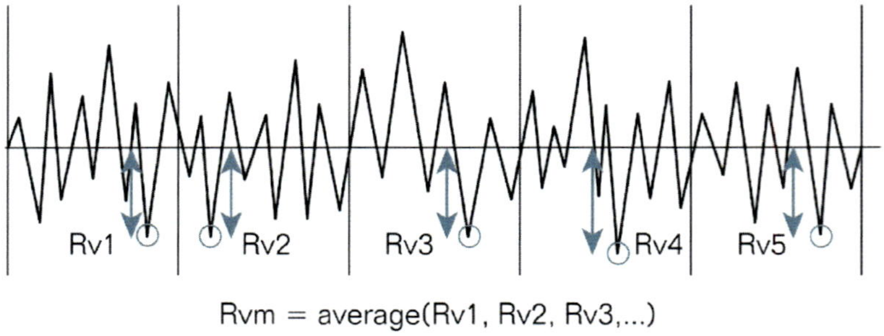

Figure 226: Illustration of Rvm (equivalent to ISO Rv)

Rvo (Pvo): The reduced valley volume from the material ratio curve, and a member of the Rk set of parameters (see the section starting on page **340**). Rvo is the area bounded by the material ratio curve, the core-to-valley transition height, and the 100% axis. On lubricated surfaces this parameter may provide a measure of the oil retention volume. Rvo has units of length because it is calculated from a two-dimensional graph with one axis being percent. Also known as A2.

Rvq (Pvq): The RMS roughness of the valley component of a stratified (plateaued) surface, derived from the material probability curve as the slope of the valley region. Additional information about the material probability curve and its associated parameters can be found on page **346**.

Rx: Maximum roughness motif depth. See page **354** for a discussion of the "motif" parameters.

Rz (Pz, Wz): Average peak-to-valley roughness based on the single highest peak and single deepest valley in each sampling length (see **Figure 227**). The meaning of Rz has varied depending on the standards organization and the release date of the standard, so this book occasionally uses the (unofficial) parameter names RzDIN and RzJIS to distinguish between the meanings. Rz is currently equivalent to RzDIN.

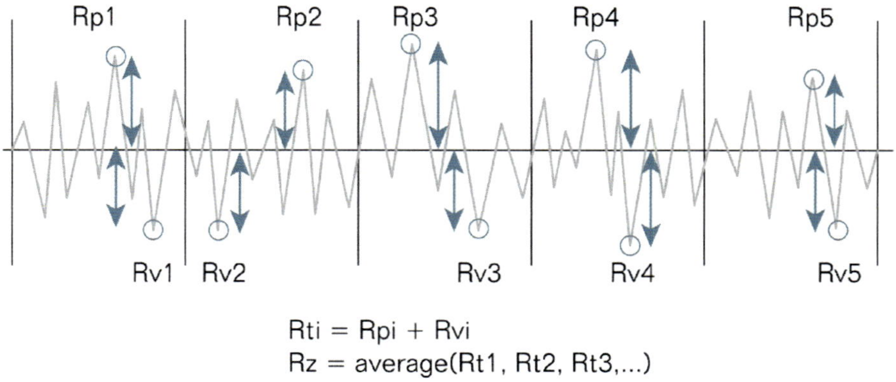

Figure 227: Illustration of Rz

RzDIN (PzDIN, WzDIN): Average peak-to-valley roughness based on the single highest peak and single deepest valley in each sampling length. These peak-to-valley distances are averaged over the evaluation length (see **Figure 227**). "RzDIN" is not an official name, but it is used here (and in some software) to distinguish it from RzJIS, the other Rz parameter (see **Figure 228** for a comparison of RzDIN and RzJIS). Rz is the second-most specified roughness parameter after Ra. It is more sensitive to changes in a surface than Ra because the maximum peaks and valleys are being considered, rather than the average of all features represented by Ra. Note that the current ISO Rz parameter is equivalent to RzDIN.

RzJIS (PzJIS, WzJIS): Average peak-to-valley roughness based on the five highest peaks and five deepest valleys in each sampling length. These "per sampling length" peak-to-valley distances are averaged over the evaluation length (see **Figure 228**). "RzJIS" is not an official name, but it is used here (and in some software) to distinguish it from RzDIN, the other Rz parameter. RzJIS will always have a smaller value than RzDIN. Note that prior to 1997, the ASME and ISO Rz parameter was equivalent to RzJIS.

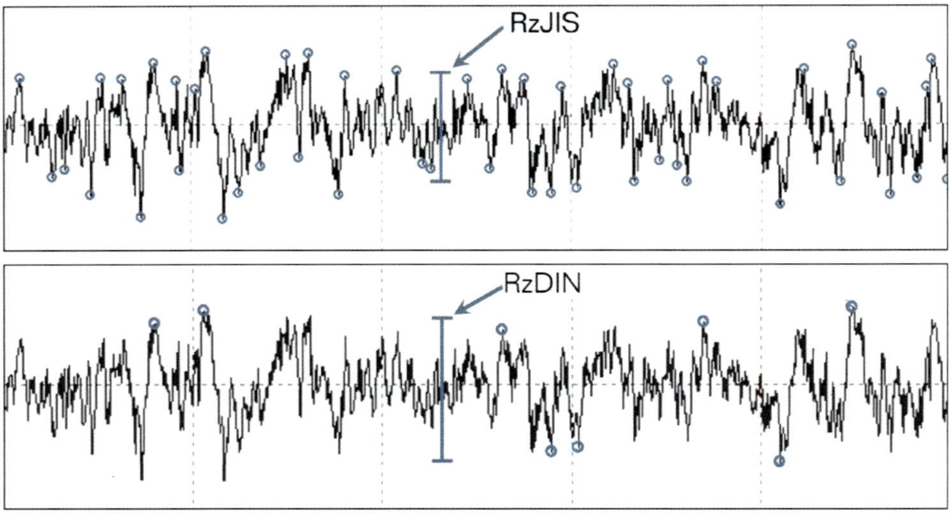

Figure 228: Comparison of RzJIS and RzDIN

Rzmax: In some instruments, the largest Rz of any sampling length within an evaluation length. It is equivalent to Rmax (Ry) or Rz1max in some instruments. The Rzmax parameter is not recognized by ISO or ASME, and can be easily confused with the ISO max-rule applied to the Rz parameter, which is also written as "Rzmax". See the section, "**What does "max" mean in a surface texture callout?**" on page **187** for information about the max-rule.

Sampling length: The length along the x-axis used for calculating some surface texture parameters (see **Figure 204** on page **296**). The roughness sampling length is designated *lr* and by default has the same numerical value as the roughness cutoff (λc). The waviness sampling length is designated *lw* and by default has the same numerical value as λf. The sampling length for the *primary* profile, *lp*, is numerically equal to the evaluation length.

Section level: The depth within a profile at which a material ratio is calculated. This depth is sometimes designated as "c" or "p"; for example, Rmr(c). The section level can be measured relative to different reference points, depending on how it is specified. For example:

1. the distance below the highest profile peak (the default method),
2. the distance below the depth to a reference material ratio (Rmr0), or
3. the height above the mean line.

Some industries use a "fraction of Rz" as the depth; for example, 20% of Rz, generally assumed to be relative to the highest peak. Some instruments automatically apply a 5% Rmr0 reference. More information can be found in the section on material ratio on page **340**.

Short-pass: A filter that lets the short wavelength/high frequency features through while attenuating the long wavelengths. Also referred to as "high pass" in some instruments.

Surface texture: The topography of a surface, consisting of a series of peaks and valleys with varying heights and spacings. Surface texture is typically defined by the four characteristics of lay, roughness, waviness, and form. This texture is created by the material's response to the manufacturing processes, is typically modified by use, and influences the function of the surface as it interacts with its environment. Surface texture is also known as surface finish.

Transmission band: The range of wavelengths included in an analysis. For a roughness parameter the range is from lambda-s to lambda-c (see **Figure 229**). For a bandpass waviness parameter the range is from lambda-c to lambda-f. Form starts at lambda-f and extends to the length of the measurement. More information can be found on page **324**.

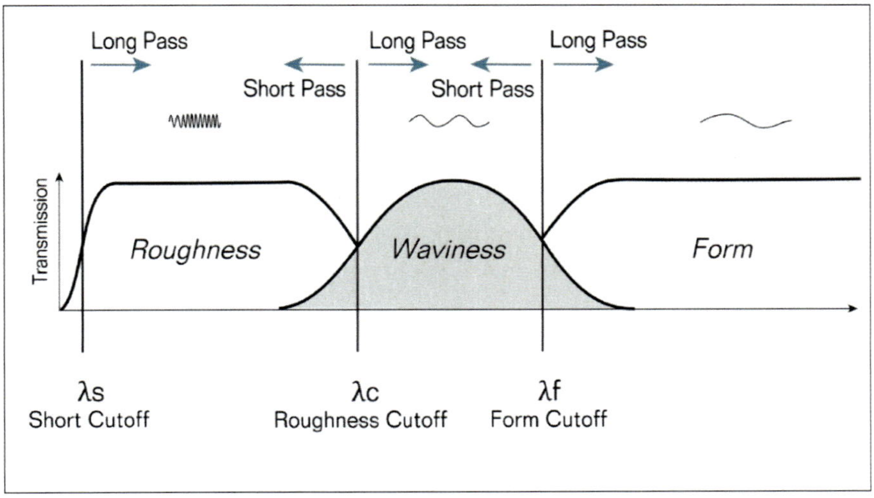

Figure 229: Illustration of transmission bands

Transmission curve: The transmission of sine wave amplitudes through a filter as a function of wavelength. **Figure 230** shows the standard short-pass transmission curves for a Gaussian filter with a 50% transmission at the cutoff wavelength. This means that, for a feature with a wavelength equal to the cutoff wavelength, half of that feature's amplitude will be included on either side of the filter (e.g., half in roughness, half in waviness). For a short-pass filter, the percentage of the amplitude passing through the filter increases with decreasing wavelength. For a long-pass filter, the percentage of the amplitude passing through the filter increases with increasing wavelength. Other types of filters can have different shapes of the transmission curve, with possibly different transmission percentages designated as the cutoff.

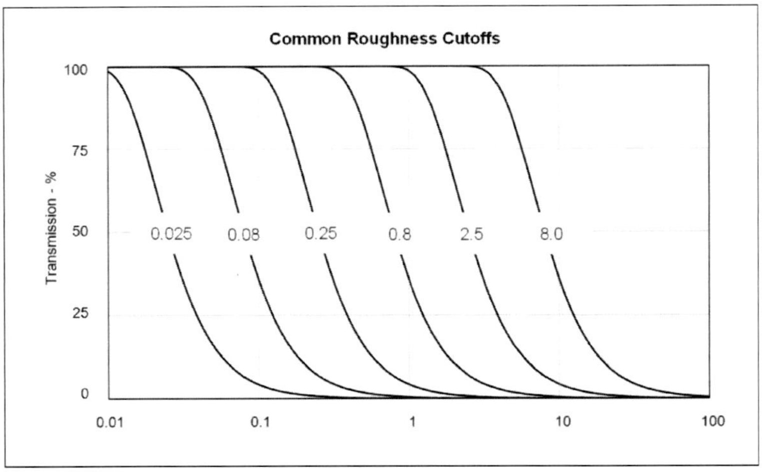

Figure 230: Standard short-pass transmission curves for a 50% Gaussian filter

Traverse length: The total distance during a measurement over which data is gathered (see **Figure 204** on page **296**). The traverse length is sometimes designated as *lt*. Also known as trace length or tracing length. Many instruments require a small amount of pre-travel, which is not typically included in the traverse length.

Valley: See **Profile valley**.

Vertical resolution: The smallest change in height (z-axis distance) that can be detected by a surface texture measurement system. See page **359** for more information.

W: Average waviness motif depth. See page **354** for a discussion of the "motif" parameters.

Waviness profile: The long wavelength dataset obtained after reference geometry or form removal and the application of the roughness filter. The *waviness* profile is bounded by the reference geometry or form cutoff (lambda-f, or λf) on the long wavelength side and by the roughness filter cutoff (lambda-c, or λc) on the short wavelength side. The "W" parameters are based on this profile.

Note that ASME terminology includes λsw as the short wavelength limit for waviness, with the assumption that this can be different than λc. While it is true that the values can be different, it is strongly recommended that λsw = λc to avoid missing or double-counting some surface wavelengths. This book assumes that λsw = λc.

Wc (Rc, Pc): The average peak-to-valley height of all profile elements in the evaluation length, calculated as the average peak height plus the average valley depth (both as positive values). When there are multiple local peaks or valleys between mean line crossings, only the highest peak or deepest valley is considered (see **Figure 231**); this is inherent in the definition of a profile element.

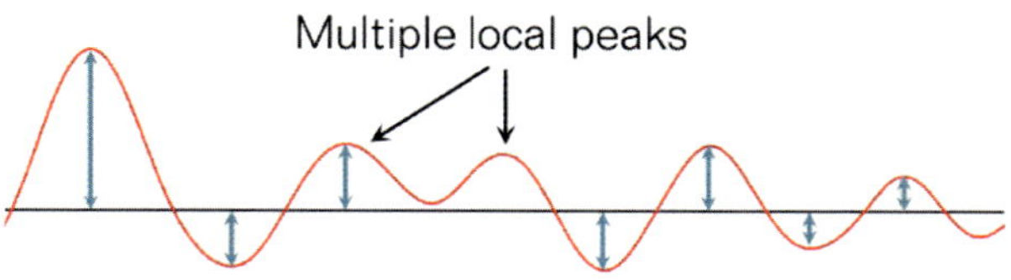

Figure 231: Illustration of Wc

The value of Wc is impacted by which waviness features are included. Numerous small waves can significantly reduce the Wc value, even though they may not be functionally important. This can be controlled by setting discrimination levels for the feature height and length; an example is shown in **Figure 232.** A proposed syntax for including these levels in engineering documents can be found on page **201**. More information on feature discrimination is given on page **352**.

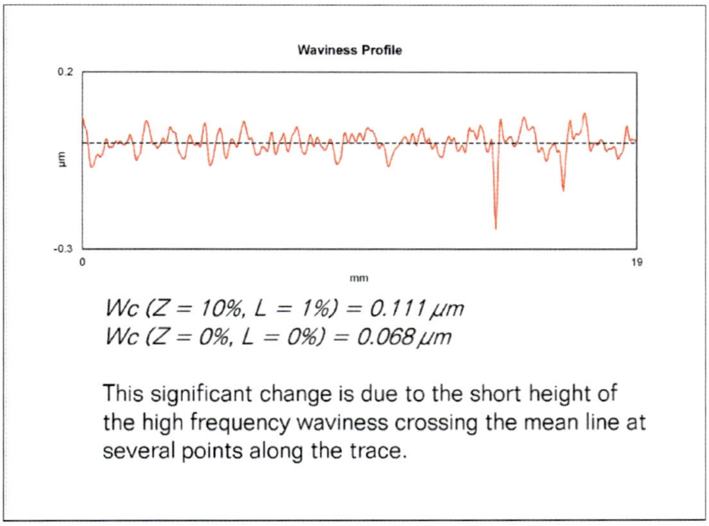

Figure 232: Example of the effect of discrimination level on Wc

Wcvx: Waviness convexity, the vertical height of the highest local convexity ("bump") in the *waviness* profile relative to the *opening* profile (see **Figure 233**). The *opening* profile is created by applying a morphological opening filter to the relevant profile (here, the *waviness* profile). The morphological filter can be visualized as rolling a disk/ball of specified radius under the profile; the surface of that disk/ball creates the *opening* profile. Wcvx is not a standardized parameter and may not be available in all commercial software.

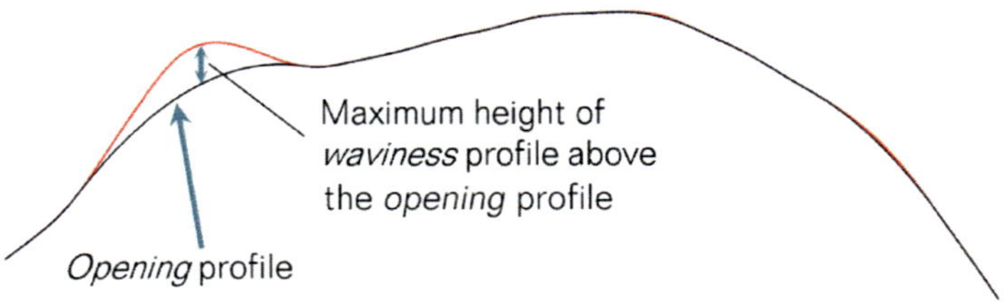

Figure 233: Illustration of Wcvx

Wcvxl: The x-axis position of the Wcvx value. Wcvxl is not a standardized parameter and may not be available in all commercial software.

WDt: The maximum peak-to-valley distance (the "t" characteristic) of the dominant wavelength in the *waviness* profile, as described in German automotive industry standard VDA 2007. The dominant wavelength is the one with the highest amplitude within the waviness domain (see **Figure 234**). WDt focuses on this wavelength while ignoring the other waviness features[74]. Other dominant waviness parameters are WDc, the mean height of the dominant waviness profile elements, and WDSm, the mean period length of the dominant waviness.

[74] If WDt is not available in your measurement system software, it can be calculated by applying a zero-width bandpass filter centered at the wavelength of interest and then multiplying the amplitudes by 4 to account for the 25% transmission from the two (combined) 50% transmission filters. For example, if the dominant wavelength is 2 mm, apply a roughness cutoff (λc) and form cutoff (λf) of 2 mm, calculate the Wt parameter, and multiply by 4.

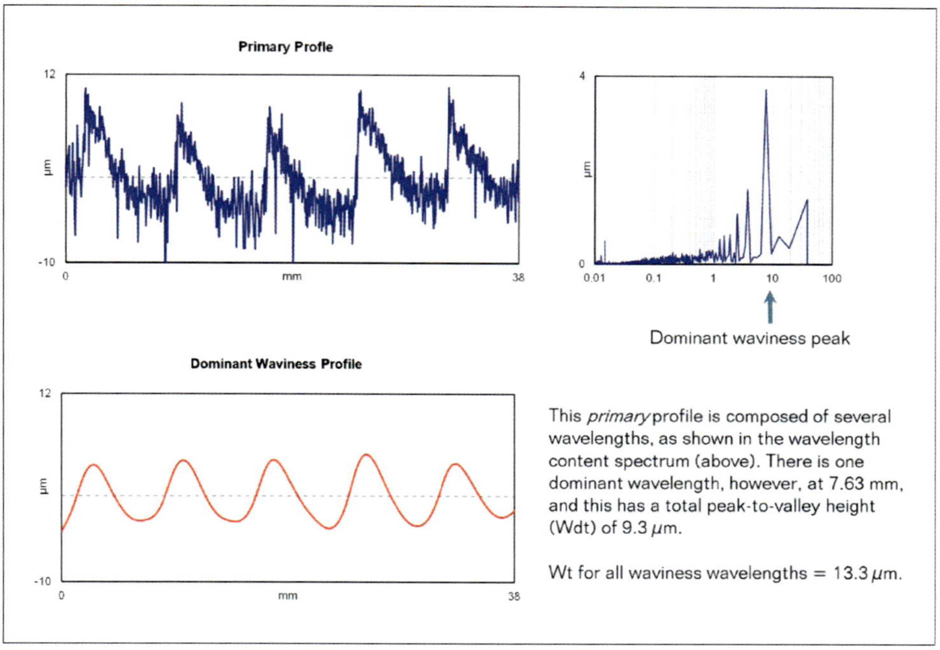

Figure 234: Illustration of WDt

Wseg: The largest local Wt (peak-to-valley distance). A window of user-specified width is moved along the *waviness* profile and the maximum peak-to-valley distance is calculated within that window (see **Figure 235**). The largest vertical distance over the evaluation length is reported as Wseg (segment waviness). The profile within the window is re-leveled using a least squares line at each location of the window as it is passed through the data. Wseg is not a standardized parameter and may not be available in all software.

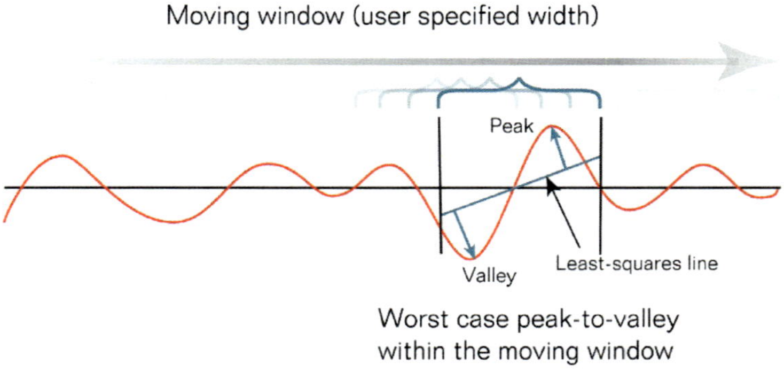

Figure 235: Illustration of Wseg

Wtc: Maximum peak-to-valley distance considering only local peaks and valleys and disregarding low or high points at the ends of the trace. Measured profiles often tend upwards or downwards at the ends, typically due to a transition such as falling off the edge of the surface or entering a fillet. The total waviness, Wt, for these types of measurements is very dependent on the starting and ending points of the trace, and is therefore not very repeatable. Wtc provides an improved repeatability by considering only the local peaks and valleys. Local peaks and valleys are those which are bounded by a falling or rising profile, respectively (see **Figure 236**). Wtc is not a standardized parameter and may not be available in all commercial software.

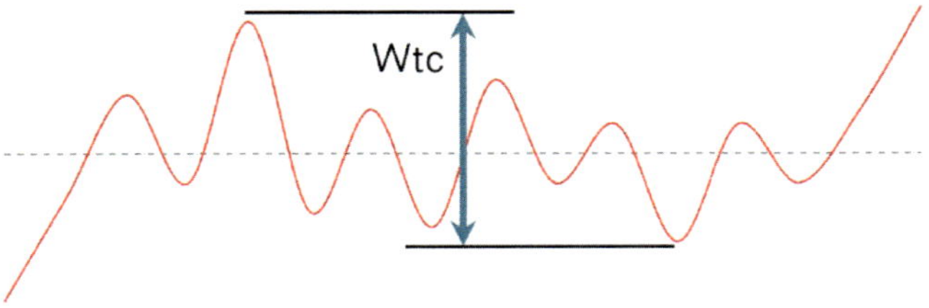

Figure 236: Illustration of Wtc

Wte: Peak-to-valley height of the waviness envelope in a motif analysis. See page **354** for a discussion of the "motif" parameters.

Wvdd: Waviness void depth, the maximum depth of the *waviness* profile relative to a *closing* profile—see **Figure 237**. The *closing* profile is created by applying a morphological closing filter to the top of the profile; this can be visualized as rolling a disk/ball of specified radius over the surface. The surface that disk creates is the *closing* profile. The section starting on page **330** contains more information about the morphological filters. An example showing the usefulness of this "virtual gasket" analysis can be found on page **243**. Wvdd is not a standardized parameter and may not be available in all commercial software.

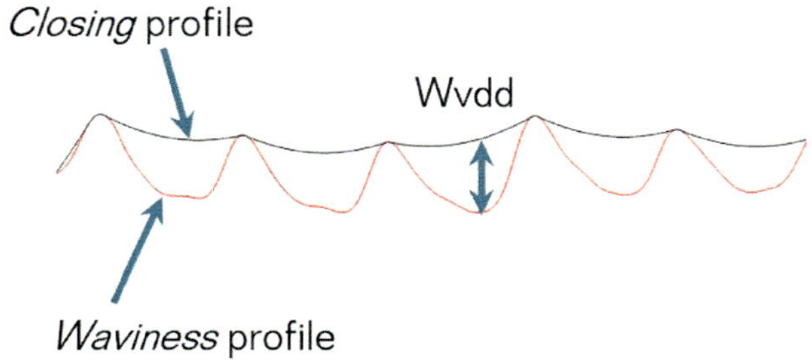

Figure 237: Illustration of Wvdd from a virtual gasket analysis

Wvoid: Waviness void area, the normalized void (leakage) area from a virtual gasket analysis (morphological disk closing filter)—see **Figure 238**. Wvoid is equivalent to the area per unit length enclosed between the *waviness* profile and the *closing* profile (the shaded areas in **Figure 238**). The units are $\mu m^2/\mu m = \mu m$. More information about the morphological filter is given on page **330**, and an example of the benefit of this analysis can be seen on page **243**. Wvoid is not a standardized parameter and may not be available in all commercial software.

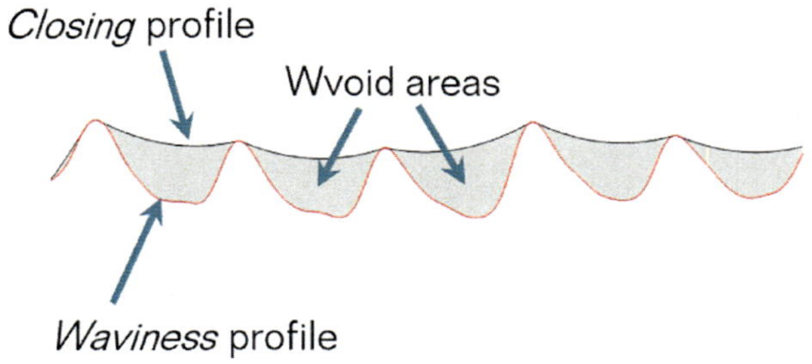

Figure 238: Illustration of Wvoid from a virtual gasket analysis

Wx: Maximum waviness motif depth. See page **354** for a discussion of the "motif" parameters.

What do R, W, and P stand for in the parameter designations?

Surface texture characteristics are derived from three wavelength ranges, or profiles: the *primary* (P) profile, the *waviness* (W) profile, and the *roughness* (R) profile. These are defined as follows, noting that all figures in this section are analyses of the same *measured* profile.

The *primary* profile is obtained from the *measured* or *direct* profile by subtracting a reference geometry and applying the short filter. The *primary* profile includes roughness, waviness, and any longer wavelength shape not removed by the reference geometry (see **Figure 239**).

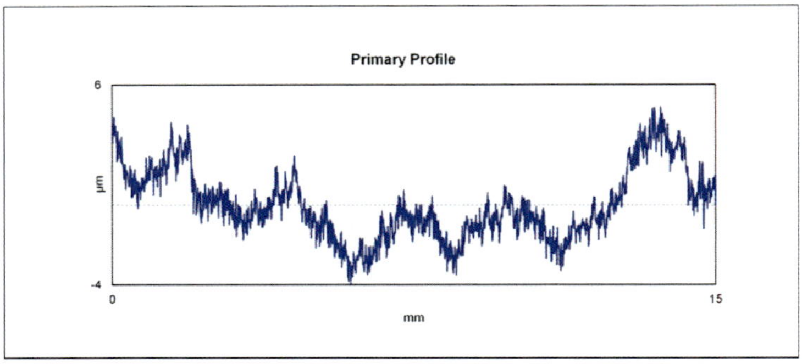

Figure 239: Example of a *primary* profile

The *waviness* profile includes the medium-wavelength components of the surface and can be of two types. If bandpass waviness is used, the *waviness* profile consists of the wavelengths between the roughness cutoff wavelength (lambda-c, or λc) and the form cutoff wavelength (lambda-f, or λf)—see the left-hand plot in **Figure 240**. If bandpass filtering is not used, waviness consists of all wavelengths longer than roughness (the right-hand plot in **Figure 240**).

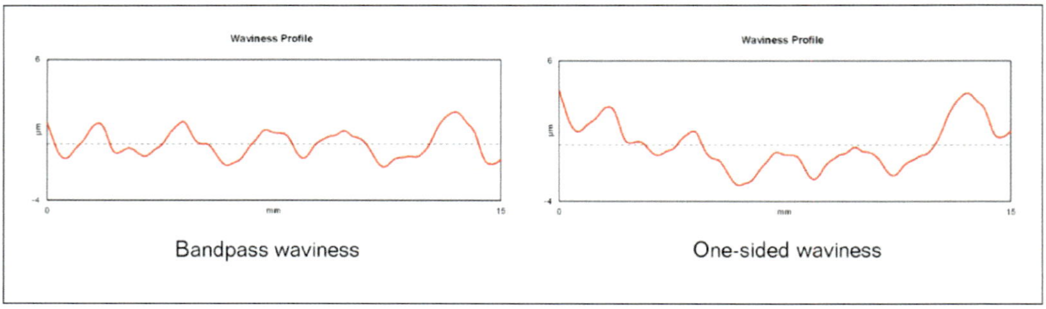

Figure 240: *Waviness* profiles from the *primary* profile shown in **Figure 239**

The *roughness* profile includes the wavelengths shorter than waviness, as defined by the cutoff filter (lambda-c, or λc) but longer than the noise defined by the short cutoff wavelength filter (λs)—see **Figure 241**. It is created by subtracting the *waviness* profile from the *primary* profile.

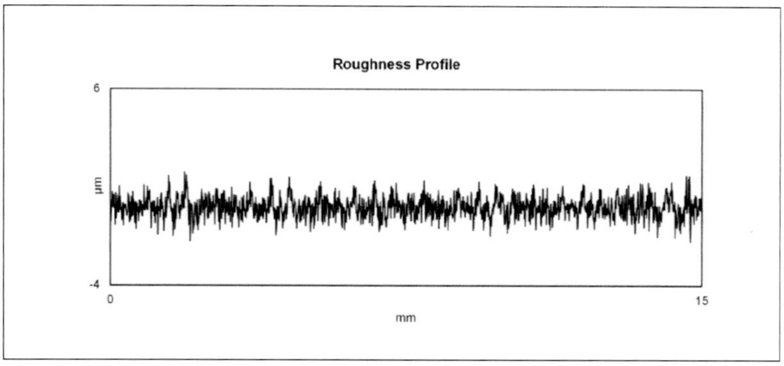

Figure 241: *Roughness* profile from the *primary* profile in **Figure 239**

When a surface texture parameter is specified (such as Pa, Wt, or Rmr) the P, W, or R designates which profile is analyzed, while the lower case letter(s) designates which attribute of that profile is reported. Therefore, Pa is the average (a) height of the *primary* (P) profile, Wt is the total height (highest peak to deepest valley) of the *waviness* profile, and Rmr is the material ratio of the *roughness* profile.

What is waviness?

The concept of surface "roughness" is common enough that most people have a reasonable idea of what it means. But 'waviness" is not as well understood. Specifically, it is often not clear which wavelengths are included in waviness. This ambiguity is increased by differences between what is written in standards and what is common in practice, both within the engineering community and within the surface texture instrument and software suppliers.

The term "waviness" is applied to those wavelengths longer than roughness. The confusion arises when deciding whether waviness consists of *all* wavelengths longer than roughness (one-sided waviness) or only those wavelengths up to some maximum value (bandpass waviness). **Figure 242** illustrates these two options. For bandpass waviness, the long-wavelength limit of waviness is lambda-f (λf); the wavelengths longer than λf are considered form.

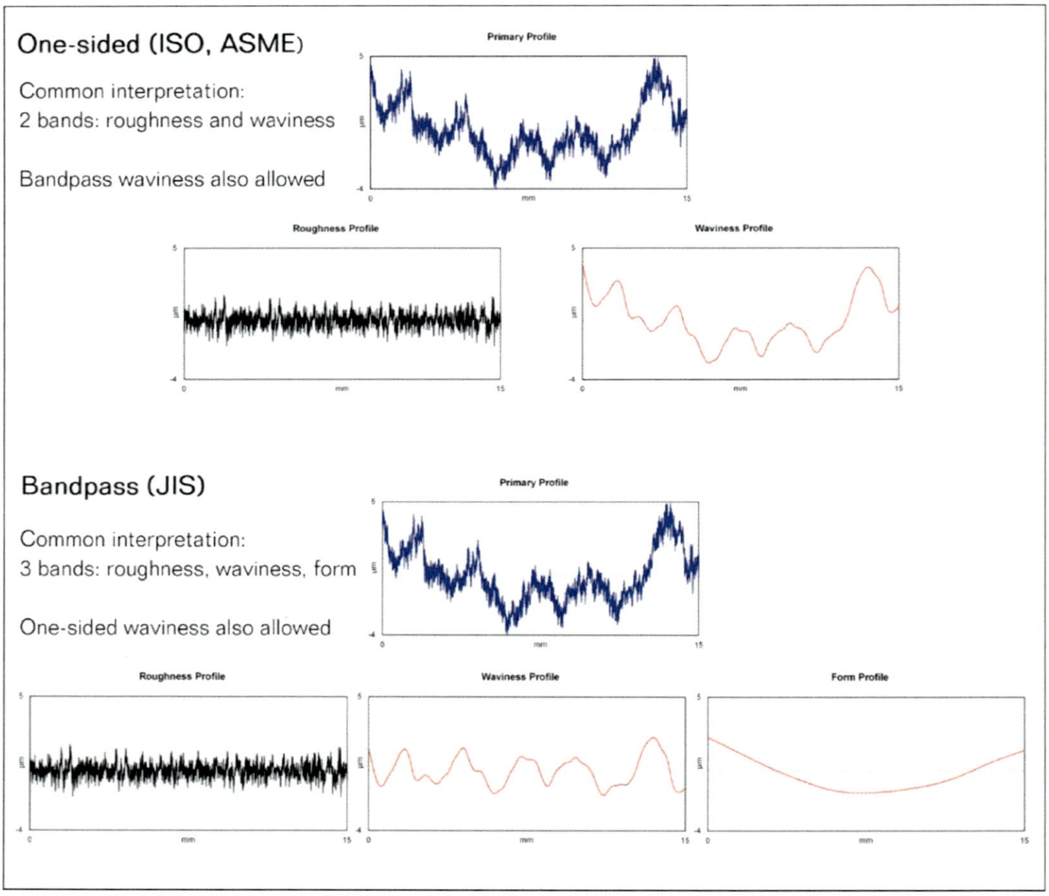

Figure 242: Two different methods of defining waviness

While ISO and ASME standards *allow* bandpass waviness, JIS documents tend to *dictate* it. Measurement equipment and software from western manufacturers usually implement a one-sided waviness—with bandpass as an option—while Japanese instruments tend to default to bandpass waviness—sometimes without saying so. It is therefore important to understand which convention your system follows.

It is also important to understand which value of λf is being applied in a bandpass analysis. Some instruments use a default value of λf = 8 mm, which may or may not be appropriate, depending on the nature of your surface and the value of λc[75]. Be sure to check your instrument's documentation for how the waviness analysis is performed.

If your surface requires a bandpass waviness analysis, the filter cutoff values must be included in the engineering documentation. One-sided waviness can also be explicitly specified in a properly-constructed specification[76].

One-sided waviness may be appropriate for surfaces that are nominally straight. In fact, in many applications, one-sided waviness is used as the measure of GD&T straightness. However, for curved surfaces it is often important to separate the waviness from the form because of their relative amplitudes (longer wavelengths commonly have higher amplitudes) and because the waviness features may influence the component's function in a manner different from the effect of the form. Even on straight surfaces, it may be important to apply tighter limits to shorter wavelengths than to the overall straightness. Bandpass waviness can facilitate this control. Further information about these effects is given in the discussion of transmission bands on page **324**. Additional information on the importance of waviness is given in the section, "**Should I be concerned with waviness?**" on page **173**.

---

[75] There is no default λf in ISO or ASME. A rule of thumb of λf = 10 · λc is often used, but the "correct" value should be determined by your component's functional requirements. Also, unlike roughness, with its 5 sampling length default evaluation length, there is no default evaluation length for waviness, so the desired length must be included in the product documentation.

[76] Bandpass transmission band limits can be included in an ISO 1302-compliant surface texture callout as follows (transmission band in bold, with λc = 0.8 and λf = 8 mm. Other parameters are given as examples only):

U "Gaussian" **0.8-8** / ln=15 / Wt 6.

For one-sided waviness, the transmission band should be specified with the λc value to the left of the hyphen and nothing to the right, as shown here:

U "Gaussian" **0.8-** / ln=15 / Wt 6.

What is a transmission band?

The wavelengths of interest must be specified whenever one refers to a surface texture parameter. This is because words like roughness or waviness don't have any inherent meaning. The moon can be described as rough, as can the ocean, a road, a piece of sandpaper, or a piece of glass. The width of the features considered as "roughness" in the above examples differ by twelve orders of magnitude. Thus, the term "roughness" only has quantitative meaning when placed in context. Similarly, when adding a surface texture specification to a technical drawing, that specification is only complete when the range of wavelengths is included (either explicitly or through a default standard).

Figure 243 is an illustration of the transmission bands for roughness, waviness, and form parameters, and **Figure 244** shows how these act to separate the features on an actual surface. The short wavelength end of roughness is the short filter cutoff, lambda-s (λs), while the long wavelength end is the roughness cutoff, lambda-c (λc). The short wavelength end of the waviness domain should be the same as the long wavelength end of the roughness domain. Waviness can be assessed over a limited band of wavelengths, with the long wavelength end designated lambda-f (λf); this is bandpass waviness. Alternatively, waviness can include all wavelengths longer than λc, incorporating the form into its value. For example, depending on your GD&T methodology, Wt without bandpass filtering can be the same as straightness, assuming the entire feature length is included in the measurement.

The ASME terms for the two wavelength limits for the waviness transmission band are λsw for the short filter and λcw for the long filter, respectively. λcw is equivalent to the ISO λf. λsw is not necessarily the same as λc, but if it isn't, some wavelengths will be double counted in your analysis ($\lambda sw < \lambda c$) or not counted at all ($\lambda sw > \lambda c$).

Page 194 provides information on how to include the transmission band limits on an ISO 1302-compliant technical drawing.

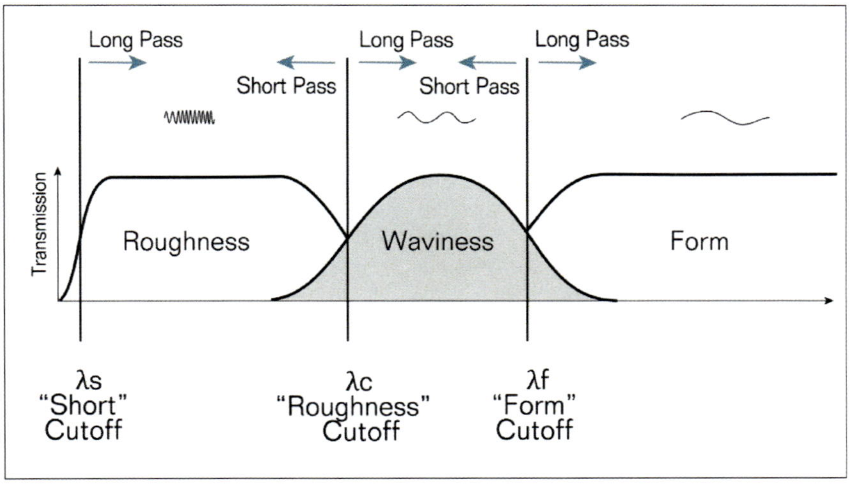

Figure 243: Illustration of transmission bands

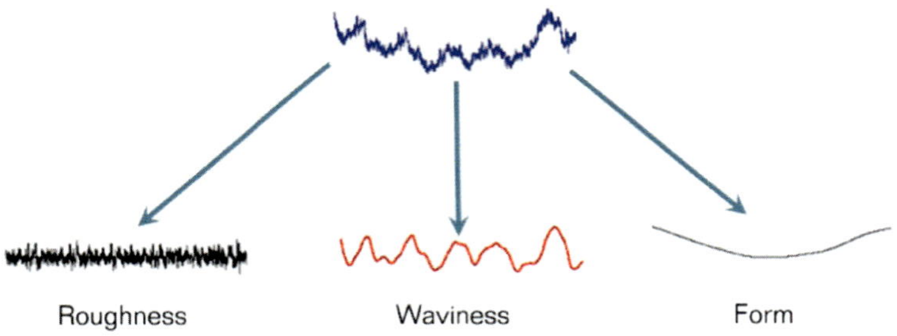

Figure 244: Separation of surface features into roughness, waviness, and form

You can also think of transmission bands in terms of the source(s) of the included features or wavelengths and their effect on function. **Table 23** lists some examples of these. An understanding of the relationship between these features and function provides the rationale for a surface texture specification and the message it communicates to the component manufacturer.

Table 23 is not meant to be comprehensive or universal. There will be situations, for example, where roughness affects clearance or form affects EHD lubrication. The point is that transmission bands, when properly selected, are more than just analysis parameters; they have a relationship to how something is made and how it is used.

Table 23: Common characteristics of surface texture domains defined by transmission bands

Characteristic name	Wavelength Range	Source	Functional impact
Noise	< λs	Instrument noise Stylus condition Surface microstructure	Considered inconsequential. If not, this should be included in roughness, with the expectation of having a suitably capable measuring instrument.
Roughness	λs to λc	Tool geometry Feeds and speeds Material structure (e.g., porosity) Process quality (e.g., non-cleanup)	Friction Wear Adhesion of coatings Visual appearance Lubrication (primarily EHD and boundary regimes) Fatigue (in extreme situations)
Waviness (sometimes combined with form to include all wavelengths longer than λc)	λc to λf	Vibration Machine ways Grinding wheel dressing Tool setup (e.g., mill cutters) Machine setup Workpiece geometry (e.g., holes in a machined surface)	Noise/vibration Sealing Local stresses Wear Lubrication
Form	> λf	Tool path Machine ways Grinding wheel dressing Workpiece clamping Workpiece geometry (e.g., holes or other discontinuities in a machined surface)	Stress Conformity to a mating surface Clearance Wear Lubrication

Are "cutoff" and "sampling length" the same thing?

"Cutoff" and "sampling length" typically have the same numerical value, so they are often assumed to be the same thing. They are not.

The "cutoff" is the wavelength at which there is a prescribed transmission through the filter. For example, when using a Gaussian filter the cutoff represents the 50% transmission point. Thus, a Gaussian roughness cutoff of 0.8 mm has 50% transmission at 0.8 mm, with 50% of the amplitude transmitted as waviness and 50% transmitted as roughness. "Cutoff wavelength" and "cutoff" are equivalent terms, with "wavelength" being implied if not stated. The short, roughness, and form filters each have a cutoff wavelength.

"Sampling length" is only used for partitioning the profile for calculating parameters such as Rz, Rpm, or Rvm. For convenience, the sampling length is set equal to the cutoff wavelength, but the similarity stops there. Sampling length identifies the portion of the evaluation length to be used for calculating a parameter. Cutoff designates the wavelength boundary between surface texture domains.

Note that the term "cutoff filter" is commonly used for the roughness filter that separates roughness from waviness. The "c" in lambda-c (or λc) stands for cutoff. It would be more correct to say, "the cutoff wavelength of the roughness filter."

What is bandpass analysis? Why would I need it?

The term "bandpass analysis" refers to the analysis of parameters within a defined range, or band, of wavelengths. For typical roughness parameters this band extends from the length of the short filter cutoff (lambda-s, or λs) to that of the roughness cutoff (lambda-c, or λc). The ratio of λc to λs is typically 300:1, except for values of λc less than 0.8 mm.

Waviness can be either single-sided or bandpass. For a single-sided analysis, waviness includes all wavelengths longer than λc. Bandpass waviness includes the wavelengths between λc and the form filter cutoff (lambda-f, or λf). This provides a separation or discrimination between waviness and the longer wavelength form. There is no standard ratio of λf to λc, but a ratio of 10 is common. This should be explicitly stated by including both λc and λf in the surface texture callout on the technical drawing.

Bandpass waviness is especially useful when evaluating waviness on a surface with a nominal shape that isn't straight. Some interfaces may be able to function acceptably with a long wavelength form error but not with higher frequency features. An example is shown in **Figure 245**, where early failures of a plain bearing system were found to correlate with the intermediate-wavelength waviness (bottom plot) but not to the longer wavelength form (top plot). A waviness specification stringent enough to control the short wavelength features would unduly restrict the manufacturer if the long wavelength form was included in the analysis. In fact, some designs require a non-straight form. In either situation, bandpass waviness is the best option for controlling the features that matter.

For additional information see the section on transmission bands on page **324**.

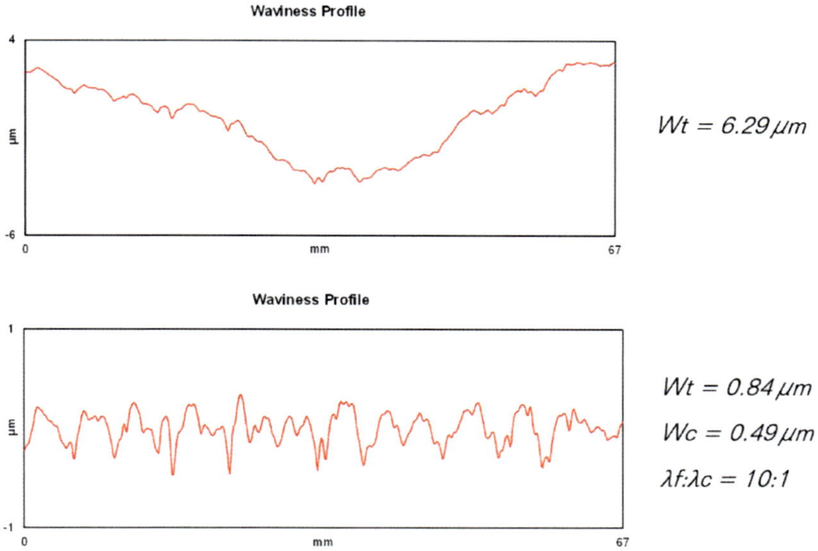

The upper graph is the single-sided waviness analysis of a workpiece with a concave form. The lower graph is the bandpass waviness analysis of this same workpiece, showing more clearly the intermediate-wavelength waviness superimposed over the longer wavelength form. Note that the bottom plot has a much smaller z-axis range.

Figure 245: Comparison of single-sided and bandpass waviness

In some industries dealing with paint and visual appearance, many different bands are specified and controlled. In these cases, the three-domain system of roughness, waviness, and form is replaced by a multi-band system, with the individual domains simply labelled by a "band name" (for example "Band A" or "Band B"). See **Figure 189** on page **267** for an example.

What is a morphological filter? What is it used for?

A morphological filter is a tool that can be used to assess the size and shape of features in a profile. It provides information not just about the height (or depth) or width of a feature, but about a combination of the two, and can therefore assess the "sharpness" (aspect ratio) of a feature. This can be useful for evaluating functional aspects of a surface such as local stress concentrators or leakage pathways.

The filter is applied by rolling a virtual shape (typically a disk or ball) over or under the profile (see **Figure 246**), creating a *closing* or *opening* profile, respectively (**Figure 247**). It is most commonly applied to the *waviness* profile. The output of the filter (the *closing* or *opening* profile) is then compared to the original profile in a morphological difference analysis to (typically) calculate either the maximum vertical distance between the two profiles or the amount of area between them. Since these output parameters are commonly derived from the *waviness* profile, their names begin with a "W". There are not yet any standard surface texture parameters that utilize a morphological filter, but the utility of some of the existing non-standardized parameters[77] will hopefully drive a broader acceptance of the concept.

Morphological filters are described in ISO 16610-40.

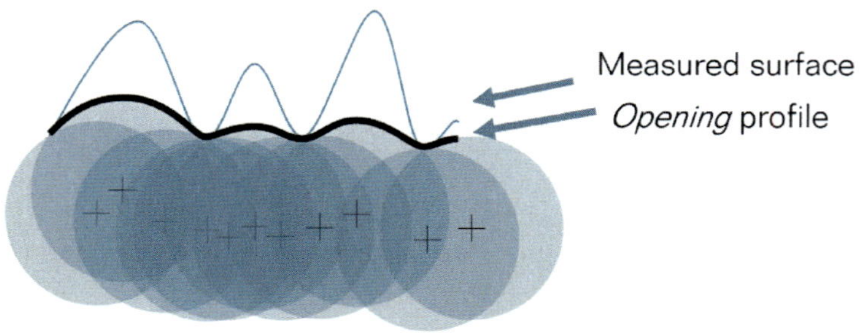

Figure 246: Schematic of a morphological opening filter for evaluating "bumps" on a surface

[77] Three such parameters included in this book are Wcvx (waviness convexity, the height of a peak or "bump" on a surface), Wvdd (waviness void depth, the depth of a valley in a surface), and Wvoid (waviness void area, the area between a closing profile and a waviness profile). More information is provided in the **Definitions** section (page **302**) and in the sections devoted to these parameters on pages **247** and **251**. These parameters may not be available in all commercial software. Other parameters can be envisioned that might be related to a surface's function, such as the number of voids or bumps, their average size, their aspect ratio, and the total void area.

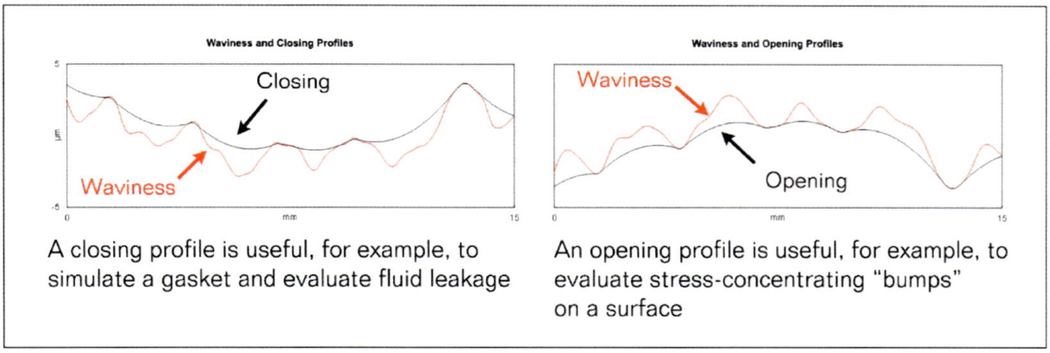

Figure 247: *Closing* and *opening* profiles

There are two key user-selectable parameters that heavily influence the output of a morphological filter when applied to a waviness profile: the virtual disk/ball radius and the λc value. These will be discussed separately.

Virtual disk/ball radius

The size of the ball helps differentiate between features of different curvature. As can be visualized from **Figure 246**, the smaller the ball, the further it will penetrate into a given feature in the *waviness* profile, and the smaller will be the resultant distance or area between profiles. This helps separate narrow "sharp" features from wider ones. This is useful, for example, when assessing the sharp, upward features of a surface as they can cause high localized stresses when contacting a mating surface.

In a similar manner, the choice of ball radius is important for assessing something like fluid leakage across a gasketed surface. This is shown in **Figure 248**, where the smaller ball reduces the amount of (shaded) area between the *closing* profile (black line) and the *waviness* profile (red line). Here the ball size is a surrogate measure of the gasket stiffness, its ability to conform to and seal a surface. More information about this application of a morphological filter is given in the section, "**Why does my gasket leak even though the roughness is correct?**" on page **243**.

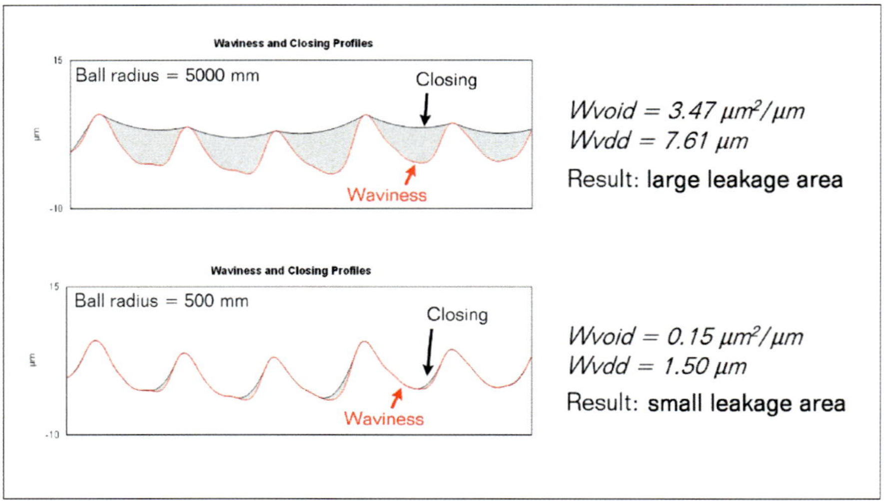

Figure 248: Illustration of the influence of ball radius on the output of a morphological filter

Lambda-c

Since morphological filters are used to evaluate the distance (or area) between the *closing* or *opening* profile and the *waviness* profile, the shape of the *waviness* profile is a key consideration. This shape is heavily influenced by the roughness cutoff wavelength (lambda-c, or λc), as has been discussed elsewhere (see, for example, **Figure 73** on page **118**). As with any choice of filter setting, the λc value must relate to both the shape and the function of the surface, creating a *waviness* profile representative of the surface and of the features on that surface that influence its function. Two illustrations are shown in **Figure 249** and **Figure 250**. The use of the Wcvx parameter to characterize the bumps shown in **Figure 249** is further discussed on page **239**.

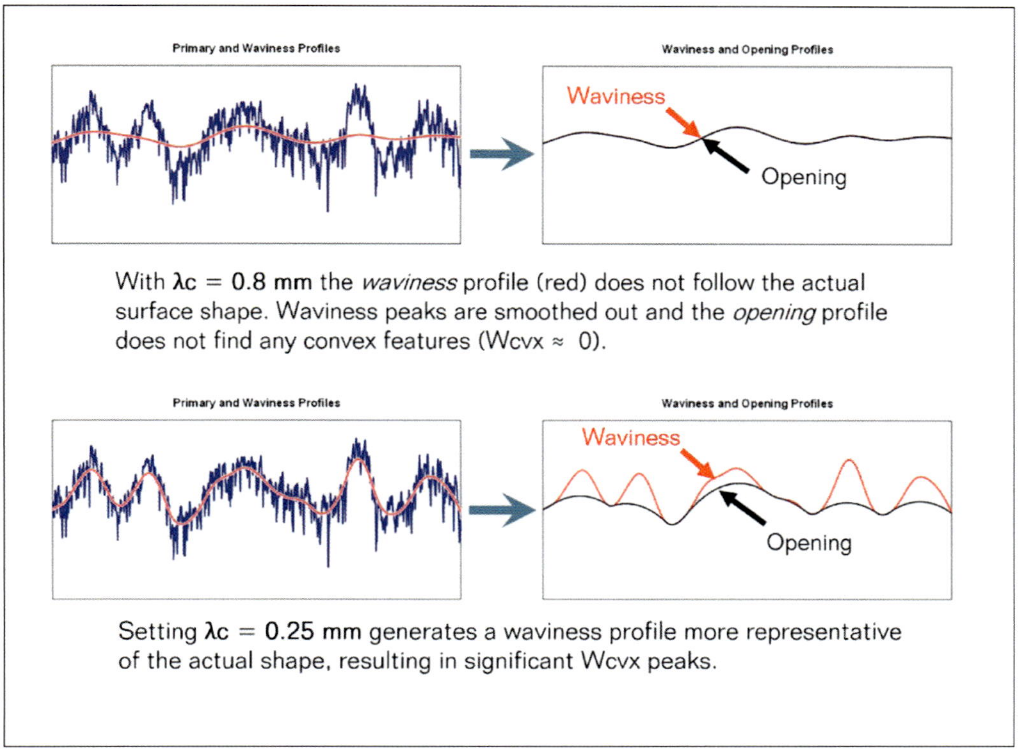

Figure 249: Effect of λc on assessing the height of a bump

The deep narrow valley shown in **Figure 250** could act as a fatigue initiation site in, for example, a highly-stressed bending fatigue application. The size of the valley can be assessed with a morphological closing filter (using, for example, the waviness void depth parameter, Wvdd), but the depth is heavily dependent on the λc value. For example, if a large λc value is chosen (e.g., 0.8 mm), the waviness profile will "bridge" the valley and thereby hide its true depth. A smaller λc value (e.g., 0.08 mm) will follow the shape of the narrow valley. This results in a feature detectable by Wvdd.

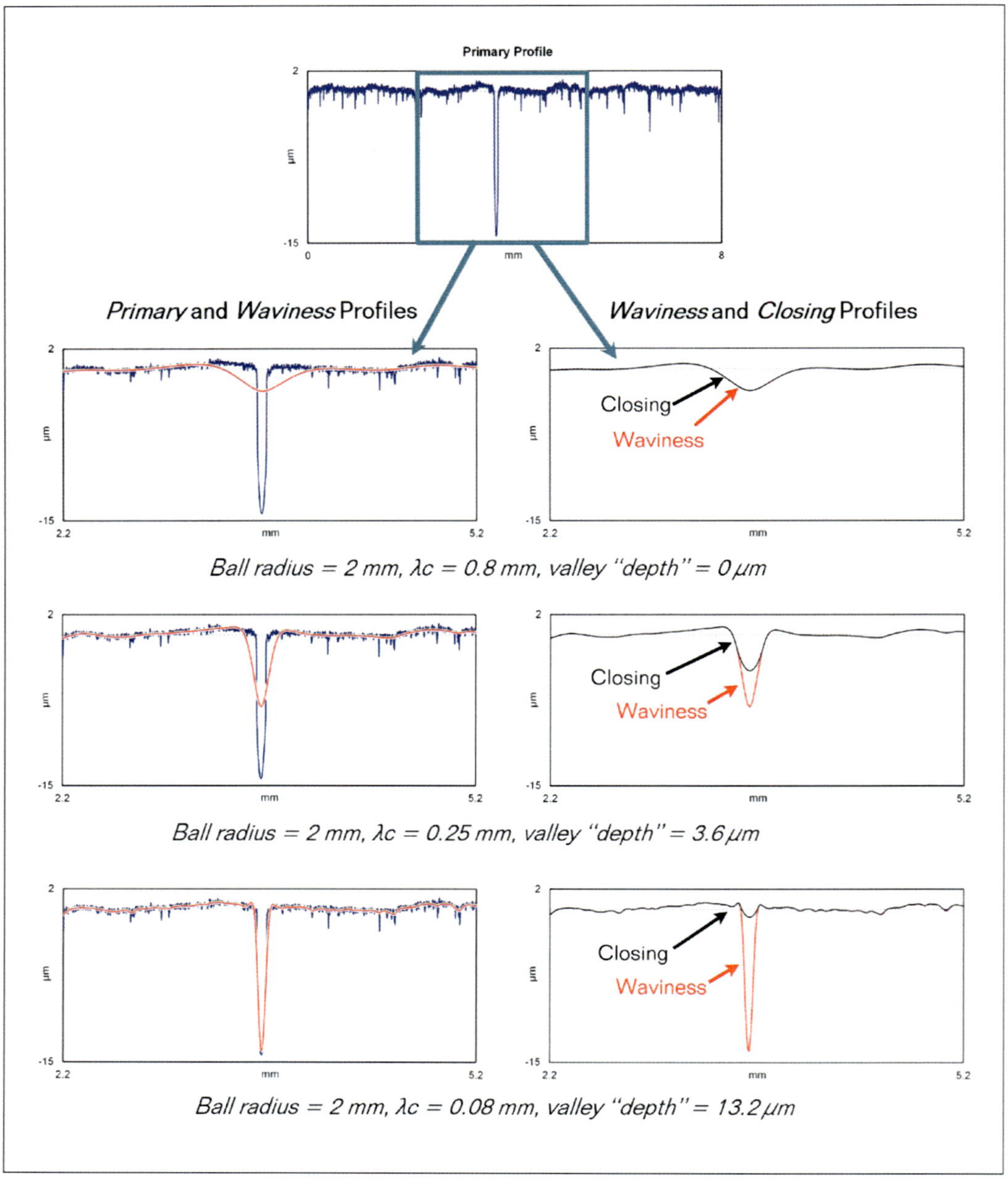

Figure 250: Effect of λc on assessing the depth of a valley

When considering two surfaces in contact, we can recognize that the finer roughness peaks can actually deform. Thus, it makes sense that we develop a *waviness* profile that accommodates this "peak deformation." This can be controlled by the selection of λc. For example, in **Figure 251** we see how the longer filter cutoff creates a *waviness* profile "below the primary peaks." This lower *waviness* profile position may better relate to a deformed surface.

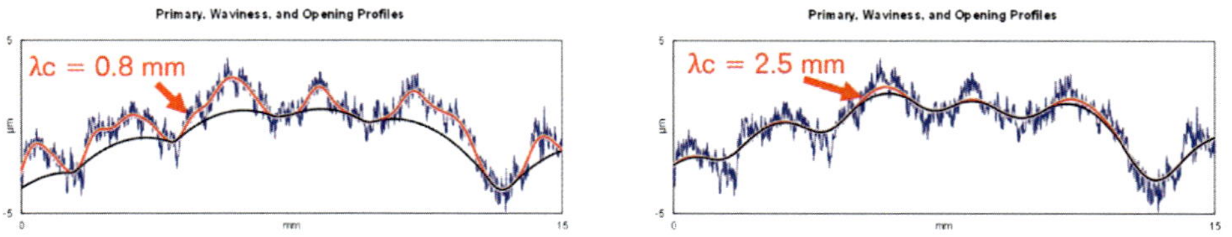

The *waviness* profile on the left closely follows the *primary* profile. The *waviness* profile on the right, slightly offset from the *primary* profile, simulates some deforming of the surface roughness peaks, which might occur under high loads or with a soft base material.

Figure 251: Using a longer λc to simulate deformation of waviness peaks

Creating a *waviness* profile that is most appropriate for a morphological filter may require a λc value different from that used to calculate roughness and other waviness parameters. For example, the depth of the valley in **Figure 250** is best captured by setting λc equal to 0.08 mm. However, a 0.8 mm λc is more appropriate for the other surface characteristics. Similarly, a 2.5 mm λc relates better to the deformation in **Figure 251** than the standard 0.8 mm value. Having two values of λc on a drawing will likely cause confusion during part inspection, so actions should be taken to minimize this confusion. Careful documentation is of course a first step. Other options include using the "normal" (in these cases, 0.8 mm) λc but specifying a larger radius for the morphological filter and setting parameter limits that accommodate an underestimation of the true feature size. Of course, whatever is done must be proven to be able to differentiate between functional and non-functional surfaces.

What is the difference between Rp and Rpm?

Rp and Rpm can be confusing because of differences between ASME and ISO definitions.

Rp and Rpm both refer to peak heights in the *roughness* profile. Historically, Rp was the highest peak in the whole evaluation length, while Rpm was the average over the evaluation length of the highest peak in each sampling length (see **Figure 252**). These are still the definitions in ASME B46.1. However, in 1997 ISO changed the definition of Rp to be the average value—the same as Rpm—and discarded the Rpm parameter. This was a consequence of the move to parameter assessments based on sampling length rather than evaluation length. As a result, you may not know if an Rp value refers to the highest peak or to the average highest peak in a measurement unless you refer to your specific software implementation. **Table 24** shows the parameter designations as they relate to the standards.

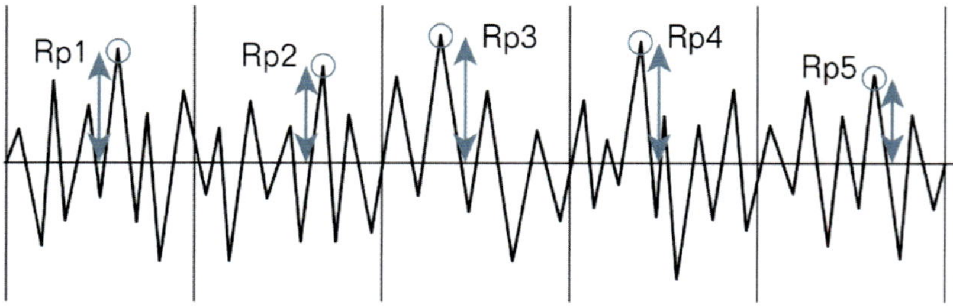

In this example:

Rp (ASME) = Rp1 (ISO) = Rp3

Rpm (ASME) = Rp (ISO) = (Rp1 + Rp2 + Rp3 + Rp4 + Rp5) /5

The analogous calculations apply to Rv, Rvm, Rt, and Rtm

Note that *Rtm = Rz (ASME) = the current Rz (ISO)*

Figure 252: Illustration of Rp, Rpm, and analogous parameters

Table 24: Parameter designations as they relate to standards

	ASME B46.1-2019	ISO 4287-1997
Single highest peak	Rp	Rp1 or Rp1max
Average of highest peak in each sampling length	Rpm	Rp

This difference in parameter definition can make a large difference in the reported parameter value, as shown with an actual surface in **Figure 253**.

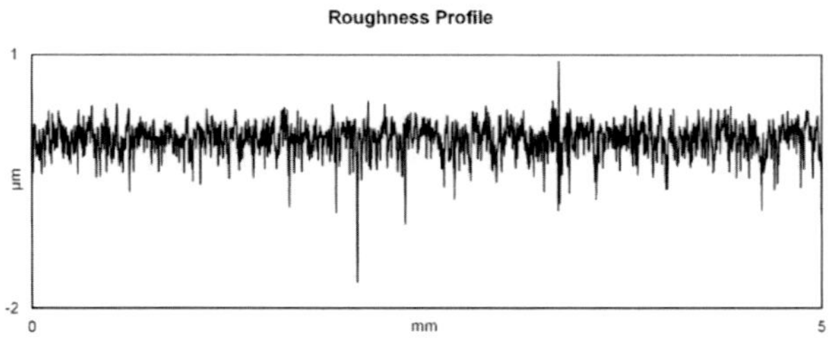

Rp (ASME) = 0.922µm due to the single peak at about 3.4 mm

Rp (ISO) = Rpm (ASME) = 0.445µm

Figure 253: Comparison of Rp and Rpm when there is one high peak

The single highest peak in a profile is extremely noisy and unstable because of the influence of dirt or vibration (that is likely the situation in **Figure 253**). It is therefore not recommended that Rp (ASME) or Rp1 (ISO) be used as specification parameters. However, the averaging parameter Rpm (or Rp according to ISO) can be a useful characteristic when dealing with surfaces that interact with each other (e.g., lubricated interfaces with thin oil films) or potentially with coatings. If Rpm is to be used, it is recommended that the technical drawing includes a note or a reference to the associated standard to provide an unambiguous definition.

Should I use Rsk or Rku to specify my surface roughness?

Rsk (skewness) and Rku (kurtosis) are parameters that describe the shape of the distribution of profile points (see **Figure 254**). They help answer questions such as, "Are there dominant peaks or valleys on my surface?" This knowledge of the surface is useful in applications such as marginally-lubricated bearings, where peaks are undesirable but valleys may be beneficial. Unfortunately, both Rsk and Rku tend to be noisy and unstable because their mathematical formulae contain the profile point amplitudes raised to the third and fourth powers, respectively (see equations **5** and **10** on pages **376 and 376**). This noisiness is illustrated in **Figure 255**, which shows traces made in the "same" location on a specific sample but in different laboratories. Occasional large valleys (or peaks) create huge swings in the values of Rsk and Rku. This makes it very difficult to demonstrate process capability or correlation between laboratories (e.g., between supplier and customer). More stable parameters, such as a combination of Rpm and Rz, are therefore recommended.

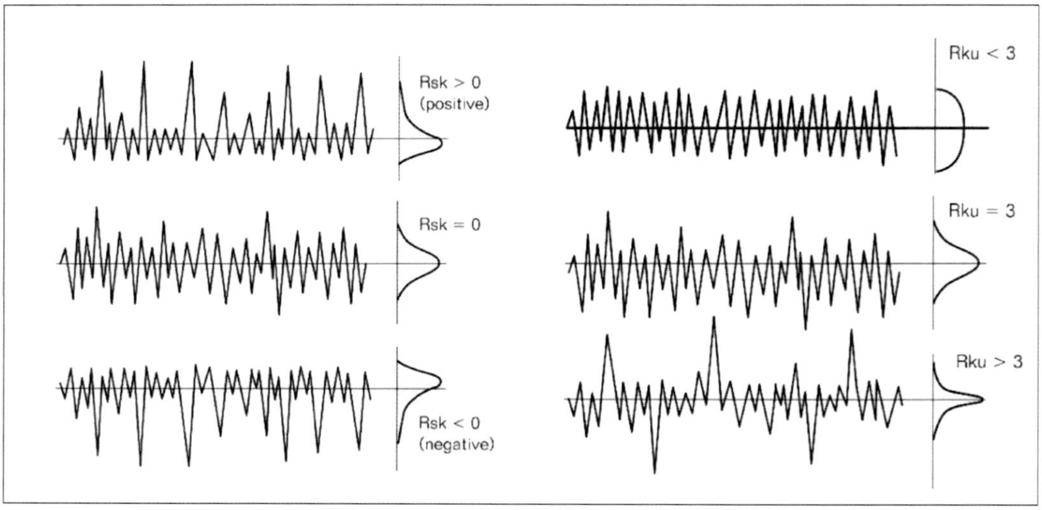

Figure 254: Graphical descriptions of Rsk and Rku

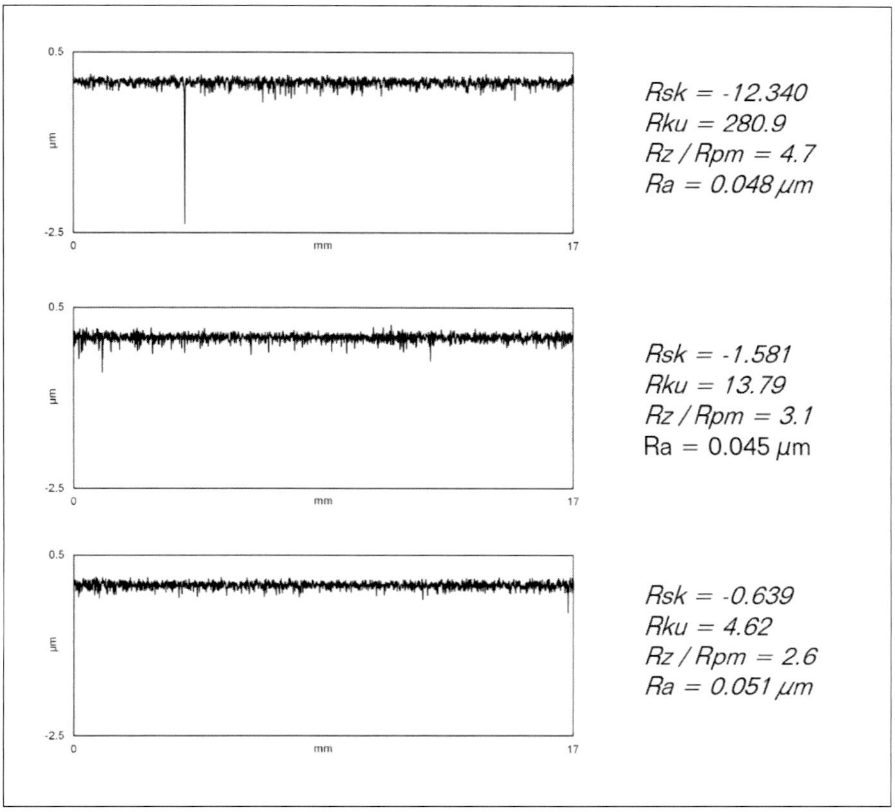

Figure 255: Example of noisiness of Rsk and Rku

What is material ratio?

Material ratio is a measure of how much material exists at a given depth below the surface. It is a concept that can help answer questions such as, "How much of my surface has to wear away before it can support the applied loads and stop wearing?" The material ratio is also sometimes called the bearing ratio or bearing area.

The material ratio is calculated by drawing a horizontal line parallel to the mean line, measuring how much of that line passes through material, and dividing that by the total line length (see **Figure 256**). If the profile under consideration is the *primary* profile, the material ratio is designated Pmr; if it is the *roughness* profile, the parameter is Rmr[78]. A plot of the material ratio as a function of depth below the surface is shown in **Figure 257** and is called the material ratio curve. It is also sometimes called the Abbott-Firestone, bearing ratio, or bearing area curve.

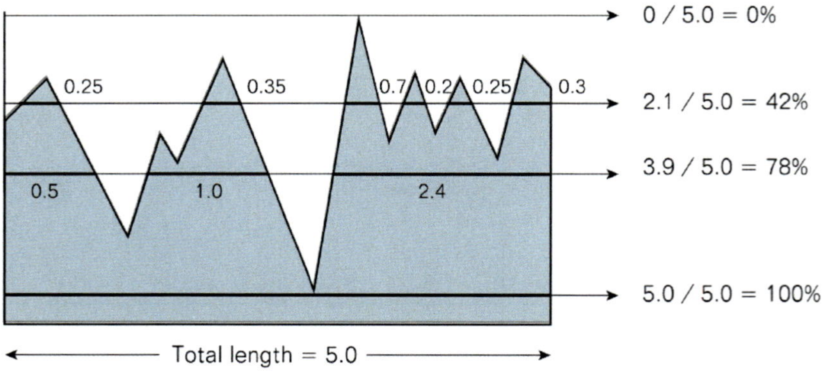

Figure 256: The calculation of material ratio at various depths below the surface

[78] The material ratio is also called tp, from the German *Traganteil Prozent* (bearing ratio percent). tpi is sometimes used instead of Rmr and tpa instead of Pmr.

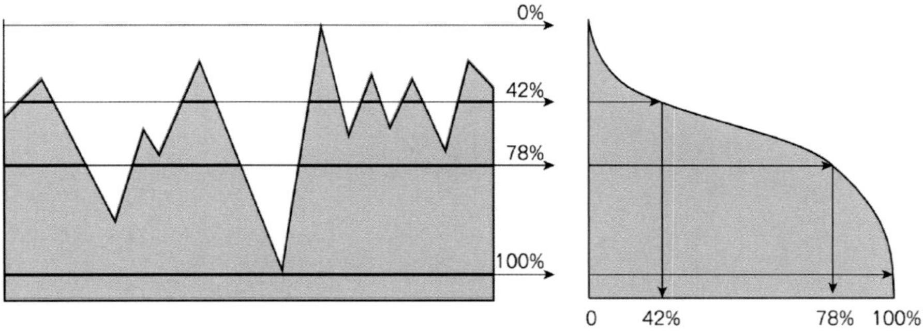

Figure 257: The derivation of a material ratio curve

A material ratio specification consists of a depth into the surface (the profile section level, designated as "c" in ISO and "p" in ASME) and the required percentage of material at that depth. The depth is typically in reference to the the highest peak or a depth relative to a specified material ratio[79]. By default, the material ratio reference point is the top of the highest peak in the profile. This tends to make the material ratio very unstable since different traces, even in the same location, will often have different highest peaks. Furthermore, the highest point on any trace may actually be debris or some other artifact. Even if that individual peak is real, its importance to the function of the component is probably small. It is therefore recommended that a material ratio specification includes a reference material ratio to ignore the top few percent of the surface and provide a more stable and repeatable measure. Five percent is a common reference value, but any number that meets the needs of the component is acceptable. The section level is then measured from the depth at which the material ratio equals this reference material ratio. **Figure 258** illustrates the difference that results from including a reference material ratio, especially on a trace with one high peak, such as the one shown.

Sample technical drawing callouts for material ratio are shown in **Figure 259**. These are roughness parameters, referencing the *roughness* profile. Pmr, referencing the *primary* profile, would be specified using the same format. Note that since Pmr includes waviness and form when calculating the material ratio, Pmr numbers will commonly (but not always) be less than Rmr (see **Figure 260**). When creating a specification, consideration must be given to factors such as interface conformability when deciding whether to specify Pmr or Rmr.

---

[79] ASME B46.1-2009 includes an additional option for the section level as a percentage of the Rt value relative to the highest peak (section1-4.3).

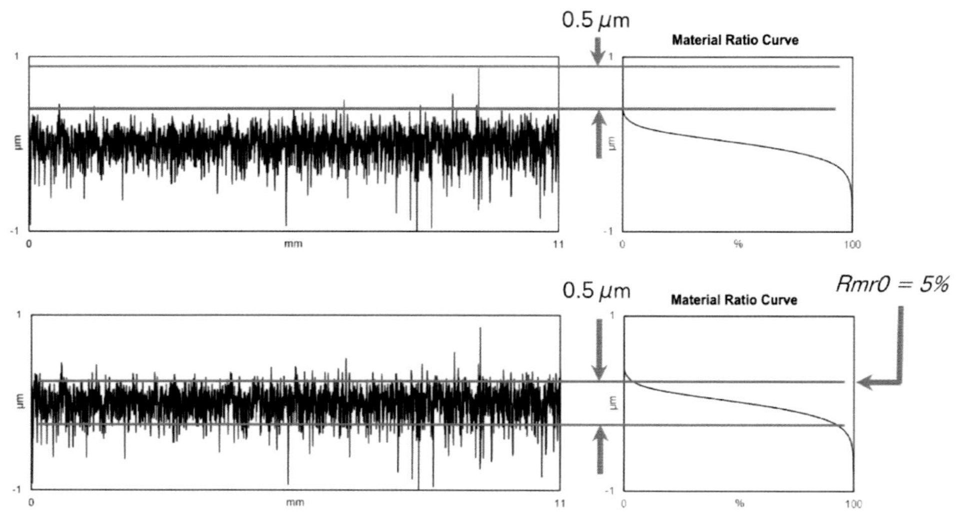

The top analysis measures from the highest peak and has a 0.07% material ratio at the 0.5 µm section level. The bottom analysis uses a 5% reference material ratio and has a 94% material ratio at 0.5 µm.

Figure 258: Illustration of the effect of a reference material ratio on material ratio stability

L "Gaussian 0.0025-0.8 / Rmr(0.5) 90%

L "Gaussian 0.0025-0.8 / Rmr0 5% / Rmr(0.5) 90%

The top callout has no reference material ratio and a 0.5 µm section level. The bottom callout has a 5% reference material ratio.

Figure 259: Sample callouts for the analyses shown in **Figure 258**

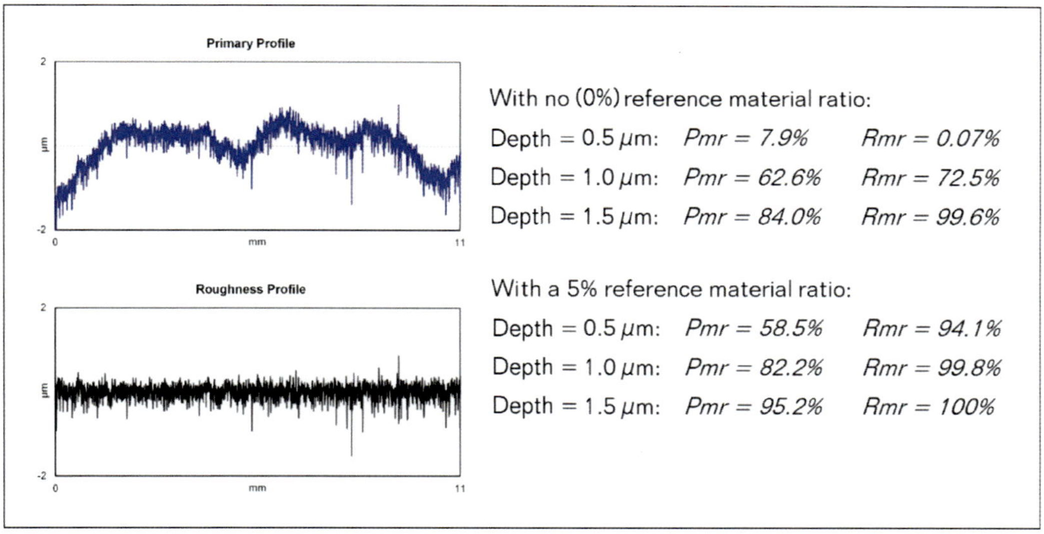

Figure 260: Comparison of Pmr and Rmr values for the same trace

Several other parameters in addition to Rmr and Pmr are derived from the material area curve. One is Rdc, also known as Htp or RHtp. Rdc is the vertical distance between the two section levels that provide two user-defined material ratios (see **Figure 261**). Rdc could be used, for example, as a measure of the material height that is expected to be removed (worn away) during the break-in of a surface.

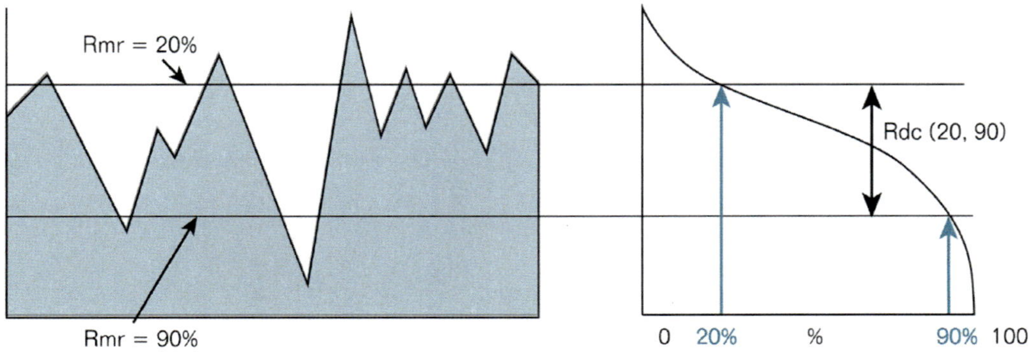

Figure 261: Illustration of Rdc

The Rk parameters, which are statndardized in ISO 13565-2, represent different regions of the material ratio curve, and are defined as follows [80]:

- Rk is the core (kernel) of the profile, the 40% section of the material ratio curve with the smallest secant slope (most horizontal). The value of Rk is the difference in height between the intersections of the slope line with the left (0%) and right (100%) axes (see **Figure 262**).
- Rpk is the reduced peak height, representing the average height of the peaks above the core roughness. It is calculated using three lines on the material ratio graph: the material ratio curve, the peak-to-core transition line (Rmr1), and the 0% material ratio axis (see **Figure 262**). Rmr1 is the material ratio at the height where the Rk line intersects the 0% axis. The "peak area" is the area bounded by these three lines. Rpk is then defined as the height of a right triangle with a base of width Rmr1 and an area equal to this peak area.
- The peak area represents the material that may be removed by wear during running in. This area is given the parameter name A1, with units of length.
- Rvk is the reduced valley depth, representing the average depth of the valleys below the core roughness. It is calculated in the same manner as Rpk, using the "valley area" bounded by the material ratio curve, the core-to-valley transition (Rmr2), and the 100% material ratio lines.
- The valley area (a valley volume in 3D) can be of interest as an oil retention zone in marginally lubricated systems. It may also capture debris. This area is given the parameter name A2 (or Rvo), with units of length.

The Rk parameter set is often applied to stratified or plateaued surfaces. However, for surfaces that are distinctly two-process, the Rpq, Rvq, Rmq parameters, based on the material probability curve, may be preferable. More information about the material probability curve and the probability parameters is given on page **346**.

---

[80] The material presented here defines the Rk parameters, but the specifics of how these parameters are calculated differ between instruments and software packages. This can result in significant differences between the reported parameter values. Variations in calculation technique include:

- Some instruments automatically use the ISO 13565-1 filter when calculating the Rk parameters, regardless of the user settings. This filter has some undesirable side effects which impact these parameters (see page **141**). Note: 13565-1 is being considered for removal from the ISO standards.
- The amount of data discarded at the ends of the trace during filtering can vary and often cannot be controlled. This, too, can impact the Rk parameter values.

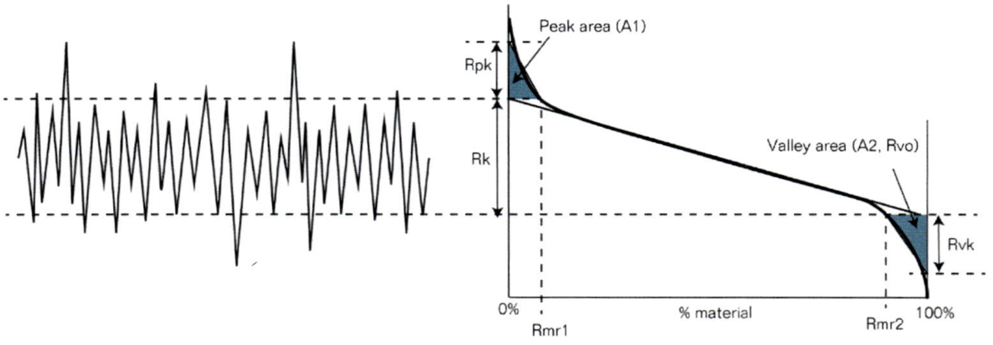

Figure 262: Rk, Rpk, Rvk, A1, and A2 (Rvo) as related to the material ratio curve

What is the material probability curve and what is it used for?

Some tribological systems utilize surfaces consisting of deep valleys overlaid by small peaks, commonly called stratified or plateaued surfaces. An example is the lubricated sliding interface in some piston ring—cylinder bore systems. These surfaces are manufactured using two separate processes—one relatively coarse, creating high peaks and deep valleys, and a second that is relatively fine to remove the high peaks and leave a smoother "plateau" on top of the remaining valleys (see **Figure 263**). Other examples of stratified surfaces might include powdered metal or ceramic surfaces. If the material ratio from such a two-process surface is plotted on probability paper, the two parts of the surface appear as (approximately) straight lines which can be characterized using parameters similar to Rq. This plot is called the material probability curve, documented in ISO 13565-3[81].

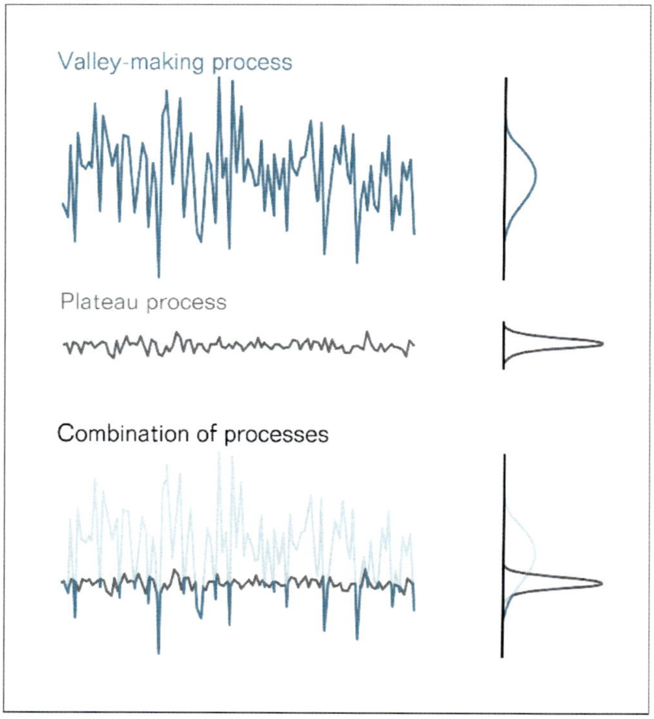

Figure 263: Combination of two processes to create a stratified surface

Figure 264 illustrates the derivation of Rvq, Rpq and Rmq parameters for a stratified surface created using the processes shown in **Figure 263** (compare the amplitude distribution functions in **Figure 263** and **Figure 264**).

[81] ISO 13565-3 defines the Rpq, Rvq, and Rmq parameters, but different instruments and software packages use different approaches for calculating the material probability curve. This can significantly impact the parameter values. See also footnote **80** on page **356**.

- Rpq is the slope of the upper part of the material probability curve (the dark gray line in **Figure 264**), and represents the Rq roughness of the plateau region (i.e., the roughness of the surface if the deep valleys were not present).
- Rvq is the slope of the lower part of the curve, and represents the Rq-roughness of the valley region (i.e., a prediction of the Rq roughness of the surface prior to the final plateauing operation).
- Rmq is the material ratio at the intersection of the two curves.

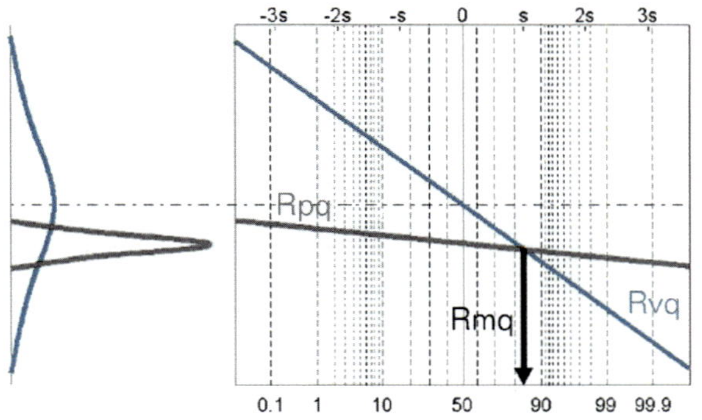

Figure 264: Derivation of Rvq, Rpq, and Rmq from a stratified surface (see **Figure 263**)

While Rvq, Rpq, and Rmq provide numerical descriptions of a functional surface for a lubricated sliding system, they are also useful for manufacturing process control. Rvq and Rpq provide independent feedback, respectively, for the rough and finishing processes (commonly honing). Rmq provides guidance on the stock removal during the final procedure, signaling when to stop the final honing.

The Rk parameter set is often applied to plateau-honed surfaces. However, for extremely pleateaued surfaces, it does not provide as good a model as the probability parameters, especially in the valley region[82] (see **Figure 265**).

[82] The green triangle on the material ratio plot (right hand plot) is used to calculate Rvk, and is meant to represent the valley portion of the material ratio curve. In this example the triangle is a very poor representation of that curve.

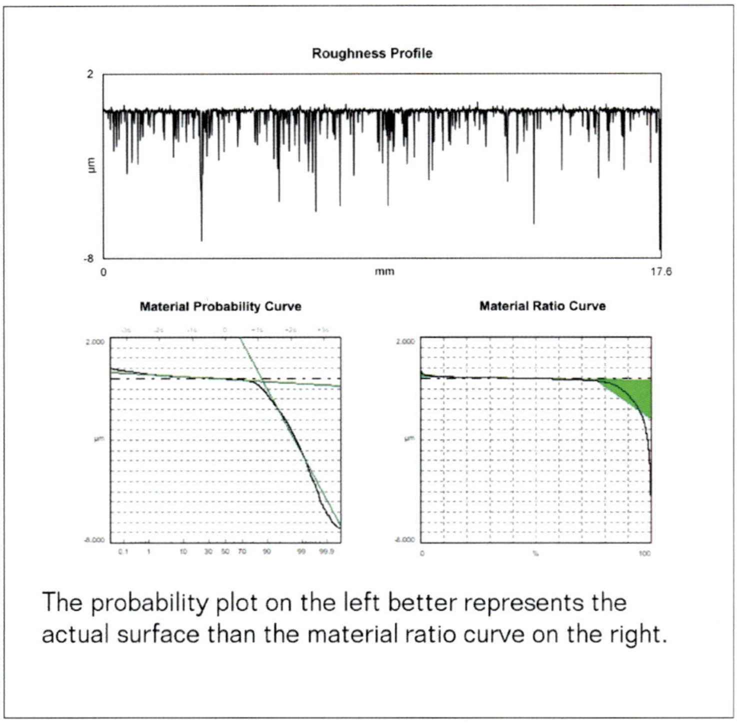

Figure 265: Comparison of a probability analysis and an Rk analysis of a plateaued surface

It should be noted that non-robust filters (e.g., Gaussian or spline) can cause distortions of the profile peaks, with false indications of raised metal (see **Figure 266**). A robust filter should be used to avoid this; see page **134** for more information about the robust filter.

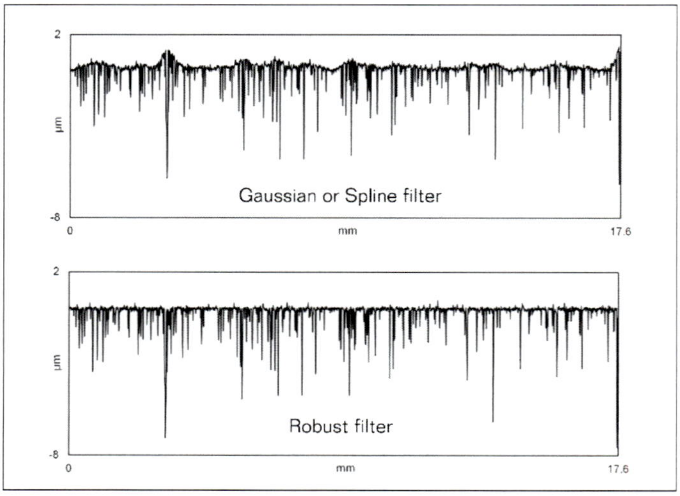

Figure 266: Distortion in a *roughness* profile corrected with a robust filter

Why can't I find the tp parameter on my system?

"tp" is the ASME term for material ratio, presumably derived from the German terminology[83]. It is sometimes used generically for the material ratio independent of the profile and can therefore be ambiguous. However, in most cases the term applies to the *roughness* profile, which in ISO 4287 is designated Rmr—the material ratio (mr) of the *roughness* profile (R). The material ratio is defined at some section depth, designated "p" in ASME and "c" is ISO (i.e., Rmr(c)).

When the material ratio is evaluated on the *primary* profile the ISO term is Pmr. In some systems tpa is used for Pmr and tpi is used for Rmr.

If you can't find "tp" on your system, look for Rmr (or Pmr) or for tpi (or tpa).

[83] *Bearing ratio percent* in German is *Traganteil Prozent.*

Which parameters measure width or spacing rather than height?

Several parameters provide a direct or indirect measure of feature spacing:

- Rsm, Psm, and Wsm, are the average spacing between "profile elements,"which are defined as peak-valley or valley-peak pairs. This is most easily visualized as the spacing between positive mean line crossings (see **Figure 267**). Positive mean line crossings are those that occur when the profile is headed upward. Discrimination levels control which features are included in the average (see below).

- Rs, Ps, and Ws are the average distance between peaks, with a peak defined as the highest point between two mean line crossings. All features in the relevant profile are counted, regardless of height or width.

- Rpc, Rhsc, and Rvc are peak or valley counting parameters that provide indirect spacing information. They are reported as the number of features over some distance[84], the inverse of average spacing. Feature sizes are controlled by specifying thresholds which the feature height/depth must exceed. Additional information is provided in the **Definitions** section starting on page **292**. Caution should be taken with the parameters that are counts over the evaluation length (e.g., Rhsc and Rvc). Different measurement systems may use different evaluation lengths, so the counts may vary.

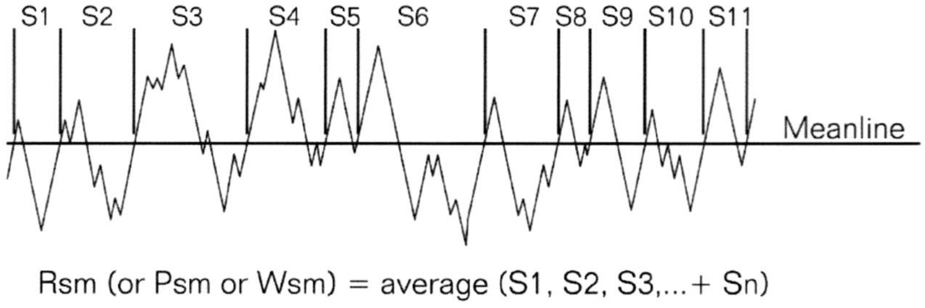

Figure 267: Illustration of the calculation of the Rsm, Wsm, or Psm parameter

A key consideration when calculating the Rsm, Wsm, and Psm parameters is determining which features are to be included. Small "wiggles" that are short in the x- and/or z-axes (see **Figure 268**) can dramatically reduce the Rsm, Wsm, or Psm value, but they may not be functionally significant. Both ASME B46.1 and ISO 4287 state that, unless otherwise specified, a feature must meet both of the following conditions in order to be included in the Rsm, Wsm, or Psm analysis:

---

[84] Rpc is evaluated as the number of features per unit length. Rhsc and Rvc are the number of features in the evaluation length (see **Table 22** on page **314**).

- Feature width ≥ 1% of the sampling length (sampling length = λc for Rsm, λf for Wsm, or the complete evaluation length for Psm or for Wsm if no λf is specified). The feature width is defined by successive positive meanline crossings.
- Feature height ≥ 10% of Pz (for Psm), Wz (for Wsm), or Rz (for Rsm). The feature height requirement applies to peaks and valleys individually.

Further information about feature discrimination can be found in the section, "**What are height and width discrimination and what are they used for?**" on page **352**.

The Rc, Wc, and Pc parameters are also affected by feature discrimination. See the comments about Wc in the ***Definitions*** section starting on page **292**.

Note that, by convention, spacing parameter values are stated in millimeters rather than micrometers, consistent with the axis to which they refer (the length axis rather than the height axis).

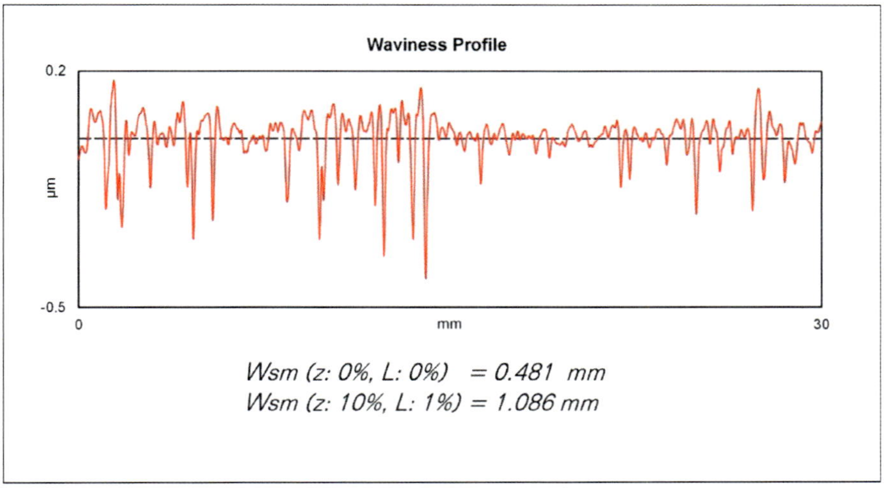

Figure 268: Example of the effect of discrimination level on the Wsm spacing parameter

What are height and width discrimination and what are they used for?

Consider the waviness profile in **Figure 269**. Your design might be sensitive to the higher-amplitude waviness seen left of center, so you may wish to control that waviness using Wc, the mean peak-to-valley distance. However, the many small peaks and valleys will influence the Wc result. This will result in a a much smaller average value than is indicated by the one region of high amplitude. Height and width discrimination are useful for situations such as this.

Wc, Wsm (the average spacing), and the corresponding roughness profile parameters Rc and Rsm, utilize feature discrimination to discard from the parameter calculation any small features that will likely have only minimal influence on a surface's function. "Small" applies to both length and height. The standard thresholds are 10% of the profile's Rz, Wz, or Pz value (for height discrimination) and 1% of the sampling length (for width discrimination). This means that, for a feature to be included, it must be longer than 1% of the relevant sampling length, and both the peak height and the valley depth must be greater than 10% of the corresponding "z" parameter (e.g., Wz for Wc and Wsm)[85]. Features are identified by the mean-line crossings[86]. Applying these discrimination levels to the profile in **Figure 269** produces a Wc value of 0.78 µm, more than double the value of 0.34 µm that results when no discrimination is used.

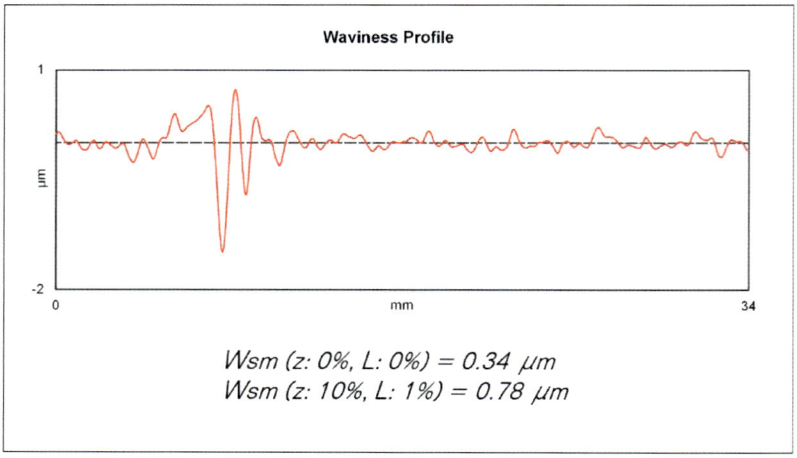

Figure 269: Effect of discrimination on Wc and Wsm for a profile dominated by small features.

The importance of this discrimination for Wc and Wsm is dependent on the nature of the surface, the choice of roughness filter cutoff (lambda-c, or λc), and whether or not the waviness is one-sided

[85] For Wc and Wsm with bandpass waviness, sampling length and Wz have their normal definitions. For single-sided waviness (no λf), sampling length equals the evaluation length and, since there is only one sampling length for calculating Wz, Wz is numerically equivalent to Wt.
[86] The peaks and valleys in Wc and Rc are defined as the highest and lowest points within a profile element, and the spacings in Wsm and Rsm are the distances between those profile elements.

or bandpass. The relationship is complex. A smaller roughness cutoff (e.g., 0.25 mm instead of 0.8 mm) will include more small features in waviness, but it also tends to increase the height of those features (the *waviness* profile will more closely follow the *primary* profile). With bandpass waviness, longer wavelengths are filtered out of the profile, tending to bring the waviness features closer to, and more likely to cross, the mean line. More features may therefore be included in the Wc and Wsm calculations, and those features will tend to have smaller amplitudes due to the removal of the form. Because of this complexity, the default discrimination levels should always be used instead of trying to rationalize different values based on the nature or function of the surface (but see below).

ISO standardized the discrimination algorithm in 2013[87]. Prior to that time, some software may not have applied any discrimination when calculating these parameters. If you have functional parameter limits based on older historical data, you need to determine how those limits were calculated. If the discrimination thresholds were different from the standard 10% and 1% values, those original thresholds must be stated in your design documentation, or you must develop new functional limits. There is no universally-accepted method for including the threshold values in a surface texture callout, but the following has been proposed (note that the numbers shown here assume that no discrimination was used to establish the 1.6 μm Wc limit):

U "Spline" 0.8-8.0 / ln=24 / Wc(z=0%, L=0%)max 1.6

If no thresholds are specified, the default values of z = 10%, L = 1% will be used.

Note that you cannot accurately derive a parameter limit for one discrimination level based on parameter values calculated using different discrimination levels.

[87] Note that as this is being written, additional changes to the algorithm are being considered by both ISO and ASME.

What are the motif parameters?

The "motif" family of parameters was developed by a French consortium of automakers based on a study of the correlation between parameter value and component function. The parameters were designed to help assess the contact between two surfaces by creating surface "envelopes" that join the highest peaks of the *primary* profile. This can be useful in applications such as lubricated or gasketed interfaces. While these parameters aren't as common outside of French industries, the goal of relating specification to function is noteworthy and commendable.

ISO 12085 contains the parameter definitions and the calculation procedure. It also contains a table providing (for information only) recommendations on which motif parameters are appropriate for specifying various surface functions.

Calculation of the motif parameters involves the detection and analysis of "motifs," which can be thought of as the "V" shaped profile features of a valley and its two adjacent peaks. Once identified, the width and two heights (left and right) of each motif can be calculated (see **Figure 270**). ISO 12085 provides standardized rules for combining any minor local peaks and valleys that lie along the sides of the "V". These rules are based on the relative heights of the peaks and a characteristic motif width—"A" for roughness and "B" for waviness. These characteristic widths can be thought of as a kind of minimum motif width, although ISO 12085 allows for wider motifs, depending on other conditions. Recommended A and B values are included in ISO 12085 and are listed in **Table 25**. If A and B are not specified the default values are A = 0.5 mm and B = 2.5 mm.

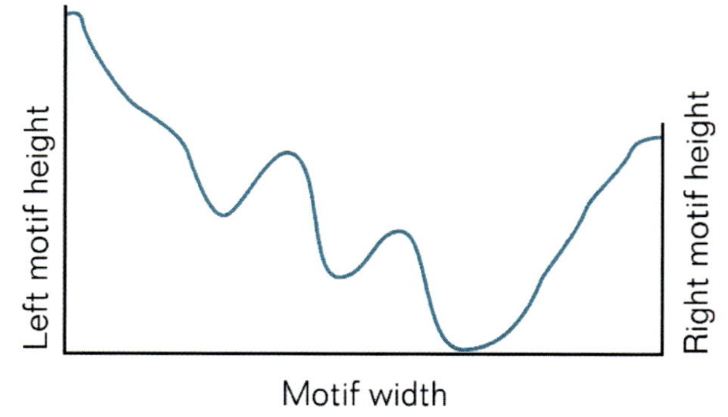

Figure 270: Identification of motif width and heights

Table 25: Recommended values for motif widths A and B

A (mm)	B (mm)	Evaluation Length (mm)	Short Cutoff (µm)	Max Tip Radius (µm)
0.02	0.1	0.64	2.5	2
0.1	0.5	3.2	2.5	2
0.5	2.5	16	8	5
2.5	12.5	80	25	10

If not otherwise specified, the default values are $A = 0.5$ and $B = 2.5$ mm

The seven motif parameters (R, Rx, AR, W, Wte, Wx, and AW) are derived from the *primary* profile as follows:

1. Detect all local peaks and valleys along the *primary* profile (see **Figure 271**).

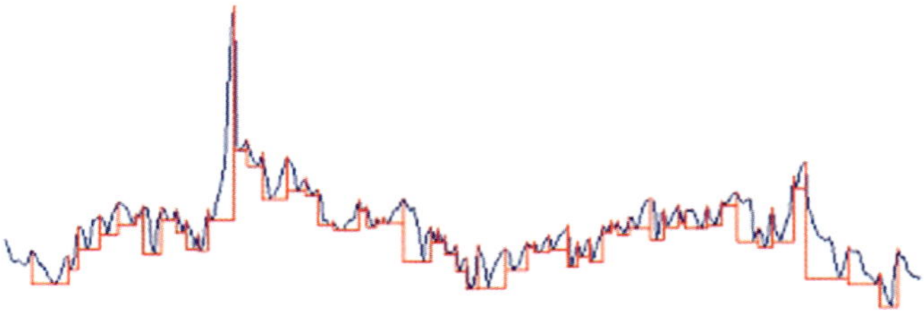

Figure 271: Identification of profile peaks and valleys

2. Remove small, local peaks and valleys using the motif combination rules described in ISO 12085, utilizing the roughness "A" characteristic and height discrimination rules. The Rx parameter is reported as the largest motif depth, which is the same as the highest motif side height (see **Figure 272**).

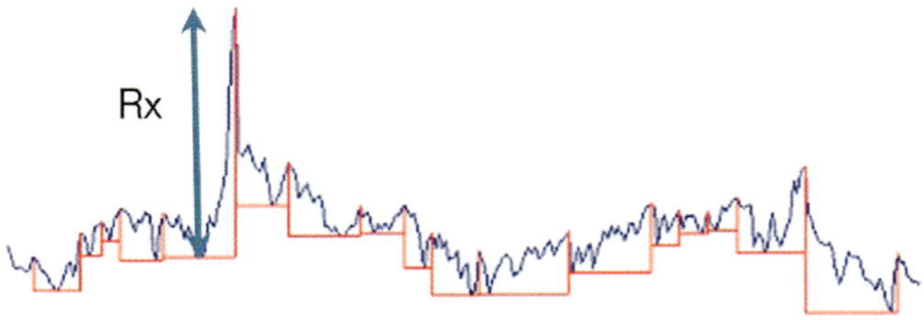

Figure 272: Motif combination and determination of Rx

3. The motif positions at extreme peaks and valleys are reduced per ISO 12085, resulting in the "height adjusted roughness motifs" (see **Figure 273**). The "R" value is reported as the average motif side height (with each motif having two heights). The "AR" value is reported as the average motif width.

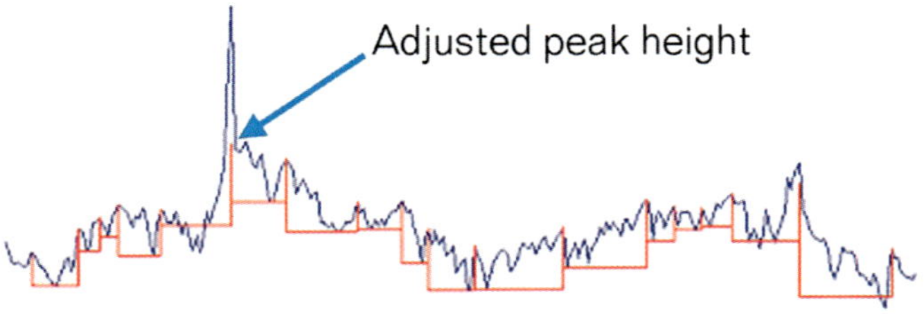

Figure 273: Height-adjusted roughness motifs

4. The motif waviness analysis begins with the generation of a "waviness envelope profile". This profile is derived by connecting the edges (peaks) of the height adjusted roughness motifs (see the green line in **Figure 274**).

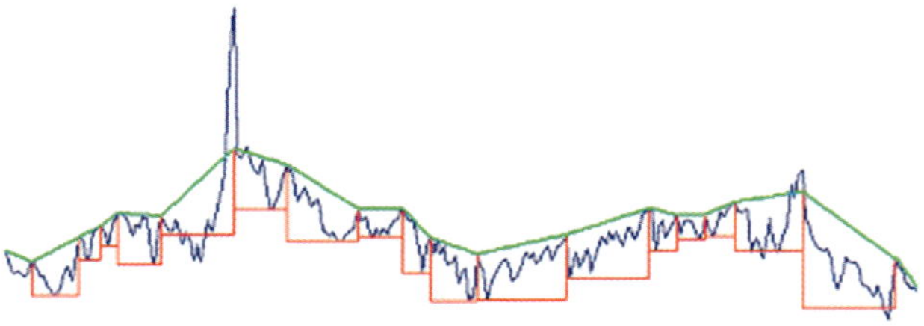

Figure 274: Creation of the waviness envelope profile

5. Peaks and valleys are detected along the waviness envelope to generate waviness motifs. These motifs are combined according to ISO 12085 using the waviness "B" characteristic and the height discrimination rules, creating the "waviness motifs". The Wte parameter is reported as the height of the top of the highest waviness motif peak relative to the bottom of the deepest motif valley bottom, not necessarily on the same motif. The Wx parameter is the largest motif depth for any single motif. "W" is the average motif side height (using two sides per motif). "AW" is the average motif width.

Figure 275: Creation of waviness motifs

The motif parameters have not been used very extensively outside of France. This may partly be due to a lack of familiarity. However, the following observations about their behavior could also contribute:

- The motif roughness parameters do a poor job of distinguishing between roughness peaks and valleys, as shown in **Figure 276**. This is curious, since the parameters were supposedly developed to, in part, address interactions between surfaces, where peaks are usually of much more interest than valleys.
- Profile inversion creates a change in the motif waviness parameters (**Figure 276**), in contrast to the analogous ASME/ISO parameters Wa and Wt, which show no change. The width/spacing parameters AW and Wsm also react differently.
- The motif analysis treats the profile in a discrete sense which is different from the more continuous sense of the other analysis systems.

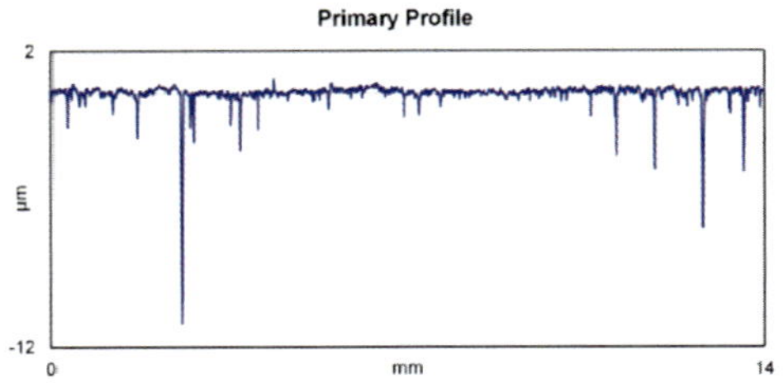

R = 0.340	*W* = 0.741	*Wa* = 0.103
Rx = 11.452	*Wte* = 8.667	*Wt* = 1.083
AR = 61.86	*AW* = 516.07	*Wsm* = 2.38

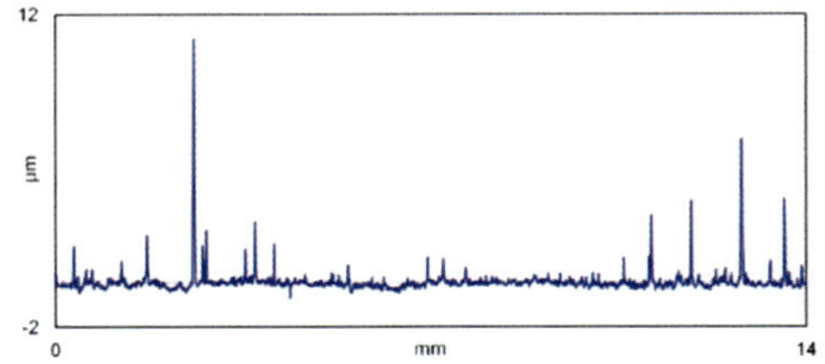

R = 0.331	*W* = 0.883	*Wa* = 0.103
Rx = 11.452	*Wte* = 3.954	*Wt* = 1.083
AR = 46.28	*AW* = 496.25	*Wsm* = 2.38

A comparison of the motif roughness parameters for a profile and for its inverted form showing a lack of discrimination between peaks and valleys. The motif waviness parameters do change with inversion, unlike the ASME/ISO waviness values. The motif widths also change, in contrast to the constant "sm" values. Units are μm, except Wsm is mm.

Figure 276: Comparison of motif parameters for inverted profiles

What are fractal methods in the context of surface analysis?

Fractal-based analysis methods have been successfully applied to surface analysis, particularly in areas such as adhesion or other applications in which an "effective surface area" is useful[88]. Fractal (also referred to as "scale sensitive") methods are commonly based on assessing a surface across a range of scales and determining the effective surface length, or effective surface area, at each scale.

Figure 277 demonstrates a simple case of the "length scale" or "compass method" fractal analysis of a surface profile. In **Figure 277** we see a profile of length 8.8 mm. However, if we were to walk along the surface at a given step length, we would travel more than 8.8 mm. A 2.4 mm step size along the example profile would result in a total distance travelled of 9.6 mm. Similarly, a 1.0 mm step size would result in a total distance traveled of 10.0 mm. As the step size decreases, we end up travelling more distance along the surface. This method has also been referred to as the "coastline" method as it relates to the distance one would have to walk along a coastline.

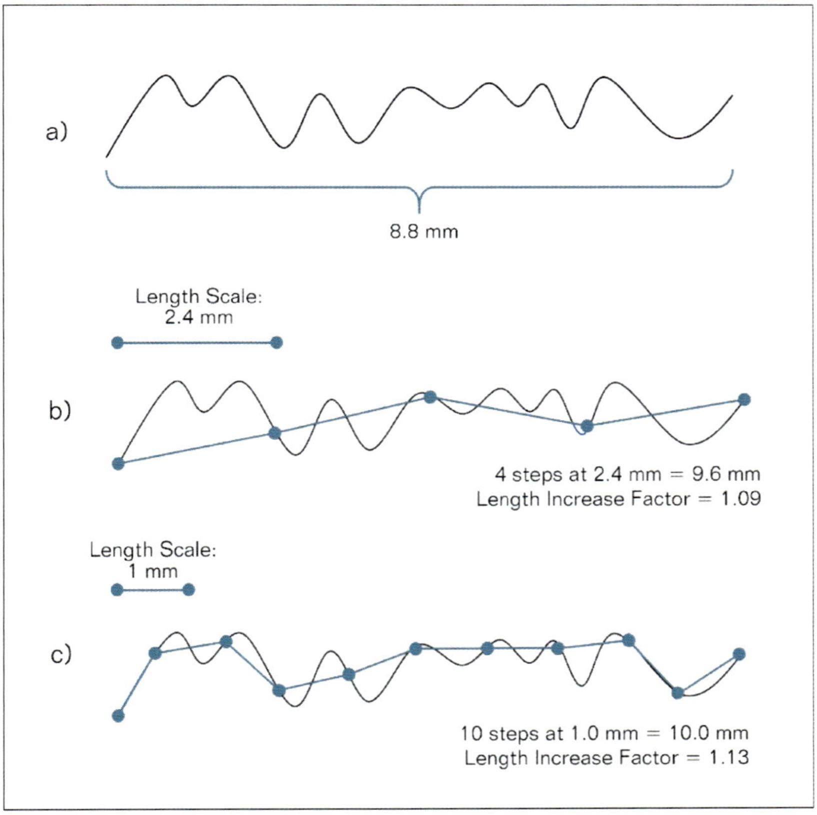

Figure 277: Decreasing step sizes result in increased distance travelled

[88] Brown, C. A. Siegmann, S., "Fundamental Scales of Adhesion and Area-Scale Fractal Analysis," *International Journal of Machine Tools and Manufacture*, Vol. 41 (2001): 1927–1933

As multiple scales are explored, a curve can be developed to show the relationship between scale (e.g., step size) and the effective length. This is represented in **Figure 278**.

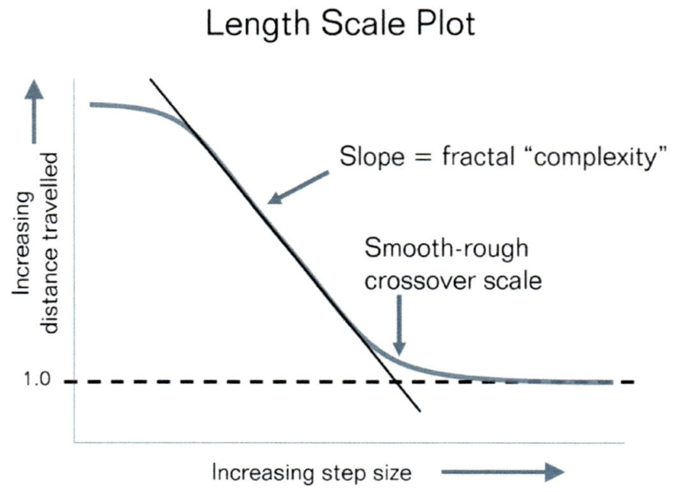

Figure 278: The "Length Scale Plot" shows relationship between step size and surface length

The length scale plot (or "area scale plot" in three-dimensional analysis) gives a representation of all scales. Specific surface functionalities may relate to specific length scales (i.e., specific step sizes) based on the physics of the interactions.

What are "vertical range" and "resolution"?

In the context of surface texture measurement, "range" typically refers to the total available vertical travel of the stylus probe tip. This is the maximum height over which usable data can be obtained. "Vertical resolution" is the smallest vertical height difference that can be discerned. The two terms are typically reported together because they are often linked in the operation of surface texture instruments. For a given instrument, an increase in range often results in a decrease in resolution. For example, if you increase the length of the probe to increase the range, you will typically reduce the resolution (the minimum resolvable feature size becomes larger).

Laboratory instruments are available with resolutions as small as 0.2 nm over a range of 25 mm or more. Even reasonably-priced systems can have resolutions as small as 16 nm with 1 mm ranges.

Poor resolution will appear as a stepped appearance in the data—see **Figure 279**.

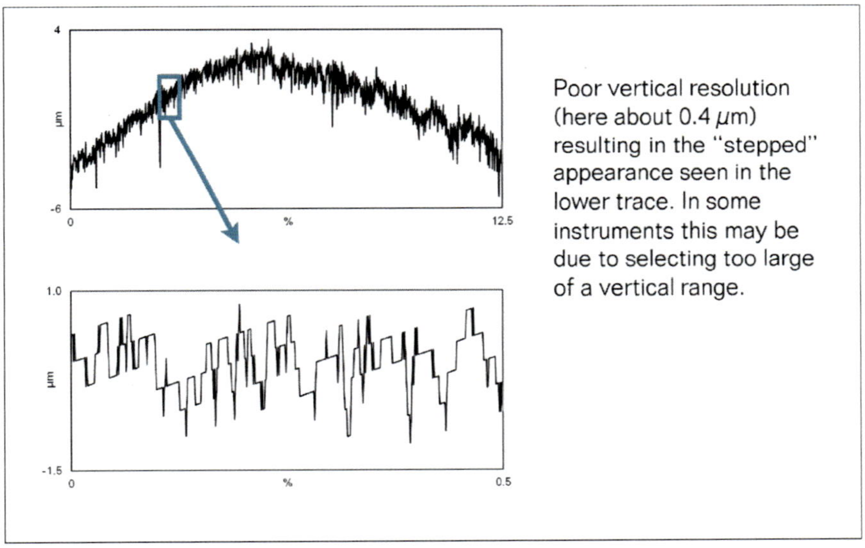

Figure 279: Poor vertical resolution

What is the autocorrelation function?

The autocorrelation function (ACF) is used to determine the periodicity of a surface by evaluating the similarity between the profile and a copy of the profile shifted along the x-axis. With zero shift the copy of the profile exactly matches the original and the ACF equals 1.0. As the copy is shifted, the amount of agreement decreases. ASME B46.1 defines the correlation length (labeled "Rcl") as the shift distance at which the autocorrelation drops below 37% (1/e) (see **Figure 280**). It should be noted that some systems use different thresholds such as 0.1 (10%) or 0.5 (50%). Thus, care should be taken when comparing results from different instruments and software.

With continued shifting, a periodic surface will produce an ACF with obvious peaks, while a random surface will produce an ACF with small fluctuations around zero. One way to visualize the ACF for a periodic profile is to places your two hands together, palm to palm, with fingers spread. If you then shift one hand relative to the other, at some point you will no longer be able to feel your fingers with the other hand—this distance is analogous to the correlation length. With continued shift you'll sense the next finger, but it won't be an exact match; instead you'll sense the index finger with, say, the middle finger or the pinkie.

On a periodic surface the ACF provides a measure of the width or sharpness of the periodic features. **Figure 281** shows two periodic profiles, with the regularity apparent in the *roughness* profiles as well as in the ACF graph. The relative sharpness of both profiles is also visually apparent and is captured numerically by the Rcl values: 11.24 μm for the first profile and 55.25 μm for the second. For comparison purposes, **Figure 281** also includes a random ground surface with an ACF that rapidly approaches, and then hovers around, zero.

The correlation length can be combined with a height parameter such as Rq to represent peak curvature, a parameter of interest in disciplines such as tribology. Unlike some of the other parameters used for surface modelling (e.g., the spectral moment parameters or Rdq), Rcl is not sensitive to data spacing.

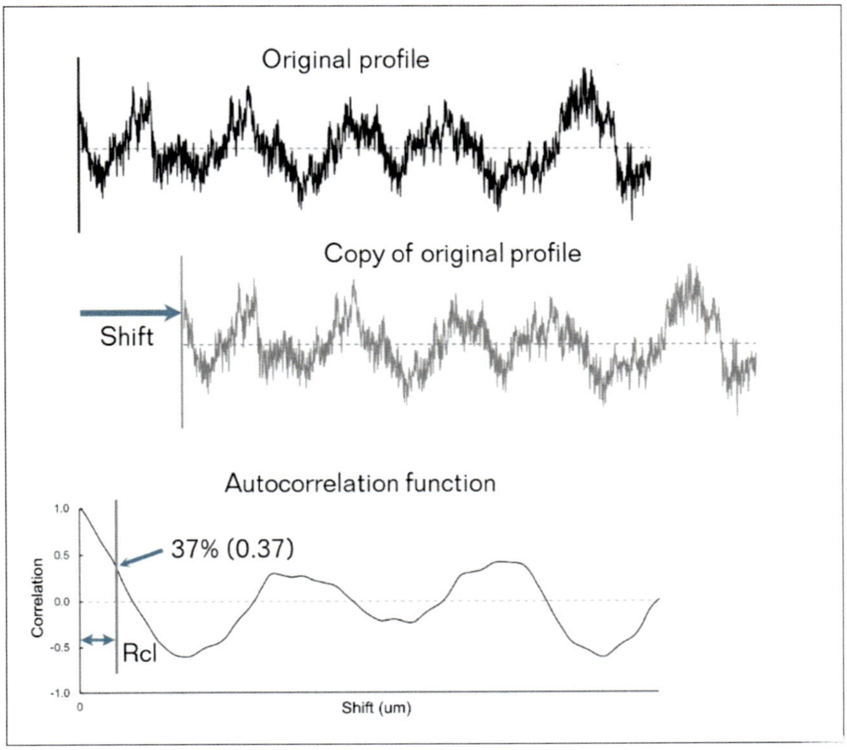

Figure 280: Illustration of the autocorrelation function and correlation length (Rcl)

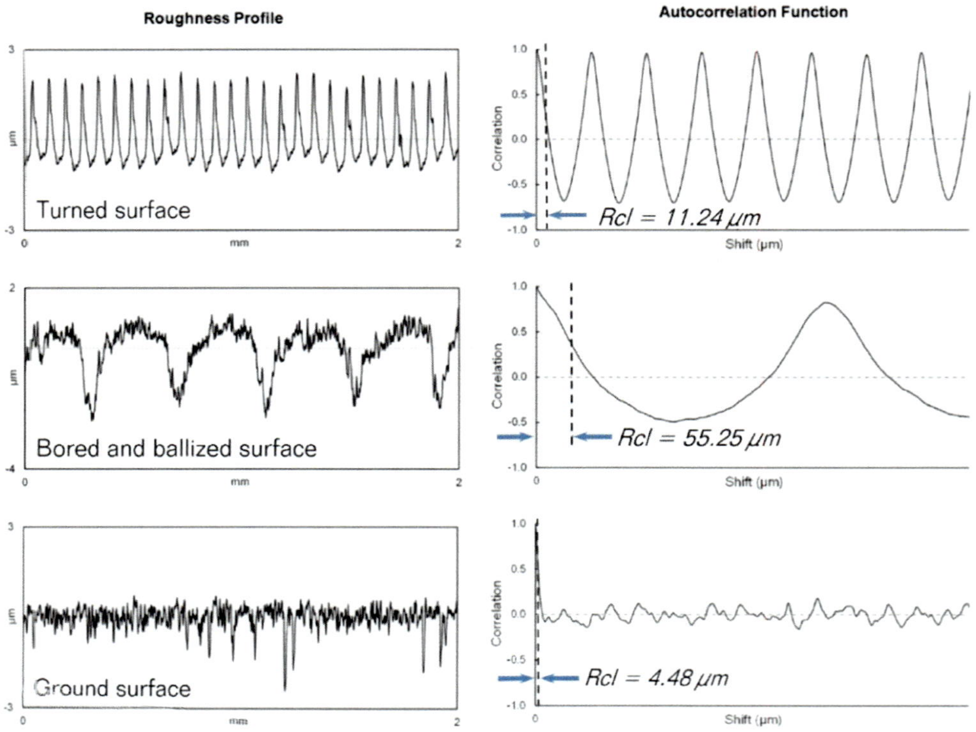

Note that the *roughness* profiles on the left have the normal x-axis units of millimeters. The autocorrelation function plots on the right show the dependence of the ACF on the amount of shift, so the x-axis units are microns. All three ACF plots have the same scale, with a maximum shift of 600 µm.

Figure 281: Examples of *roughness* profiles and their autocorrelation functions

How do I know which wavelengths are in my surface data?

While there are standard parameters for describing the spacing of features in a surface profile (e.g., Psm, Wsm, and Rsm), most surfaces texture is comprised of multiple spacings. Furthermore, there are often cases where random texture can mask the regularly-spaced features. As an example, the regularly spaced tool marks from a turning operation may be masked by a subsequent grinding operation. In other cases, a measured profile may contain periodic features or noise due to vibration or noise in a measurement itself. In these cases it's useful to be able to pull apart the different aspects of the data.

Two tools are available to display the regular spacings, or wavelengths, present within surface texture: the "power spectral density" (PSD) plot, and the "wavelength content" graph.

The power spectral density plot (often called the "power spectral density function") represents the amount of power that is present at each frequency. If there is a strong sine wave present in the data, there will be significant power at the associated frequency. **Figure 282** shows the PSD for a turned surface with a spacing (i.e., feedrate) of approximately 0.075 mm.

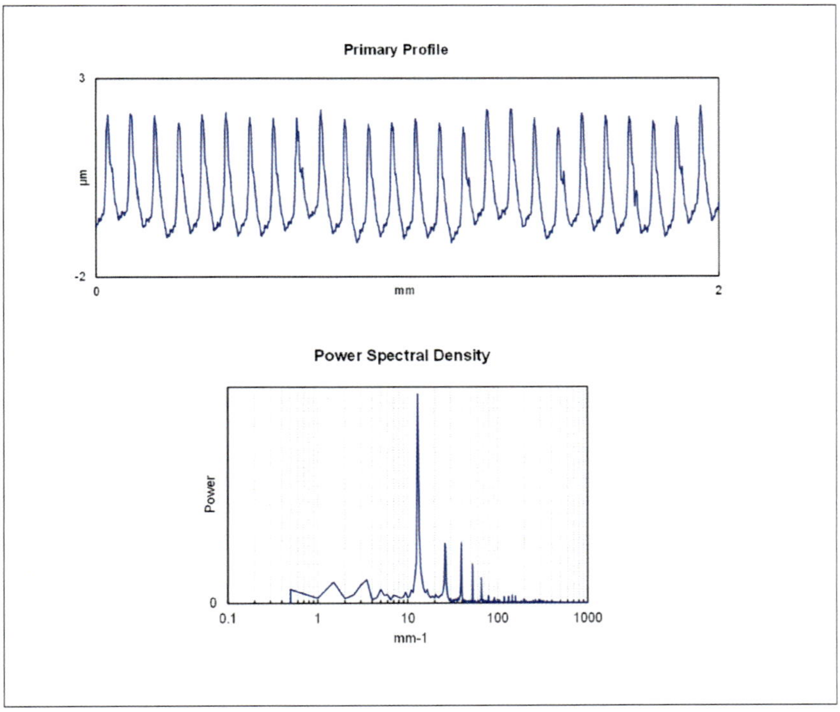

Figure 282: Power spectral density plot for a turned profile

The PSD is plotted with "power" on the y-axis versus the frequency (features per unit length) on the x-axis. For the data in **Figure 282**, the strong wavelength of 0.075 mm translates to an x-axis frequency of 13.33 cycles per millimeter. Since the surface is not a pure sine wave, we can also see smaller peaks in the PSD at higher frequencies. It should be noted that the units and calculation of the power is not well standardized in the context of surface texture and can vary from instrument to instrument. Thus, the usefulness of the PSD is often limited to a visual interpretation of the location and relative heights of the peaks.

The wavelength content graph provides the amplitude (i.e., height) at each wavelength in the surface profile. **Figure 283** compares the wavelength content graph to a PSD plot for a turned profile. In this wavelength content plot the dominant "spike" appears at the 0.075 mm wavelength. This ability to directly read a wavelength (without having to convert from frequency) can be much more intuitive for describing surface features.

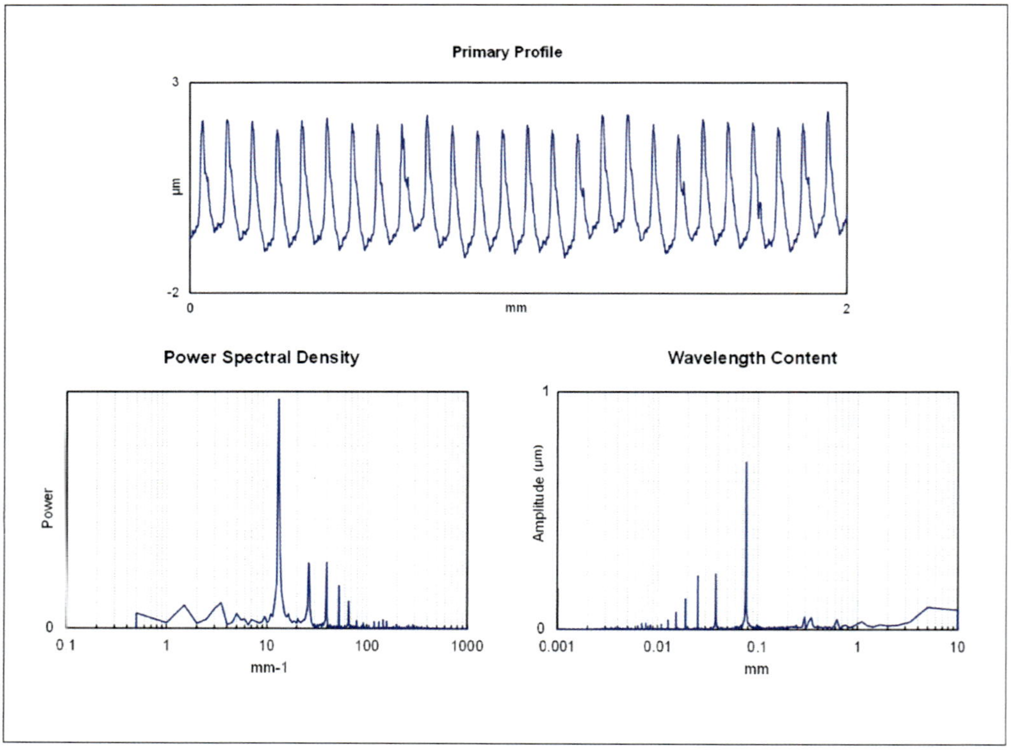

Figure 283: Power spectral density and wavelength content plots for a turned profile

The height of each peak is the (mean-to-peak) amplitude at that frequency. An easy way to demonstrate this is to plot the wavelength content graph for a single sine wave, as shown in **Figure 284**.

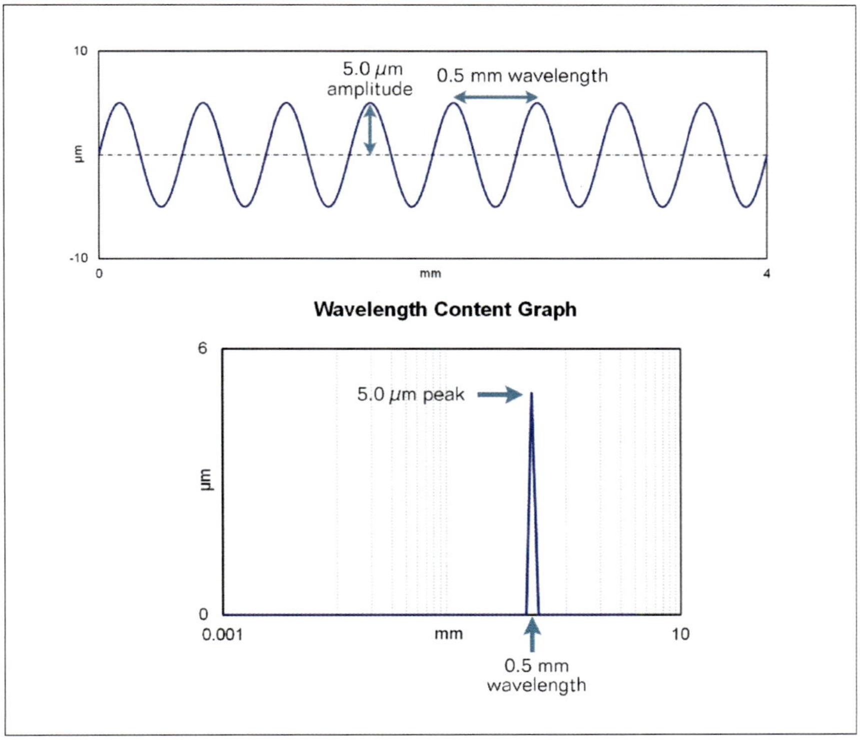

Figure 284: The wavelength content graph for a sine wave.

The wavelength content graph can provide valuable information about a manufacturing process. This information can be used either to reverse engineer a process or to trouble-shoot a problem. For example, a measurement of a turned surface will have a major peak at the wavelength defined by the tool feedrate and the rotation rate of the workpiece or tool (see **Figure 283** above). An unintentional manufacturing input, such as a bad spindle bearing or vibration-induced chatter, will also leave a periodic signature, typically different from the feed plus speed, and this can suggest the source of the problem.

Figure 285 presents another example showing a ground surface profile and its wavelength content graph. In this case, the dressing of the grinding wheel imparted a 1.0 mm periodic shape into the wheel surface which was then translated to the workpiece. This surface error is often undetected as the 1.0 mm wavelength is partially transmitted as roughness and partially transmitted as waviness when using the common roughness cutoff wavelength of 0.8 mm. However, this wavelength is easily detected in the wavelength content graph.

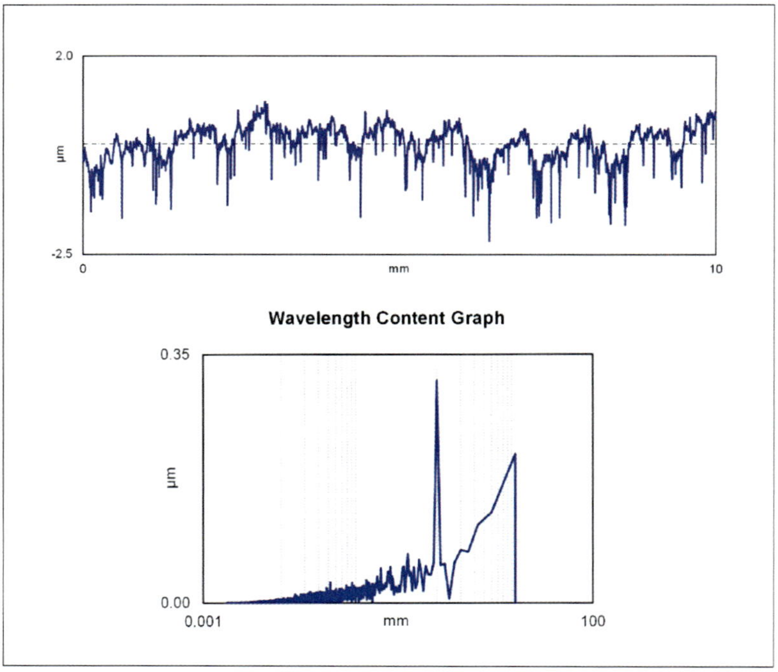

Figure 285: Wavelength content graph showing 1.0 mm period due to a grinder dressing process

The wavelength content graph is also a powerful tool for assessing the measuring instrument itself. **Figure 286** gives the profile of a measured optical flat. The Ra is much higher than expected (Ra = 0.123 µm versus a typical value of 0.01 µm or less), but the cause is not visually obvious in the *primary* profile. However, the wavelength content graph shows a prominent peak at 0.0083 mm. This equates to a frequency of 60 Hz at the traverse speed of 0.5 mm/sec, and was later determined to be related to unisolated vibration.

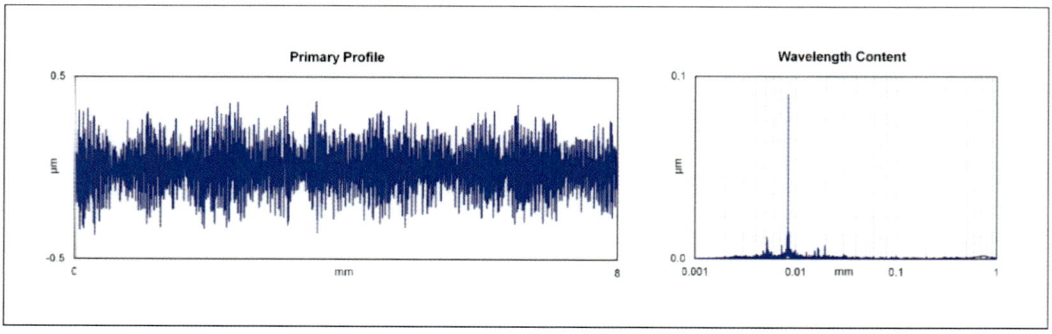

Figure 286: Peak in the wavelength content graph at a noise wavelength

REFERENCES AND RESOURCES

Bibliography

Badami, Vivek G., et al. "Evaluation of the measurement performance of a coherence scanning microscope using roughness specimens," *Proceedings of ASPE Annual Meeting* (2011).

Brown, C.A., et. al., "Multiscale analyses and characterizations of surface topographies," *CIRP Annals—Manufacturing Technology*, 67(2018): 839-862

Brown, C. A. Siegmann, S., "Fundamental Scales of Adhesion and Area-Scale Fractal Analysis," *International Journal of Machine Tools and Manufacture*, Vol. 41 (2001): 1927–1933

Clarke, J., S.J. Dixon, L. Damm, and M.J. Carré, "The effect of normal load force and roughness on the dynamic traction developed at the shoe-surface interface in tennis," *Sports Engineering* (2013): 165-171

Cohen, D.K., R. P. Garibary, J. McNeill, and J. Huang, "Correlation of Optical Surface Profilometry with Contact Profilometry on Precision Peened Surfaces," *SME Conference on 3D Gaging and Technology* (March 4, 1997) Detroit, MI,

DeHart, A. and Smiley, J., "Imperfect Journal Geometry—Its Effect on Sleeve Bearing Performance," SAE Technical Paper 640837 (July 1964)

Malburg, M.C., "Surface Profile Analysis for Conformable Interfaces," *J. Manuf. Sci. Eng.* 125(3) (Jul 23, 2003): 624-627

Malburg, M.C., Chetwynd, D.G., and Raja, J, "Local Slope Analysis in the Stylus Based Assessment of Surface Integrity," *Tribology International*, 30(7) (July 1997): 527-532

McCool, J. I., "Comparison of models for the contact of rough surfaces," *Wear*, 107(1) (Jan 1986): 37-60

Rhee, H.G., T.V. Vorburger, J.W. Lee and J. Fu, "Discrepancies between roughness measurements obtained with phase shifting interferometry and white-light interferometry," *Applied Optics*, 44(2005):5919.

Roper, H.J., et. al., "The Effect of Peak Count of Surface Roughness on Coating Performance," *Journal of Protective Coatings and Linings* (June 2005) (Available at www.paintsquare.com)

Tabenkin, Alex. "The ABCs of Rz." Quality 46, no. 11 (2007): 40-43.

Thomas, T.R., *Rough Surfaces*. London: Imperial College Press, 1999, 2nd edition

Todhunter, L.D., R.K. Leach, S.D.A. Lawes, and F. Blateyron, "An analysis of Type F2 software measurement standards for profile surface texture parameters," *Meas Sci Technol*, 28(6)(2017):5017

Vorburger, T.V., A. Zheng, T. B. Renegar, J-F. Song, and L. Ma. "An iterative algorithm for calculating stylus radius unambiguously," *Journal of Physics*: Conference Series, 311(2011): 012029. IOP, 2011

Williamson, J.B.P., "Microtopography of Surfaces," *Proceedings of the Institution of Mechanical Engineers*, Vol. 182, Part 3K(1967-68): 21-30

Background reading material – books and journal articles

Blunt, L., Jiang, X., Advanced Techniques for Assessment Surface Topography, Butterworth-Heinemann, 2003.

Leach, R., *Characterisation of Areal Surface Texture*, Springer-Verlag Berlin Heidelberg, 2013.

Malburg, M.C., "A Unified Methodology for the Application of Surface Metrology," PhD thesis, University of Warwick, Coventry, UK

Muralikrishnan, B., Raja, J., *Computational Surface and Roundness Metrology*, Springer-Verlag London, 2009.

Song, J.F., and Vorburger, T.V., "Surface Texture," in *ASM Handbook*, vol 18, Friction, Lubrication, and Wear Technology, 1992, ASM-International, p 334 ff

Thomas, T.R., *Rough Surfaces*, Imperial College Press, 1998.

Vorburger, T.V., and Raja, J., "Surface Finish Metrology Tutorial," NISTIR 89-4088, 1990.

Whitehouse, D.J., *Surfaces and Their Measurement*. Butterworth-Heinemann, 2012.

Whitehouse, D.J., *The Handbook of Surface Metrology*. CRC Press, 1994

Selected standards

ASME B46.1, Surface Texture (Surface Roughness, Waviness, and Lay)

ASME Y14.36, Surface Texture Symbols

ASME Y14.5, Dimensioning and Tolerancing

ISO/IEC Guide 98-3: Uncertainty of measurement—Part 3: Guide to the expression of uncertainty in measurement

ISO 1101, Geometric tolerancing: Tolerances of form, orientation, location, and run-out

ISO 1302, Indication of surface texture in technical product documentation

ISO 2632-1, Roughness Comparison Specimens—Part 1: Turned, ground, bored, milled, shaped and planed

ISO 3274, Surface texture: Profile method; Nominal characteristics of contact (stylus) instruments

ISO 4287, Surface texture: Profile method; Terms, definitions, and surface texture parameters

ISO 4288, Surface texture: Profile method; Rules and procedures for the assessment of surface texture

ISO 5436-1, Surface texture: Profile method; Measurement standards—Part 1: Material measures

ISO 8785, Surface imperfections: Terms, definitions and parameters

ISO 11562, Surface texture: Profile method—Morphological characteristics of phase correct filters

ISO 12085, Surface texture: Profile method—Motif parameters

ISO 13565-2, Surface texture: Profile method; Surfaces having stratified functional properties—Part 2: Height characterization using the linear material ratio curve

ISO 13565-3, Surface texture: Profile method; Surfaces having stratified functional properties—Part 3: Height characterization using the linear material probability curve

ISO 16610-21, Filtration—Part 21: Linear profile filters: Gaussian filters

ISO 16610-22, Filtration—Part 22: Linear profile filters: Spline filters

ISO 16610-31, Filtration—Part 31: Robust profile filters: Gaussian regression filters

ISO 21920 Series: Profile-based surface texture

ISO 25178 Series: Areal surface texture

Websites

The following websites are provided by various surface measurement equipment and software suppliers, measurement laboratories, and academic institutions, and contain definitions, illustrations, journal articles, and other useful information. These are listed for information only; the authors neither promote specific sites nor verify the accuracy of the information contained therein. All of these sites were available at the time this is being written. Some may require the creation of an account before access is granted. A Google search will find additional general and specific information.

digitalmetrology.com—A site with information on software, consulting, and training services from Digital Metrology Solutions, Inc. A blog provides useful information and commentary on selected metrology topics.

digitalsurf.fr/en/guide.html - A surface metrology guide from Digital Surf. This contains a lot of information on areal (3D) measurements as well as some useful 2D material.

jenoptik.com/en-roughness-measurement-and-contour-measurement - A basic guide, a poster, and measurement equipment information from Hommel-Etamic.

mahr.com/ - Information on a range of metrology equipment and techniques, including surface metrology.

michmet.com - A selection of articles and reference material, including 3D measurements, from this supplier of laboratory services.

mitutoyo.com/wp-content/uploads/2012/11/1984_Surf_Roughness_PG.pdf - A short guide on surface roughness measurements from this instrument manufacturer.

olympus-ims.com/en/knowledge/metrology/roughness/ - Several pages on 2D and 3D measurement equipment and terminology.

pml.nist.gov/VSC/jsp/About.jsp - The NIST Surface Metrology Algorithm Testing Service (SMATS). This is intended for testing the accuracy of analysis software, thus providing traceability to this national standard. The site can also be used to analyze the user's raw data files—both 2D and 3D. The service is described in the article: S.H. Bui and T.V. Vorburger, Surface Metrology Algorithm Testing System, Prec. Eng. 31, 218-225 (2007).

predev.com/smg/surface_metrology_guide.php - A surface metrology guide from the instrument manufacturer Precision Devices.

taylor-hobson.com/ - A wide range of guides and training materials from the instrument manufacturer Taylor Hobson.

zeiss.com/ - Information on a range of metrology equipment and techniques, including surface metrology.

Equations

This section provides the mathematical equations for several of the parameters, and is intended for reference only. Some parameters are calculated according to specific algorithms and cannot be represented by a simple equation. These algorithms can often be found in the relevant ISO standard; e.g., ISO 13565-2 for the Rk parameter set.

The equations use the following nomenclature:

- z_i is the distance from the mean line of the ith data point in the relevant profile
- z_{pi} is the height above the mean line of the ith peak in the relevant profile
- z_{vi} is the depth below the mean line of the ith valley in the relevant profile
- z_{ti} is the sum of the absolute values of z_{pi} and z_{vi} for a peak and the associated valley in the relevant profile
- dx is the distance between the individual data points along the x-axis
- s_i is the x-axis width of the relevant feature
- l is the evaluation length
- m is the number of relevant features in the evaluation length
- n is the number of data points in the evaluation length

Formulae are given in terms of the roughness parameter. The same characteristic is commonly available for the *primary* and *waviness* profiles, and can be calculated in a manner analogous to the roughness parameter.

Ra Average amplitude

$$Ra = \frac{1}{l}\int_0^l |z(x)|dx \qquad \text{(Eqn 1)}$$

Rc Average peak height plus valley depth

$$Rc = \frac{1}{m}\sum_{i=1}^{m} |z_{ti}| \qquad \text{(Eqn 2)}$$

Rda Average slope

$$Rda = \frac{1}{n-6}\sum_{i=4}^{n-3} |\Delta_i| \qquad \text{(Eqn 3)}$$

where

$$\Delta_i = \frac{1}{60dx}(z_{i+3} - 9z_{i+2} + 45z_{i+1} - 45z_{i-1} + 9z_{i-2} - z_{i-3})$$

Rdq	RMS slope	$$Rdq = \left(\frac{1}{n-6}\sum_{i=4}^{n-3}\Delta_i^2\right)^{\frac{1}{2}}$$ where $$\Delta_i = \frac{1}{60dx}(z_{i+3} - 9z_{i+2} + 45z_{i+1} - 45z_{i-1} + 9z_{i-2} - z_{i-3})$$	(Eqn 4)
Rku	Profile kurtosis	$$Rku = \frac{1}{Rq^4}\left[\frac{1}{l}\int_0^l z^4(x)dx\right]$$	(Eqn 5)
Rm0	Zeroth spectral moment	$$Rm0 = \frac{1}{l}\int_0^l z^2(x)dx$$	(Eqn 6)
Rm2	Second spectral moment	$$Rm2 = \frac{1}{l}\int_0^l \left(\frac{dz(x)}{dx}\right)^2 dx$$	(Eqn 7)
Rm4	Fourth spectral moment	$$Rm4 = \frac{1}{l}\int_0^l \left(\frac{d^2z(x)}{dx^2}\right)^2 dx$$	(Eqn 8)
Rq	RMS amplitude	$$Rq = \sqrt{\frac{1}{l}\int_0^l z^2(x)dx}$$	(Eqn 9)
Rsk	Profile skewness	$$Rsk = \frac{1}{Rq^3}\left[\frac{1}{l}\int_0^l z^3(x)dx\right]$$	(Eqn 10)

Rsm Mean spacing

$$Rsm = \frac{1}{m}\sum_{i=1}^{m} s_i$$ (Eqn 11)

Autocorrelation function

$$ACF(\Delta x) = \left(\frac{1}{Rq^2}\right)\left(\frac{1}{n-k}\right)\sum_{i=1}^{n-k} z_i z_{i+k}$$ (Eqn 12)

where

Δx = the distance the profile is shifted ($k \cdot dx$)

k = the number of points the data is shifted

INDEX

Made in United States
Orlando, FL
04 November 2026

90208243R00240